Knowledge and Skill Chains in Engineering and Manufacturing

Information Infrastructure in the Era of Global Communications

IFIP – The International Federation for Information Processing

IFIP was founded in 1960 under the auspices of UNESCO, following the First World Computer Congress held in Paris the previous year. An umbrella organization for societies working in information processing, IFIP's aim is two-fold: to support information processing within its member countries and to encourage technology transfer to developing nations. As its mission statement clearly states,

> *IFIP's mission is to be the leading, truly international, apolitical organization which encourages and assists in the development, exploitation and application of information technology for the benefit of all people.*

IFIP is a non-profitmaking organization, run almost solely by 2500 volunteers. It operates through a number of technical committees, which organize events and publications. IFIP's events range from an international congress to local seminars, but the most important are:

- The IFIP World Computer Congress, held every second year;
- Open conferences;
- Working conferences.

The flagship event is the IFIP World Computer Congress, at which both invited and contributed papers are presented. Contributed papers are rigorously refereed and the rejection rate is high.

As with the Congress, participation in the open conferences is open to all and papers may be invited or submitted. Again, submitted papers are stringently refereed.

The working conferences are structured differently. They are usually run by a working group and attendance is small and by invitation only. Their purpose is to create an atmosphere conducive to innovation and development. Refereeing is less rigorous and papers are subjected to extensive group discussion.

Publications arising from IFIP events vary. The papers presented at the IFIP World Computer Congress and at open conferences are published as conference proceedings, while the results of the working conferences are often published as collections of selected and edited papers.

Any national society whose primary activity is in information may apply to become a full member of IFIP, although full membership is restricted to one society per country. Full members are entitled to vote at the annual General Assembly, National societies preferring a less committed involvement may apply for associate or corresponding membership. Associate members enjoy the same benefits as full members, but without voting rights. Corresponding members are not represented in IFIP bodies. Affiliated membership is open to non-national societies, and individual and honorary membership schemes are also offered.

Knowledge and Skill Chains in Engineering and Manufacturing

Information Infrastructure in the Era of Global Communications

Proceedings of the IFIP TC5 / WG5.3, WG5.7, WG5.12 Fifth International Working Conference of Information Infrastructure Systems for Manufacturing 2002 (DIIDM2002), November 18-20, 2002 in Osaka, Japan

Edited by

Eiji Arai
Osaka University
Japan

Jan Goossenaerts
Eindhoven University of Technology
The Netherlands

Fumihiko Kimura
The University of Tokyo
Japan

Keiichi Shirase
Kobe University
Japan

Springer

Eiji Arai
Osaka University
Japan

Jan Goossenaerts
Eindhoven University of Technology
The Netherlands

Fumihiko Kimura
The University of Tokyo
Japan

Keiichi Shirase
Kobe University
Japan

Library of Congress Cataloging-in-Publication Data

A C.I.P. Catalogue record for this book is available from the Library of Congress.

Knowledge and Skill Chains in Engineering and Manufacturing: Information Infrastructure in the Era of Global Communications / Edited by Eiji Arai, Jan Goossenaerts, Fumihiko Kimura, Keiichi Shirase

p.cm. (The International Federation for Information Processing)

ISBN: (HC) 0-387-23851-4 / (eB00K) 0-387-23572-2 Printed on acid-free paper.

Printed in the United States of America.

9 8 7 6 5 4 3 2 1 SPIN 11350422 (HC) / 11350460 (eBook)

springeronline.com

TABLE OF CONTENTS

Preface

Since the first DIISM conference, which took place 9 years ago, the world has seen drastic changes, including the transformation of manufacturing and engineering software, and the information and communication technologies deployed. The conditions for manufacturing and engineering have changed on a large scale, in terms of technology-enabled collaboration among the fields of design, engineering, production, usage, maintenance and recycling/disposal. These changes can be observed in rapidly-growing fields such as supply chain management. As for production technologies at factory floors, new visions on human-machine co-existing systems involve both knowledge management and multi-media technologies. Therefore, because of these changes, the importance of information infrastructure for manufacturing has increased, stunningly. Information infrastructure plays a key role in integrating diverse fields of manufacturing, engineering and management. This, in addition to its basic role, as the information and communication platform for the production systems. Eventually, it should also serve the synthetic function of knowledge management, during the life cycles of both the production systems and their products, and for all stakeholders.

Over the past decade, the conference objectives have reflected changes of the engineering, manufacturing and business processes due to the advancements of information and communication technologies. The Fifth International Conference on Design of Information Infrastructure Systems for Manufacturing (DIISM 2002) held November 18-20, 2002 at Osaka University, in Osaka carried the theme: "Enhancing Engineering and Manufacturing Knowledge and Skill Chains in the era of Global Communications". The theme expresses both the wide scope and the technical depth that we are faced with in designing the information infrastructure for manufacturing. Yet, the globality and connectedness of the economic fabric and its problems obliges us to contain it... a mission impossible? Yes, if we stick to the traditional divide of mono-disciplinary academia and product-by-product industry. But do we have an alternative? Let us recall Hiroyuki Yoshikawa's vision of technical cooperation transcending cultural differences (among nations and among industry and academia) as set out in his keynote address to the 1st DIISM in Tokyo, November 1993. This vision has been guiding the global research programme on Intelligent Manufacturing Systems (www.ims.org). Over its five editions the DIISM working conferences have enjoyed very valuable contributions from several industry-led IMS projects such as Globeman 21, Next Generation Manufacturing Systems, Holonic Manufacturing Systems, Gnosis, Globemen, Mission, Humacs and Prodchain. The DIISM community has been honored to include these projects' contributions, facilitating interchange of ideas within these projects and with others outside of the projects.

The information infrastructure supportive of improving the state of "manufacturing industries as a whole" as Yoshikawa described it, must draw on both academic and industrial excellence, vision, knowledge, skill and ability to execute. It must support a wide range of scenarios, and involves an ever growing variety of devices, software and knowledge.

At the conference a great number of prominent experts from both academia and industries have presented significant results, approaches, knowledge, scenarios, and prototypes. Reworked versions of most of the presented papers are grouped into four parts: Generic Infrastructure Components, External Collaboration, Factory Floor Infrastructure and Man-System Collaboration. Applying principles of architecture descriptions for evolutionary software intensive systems. An introductory paper explains the DIISM problem statement and this book's structure.
As a whole, this compilation will be a great source of information, providing guidance toward design, implementations and utilization of information infrastructure for manufacturing.

The conference was sponsored by the International Federation of Information Processing (IFIP), through Working Groups 5.3 (Computer Aided Manufacturing) and 5.7 (Computer Applications in Production Management). The working conference would not have been a success without the help and hard work of many volunteers. First, we thank the members of the Organizing Committee. Further thanks go to the authors, the members of the International Program Committee and the conference participants for their contribution to the success of the conference and this book.
In conclusion, we strongly hope that this book will have a useful shelf life, and becomes another step towards solving problems of a fabric that we all share.

The editors,
Eiji Arai,
Jan Goossenaerts
Fumihiko Kimura
Keiichi Shirase

1

ENHANCING KNOWLEDGE AND SKILL CHAINS IN MANUFACTURING AND ENGINEERING

Jan B.M. Goossenaerts[1], Eiji Arai[2], Keiichi Shirase[3], John J. Mills[4], and Fumihiko Kimura[5]
[1]*Dept. of Technology Management, Eindhoven University of Technology, the Netherlands*
[2]*Dept. of Manufacturing Science, Graduate School of Eng., Osaka Univ., Japan*
[3]*Dept. of Mechanical Eng., Kobe Univ., Japan*
[4]*Dept. of Mechanical and Aerospace Eng., The Univ. of Texas at Arlington, TX, USA*
[5]*Dept. of Precision Machinery Eng., The Univ. of Tokyo, Japan*
e-mail: j.b.m.goossenaerts@tm.tue.nl

Abstract: This introductory paper to the volume explains the DIISM problem statement and applies principles of architecture descriptions for evolutionary systems (IEEE 1471-2000) to the information infrastructure for engineering and manufacturing. In our vision, knowledge and skill chains depend on infrastructure systems fulfilling missions in three kinds of environments: the *socio-industrial domain* of society and its production systems as a whole, the *knowledge domain* for a scientific discipline, and the *sectorial domain,* which includes the operational entities (companies, organisational units, engineers, workers) in engineering and manufacturing. The relationships between these different domains are captured in a *domain paradigm*. For companies, the original scope for infrastructure systems was the *factory floor* and the engineering office. Recently the scopes of *external* collaboration and of *man-system* collaboration have gained importance. Within each of the four identified scopes a system can offer services to different operational levels: *operations*, development or *engineering*, and *research*. The dimensions of scope and service level are briefly explained in relation to the architecting of an infrastructure. Papers are grouped according to their contribution to an infrastructure scenario or to an infrastructure component.

Keywords: architecture, engineering, information infrastructure, manufacturing

1. INTRODUCTION

The context of engineering and manufacturing has witnessed a striking expansion: from the product at the workshop during the workday of the craftsman, towards the portfolio of products and services, the resource base, and the business processes of the globally operating virtual enterprise. Simultaneously, the *set of information-based tools,* supporting the knowledge and skill chain has expanded: from the paper, pen and ruler to computer-and-communications aided applications for a growing range of functions ("CCAx"), with their impacts ranging from the core manufacturing process, over intra- and inter-enterprise integration, to the supply chain and the total life time of the extended product.

Computer-and-communications applications do well support many of the engineering, manufacturing and business functions that are key to manufacturing excellence and product success. But still, the engineering and manufacturing knowledge and skill chain shows many inefficiencies and hurdles. Therefore research and technology development on information infrastructure is ongoing, addressing a.o. information architectures, methodologies, ontologies, advanced scenarios, tools and services. This research is driven by the insight that throughout an integrated life cycle of products and enterprises, the manufacturing knowledge and skill chain sources information from globally distributed offices and partners, and combines it with situational awareness, local knowledge, skills and experience to initiate decisions, learning and action. Hence the top-level objective of the information infrastructure: enhancing knowledge and skill chains.

But how to design the information infrastructure that manages knowledge, information, data, and related services and tools that are shared by the different autonomous entities collaborating in the socio-economic fabric? Because the collaborators are part of different enterprises and economies, the information infrastructure is not regarded as a long-term differentiator in the business strategy of any enterprise. The infrastructure rather is a common enabler for the globalizing enterprise networks and professionals. For these entities, the common services matter at different levels of aggregation: for the external collaboration, for the teams and machine devices working in the factory or office, and for each person working in one or more enterprises. Hence the scope of this book: information infrastructure systems and services for any level of aggregation in the engineering and manufacturing knowledge and skill chain.

2. AN INFRASTRUCTURE PROBLEM?

A series of IFIP TC5 WG 5.3/5.7 working conferences has been dedicated to the design of the information infrastructure systems for manufacturing [1,2,3,4]. At this 5th working conference, building on recent research results and the results reported at and discussed at the previous conferences, contributions demonstrated a rich combination of breadth and depth, academic focus and industrial relevance. As multiple and more capable components are being developed, the gap grows between scenarios that are possible theoretically and experimentally and their practical realization and application. Unless a sound information infrastructure gets deployed, the chaining of the new scenarios will meet problems of quality, of interoperability of data, and of the scaling and combination of knowledge. How to offer continuity of service, the ubiquitous reuse of data and knowledge, and continuous interoperability while seizing new scenarios, as companies compete, stakeholders evolve and new technologies emerge?

Contributions to this volume address components and scenarios of future knowledge and skill chains, as seen from the viewpoints of many expert researchers in engineering, manufacturing and information technology. Traditionally, in industry, the integration of such components and scenarios is performed at companies. Today, and for the future, the globality and connectedness of the economic fabric and its problems oblige the research community to also address these chains supportive of improving the state of “manufacturing industries as a whole”.

3. ARCHITECTING THE INFRASTRUCTURE

Architecture is defined in IEEE 1471-2000 [5] as “the fundamental organization of a system embodied in its components, their relationships to each other, and to the environment, and the principles guiding its design and evolution”. Every system has an architecture which can be recorded by an architectural description (AD) consisting of one or more models. The viewpoints for use selected by an AD are typically based on consideration of the concerns of the stakeholders to whom the AD is addressed.

Modelling techniques support communication with the systems stakeholders, prior to system implementation and deployment. Methodologies and tools come available for the model driven building and deploying of information systems and information infrastructures.

The relevance of architecting for the infrastructure addressed in DIISM derives from its life cycle focus: architecting is concerned with developing satisfactory and feasible systems concepts, maintaining integrity of those system concepts through development, certifying built systems for use and assuring those system concepts through operational and evolutionary phases. This is important as the domain of engineering and manufacturing is immensily complex, diverse and evolving. Where infrastructure sub-systems fulfill missions in different scopes, these systems should co-evolve and their architectures be aligned. Their AD's should be based on stable viewpoints.

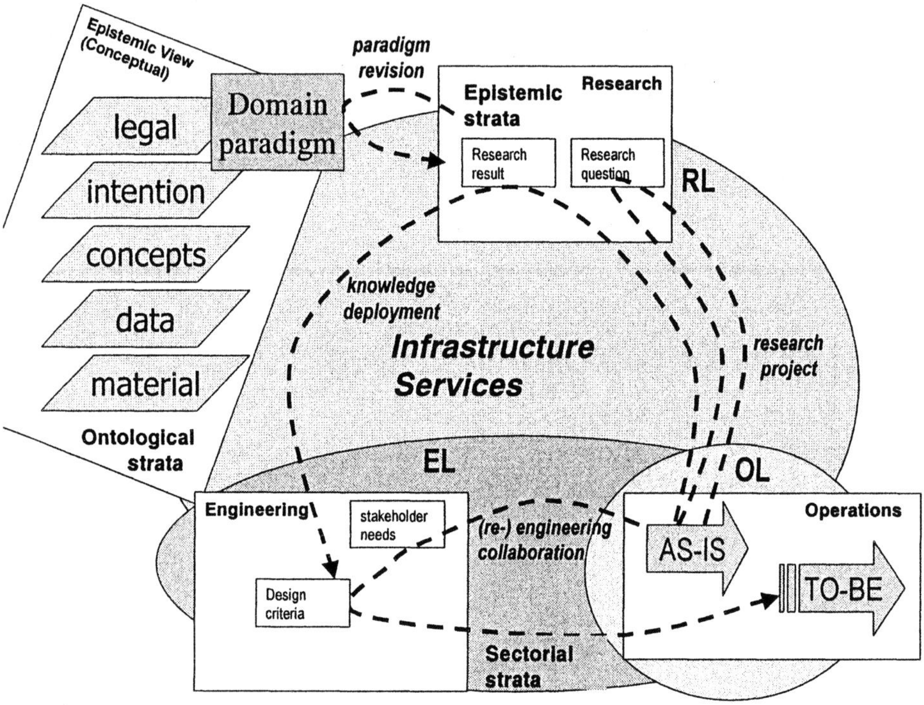

Figure 1. Three operational levels to serve

The four different scopes for which scenarios must be supported are the *natural & socio-economic domain* (DP – domain paradigm), the *external collaboration* (EC) among enterprises, the *factory floor* (FF), and the *man-system collaboration* (MS). In each scope systems evolve: problems and stakeholder needs are observed and analysed in the AS-IS, requirements analysis and design deliver an extended or new specification, development and implementation deliver the TO-BE operational system which is monitored for the occurrence of new problems.

Each of the four views in Figure 1 offers services to the above scenario of systems evolution. The *epistemic view* offers an *ontological stratification* that structures the design space within which intentions, models and operational systems evolve. The *research view* offers *epistemic stratification* (one strata per scientific discipline such as logistics, mechanics, chemistry, and ergonomy) that structures the design criteria (constraints) that must be met in modifying or creating the operational system. The *engineering view* merges constraints and contributions from ontological and epistemic strata to obtain new operational capabilities. In the *operations view* repeating tasks are performed, in accordance with the models developed; these models define operations that meet the hard laws of nature, the more soft laws of the socio-economic fabric, and the soft design criteria. Both the engineering and operations view show *sectorial stratification* which is for instance reflected in the STEP Application Protocols.

Assuming that a stable (meta-)model of the epistemic view exists, and that it rarely needs overhauls, the remaining infrastructure services are classified into three levels: *Operations Level (OL):* for the AS-IS operations (engineering or manufacturing processes); (Re-) *Engineering Level (EL):* for the (re-) engineering collaborations linking AS-IS operations and development for certain context to achieve the TO-BE operations; and *Research Level (RL):* research and the deployment of scientific knowledge pertaining to OL processes and EL collaborations.

4. INFRASTRUCTURE DESIGN AT DIISM 2002

Each infrastructure sub-system is a software intensive systems that could be developed using the widely used 4+1 view model of Kruchten[6]. The alignment of the architecture descriptions of these infrastructure sub-systems would benefit from a maximal reuse across those views, in accordance with the subsidiarity principle.

The best opportunities for such reuse are in the epistemic view which covers Kruchten's logical and process views, and in the research view. The domain paradigm would consist of universally applicable models. The domain paradigm embodies the ontological stratification of the natural & socio-economic domain, the epistemic stratification of our (scientific) knowledge, and the separation of operations, engineering and research scenarios in our activities. Part I of these proceedings contains the DIISM 2002 contributions that pertain to the epistemic view, the domain paradigm and the research view. Comparing with the present day best practice, the epistemic view and the domain paradigm could be taken into consideration when developing a 2nd generation structure for STEP's Generic Resources.

With the availability of reusable domain-level infrastructure components, the focus in the scopes of EC, FF and MS is on their differentiating aspects and scenarios. Part II, III and IV contain the DIISM 2002 contributions on

External Collaborations, the Factory Floor Infrastructure and the Man-System Collaboration. In each of these parts both Engineering Level and the Operations Level contributions are included.

4.1 Part I – Generic Infrastructure Components

This part contains the contributions that address viewpoints or services that in principle can be shared by all scopes (society, external collaboration, factory floor and man-system collaboration). It includes papers on the information infrastructure requirements, on the domain paradigm and on the epistemic viewpoint. Papers on research level services are also included because in principle, they can be shared by all scopes at operations and engineering level.

Kimura proposes basic approaches for managing life cycle support information, considering requirements such as flexible extensibility, distributed architecture, multiple viewpoints, long-time archiving and product usage information. Goossenaerts applies an architecting approach to derive specifications of a model-driven information infrastructure.

The domain paradigm is addressed from three different viewpoints. The intensification of service and knowledge contents within product life-cycles is addressed in four papers.

Shimomura and Tomiyama propose a service modelling technique that can represent services with subjective properties. Jansson and Thoben introduce the extended products paradigm and illustrate it with examples from the IMS GLOBEMEN project. Salkari et al. discuss the management of product information of process plants, complex one-of-a-kind products. Mills and Goossenaerts present the architecture of a product knowledge environment that is based on computational contexts.

Two papers focus at software intensity at the shop floor, and how to cope with it. Kanai et al. propose an object-oriented design pattern approach for the seamless modeling, simulation and implementation of distributed control systems (DCS). Matsuda et al. present an interoperability framework and manufacturing software capability profiling methodology.

External collaboration is addressed by Nienhaus et al. who propose a supply chain modelling approach which enriches the SCOR model with product-related and financial information.

The epistemic viewpoint is addressed in three papers with a complementary focus. Itoh et al. illustrate the Multi-Context Map (MCM) and Collaborative Linkage Map (CLM) and interprete these enhanced process-modelling constructs in the collaboration stratum, the workflow stratum and the state-transition stratum. Abramov et al. address ontological stratification and apply it in agent system design. Yagi et al. address behavioural aspects in their paper on logics of becoming in scheduling.

A first paper addressing research level services is by Kryssanov et al. who develop a new theory of communication to explain computer-mediated communication, which in the future will also involve products such as cars. Another paper is by Lee et al. who investigate the relationship between media choice and end-user belief on help desk service. They validate the Mediqual constructs: reliability, empathy, assurance, tangibles, and responsiveness as new belief criteria on media users' satisfaction.

4.2 Part II – External Collaboration

The papers addressing engineering level services for external collaboration cover a wide range of topics.

Two papers address inter-enterprise engineering collaboration. Kawashima et al. describe the Distributed Engineering Environment prototype that was developed as a part of the IMS Globemen project. Ratchev and Medani propose a new STEP AP224 EXPRESS based data model to facilitate the exchange of part and process data during the early design process.

Simulation in external supply chains or virtual enterprises is the topic of three papers. Mertins and Rabe describe a tested platform for performing distributed simulations using the High Level Architecture (HLA, IEEE 1516) while keeping the participating enterprise models private. Hibino and Fukuda describe and illustrate the use of an adapter and user support system between manufacturing simulators and Runtime Infrastructures based on the HLA. Nakano et al. propose a method and its tool to navigate the designers through the engineering process and generate the simulation model automatically from the design results.

The remaining five papers on engineering level services have a focus on the engineering of logistic and engineering networks and related management problems.

Zhou et al. discuss the knowledge management issues in the development of VIEWBID, a web-based system for supporting online bidding document preparation for global engineering and manufacturing projects. Laakman presents a reference model based guideline for logistics engineers. The guideline is supported by a collaborative knowledge management application. Alard et al. describe a framework for the strategic evaluation and planning of the deployment of internet-based procurement solutions for direct materials. Nishioka et al. propose the SUPREME architecture which supports web-based virtual enterprise design and collaborative planning and scheduling. Zhou et al. present a review of state-of-the-art tools and methods that can be used to manage risks in multi-site engineering projects. They then propose a risk management roadmap that can provide guidelines for project managers.

Three papers address operations level services in the context of external collaboration. Kimura et al. propose the ASSIST concept, a manufacturing support system that – for multi-vendor manufacturing systems – combines maintenance services with consulting services by engineering companies and machine tool vendors. Hartel et al. describe a model that will enable service enterprises to team up with external partners and fulfill services collaboratively. Kamio et al. discuss and illustrate a scheme that allows all parties involved in the maintenance of a chemical plant to form a service enterprise, whenever a maintenance service is necessary.

4.3 Part III – Factory Floor Infrastructure

Papers in this part address engineering and operation level services for the factory floor. The architecture of the factory floor infrastructure is addresssed in two papers. Mo and Woodman describe the development of an integrated web-based CIM environment called J-MOPS: based on the MOPS philosophy it can intelligently transform CAD information into machine programs while simplifying system requirements for the user and removing the dependency on platforms. Using the Glue Logic, Takata and Arai design a real-time data gathering system for manufacturing lines. A Scalable Intelligent Control Architecture permits expansion of the control system, in spatial dimension and in intelligence.

Two papers present advanced scenarios in process planning and CAM. Muljadi et al. propose a feature set creator that can lead to the generation of multiple process plans in support of flexibility in shop floor scheduling. Narita et al. propose and verify a two-stage strategy for automatic and interactive modification of NC program using the VMSim cutting process simulator.

Scheduling of distributed production systems is the topic of three papers. Shirase et al. use HLA to achieve a distributed scheduling simulation for dynamic work assignment and flexible work group configuration. Sashio et al. study the data handling mechanism of a distributed virtual factory that is constructed by integrating area level simulators in a manufacturing system. Iwamura et al. propose a new real-time scheduling method to select a suitable combination of resource holons and job holons to carry out the machining process.

To support operators in improving the makespan of an existing Job Shop schedule, created with a heuristic algorithm such as Tabu search, Tsutsumi and Fujimoto propose a tool for sensitivity visualization of the critical path.

Wang and Chu study the requirements of the enterprise-wide integration of managerial and automation systems in a petroleum refinery.

4.4 Part IV – Man-System Collaboration

Papers in this part address engineering and operation level services for man-system collaborations. Mizugaki et al. present a computer aided instruction system (CAI) for NC lathe programming. Multi-media objects including movies, animations, pictures and sound are used in web-browser based training procedures. Following tests with beginners, the efficiency and usefulness of the CAI system is discussed.

Fukuda et al. describe the development and test of a prototype Web-based Instruction system using the wearable computer. The web-applications include time estimation, simulator, active instruction manual system and a posture acquisition system.

Imai addresses the model-based description of human body motions of factory workers performing their work, and how to use the resulting evaluations in manufacturing process design. For the purpose of accurate posture and motion detection by multiple video cameras, Sakaki et al. describe a calibration technique for use in combination with Model based Motion Capture.

Jianhua and Fujimoto propose a Bayesian decision model for the identification of human behaviour in manufacturing on the basis of manufacturing history data. The latter data is converted to a non-parametric distribution over a feature vector by using a binary division method.

5. POST CONFERENCE GAPS

Following the DIISM 2002 conference the research and technology development challenge is to further integrate multiple advanced scenarios and components into true knowledge and skill chains. The engineering and manufacturing infrastructure should support such vertical and horizontal chains, ensuring data consistency, reuse and interoperability as operations, engineering and research proceed within the scopes of man-system collaboration, factory floor and external collaboration.

In line with good practice in software systems development, three milestones could be identified for the information infrastructure: the lifecycle objectives (LCO), the life-cycle architecture (LCA), and the initial operational capability (IOC). These milestones could structure the research and technology development activities that should also include:

- The development and validation of the Epistemic View. Overarching tasks in this validation are the development of a Domain Paradigm and the definition of Research Level services. For these tasks, the papers in part I offer a baseline from where to proceed.
- The scripting of the scenarios from Parts II, III and IV using the conceptual models of the Epistemic View, the Domain Paradigm and the Research Level services (reference models). For available components, the development of application protocol interfaces is recommended.
- The linking of scenarios and components into operational knowledge and skill chains, and their deployment in industry.

ACKNOWLEDGEMENTS

We sincerely thank all the authors, the program committee members and the conference participants for their contribution to the conference and this book. Special thanks go to the Conference and PC/OC (co-) chairs of the previous DIISM working conferences: Hiroyuki Yoshikawa, Hendrik Van Brussel, Hans Wortmann, Laszlo Nemes and John P.T. Mo for their contributions to the earlier DIISM conferences, and their continuous support.

REFERENCES

1. Yoshikawa, H. and J. Goossenaerts (eds.) (1993) Information Infrastructure Systems for Manufacturing. IFIP Transaction B-14. Elsevier Science B.V. (North Holland)
2. Goossenaerts, J., F. Kimura and J.C. Wortmann (eds.) (1997) Information Infrastructure Systems for Manufacturing, Chapman & Hall, London, UK
3. Mills, J. and F. Kimura (eds.) (1999) Information Infrastructure Systems for Manufacturing II, Kluwer Academic Publishers, Boston
4. Mo, J.P.T. and L. Nemes (eds.) (2001) Global Engineering, Manufacturing and Enterprise Networks, Kluwer Academic Publishers, Boston
5. IEEE Recommended Practice for Architectural Description of Software-Intensive Systems (IEEE Std 1471-2000) July 2000.
6. Kruchten, P., Architectural Blueprints - The "4+1" View Model of Software Architecture, IEEE Software, November 1995, 12 (6)

PART I

Generic Infrastructure Requirements and Components

2

ENGINEERING INFORMATION INFRASTRUCTURE FOR PRODUCT LIFECYCLE MANAGMENT

Fumihiko Kimura
Department of Precision Machinery Engineering
The University of Tokyo
Hongo 7-3-1, Bunkyo-ku, Tokyo 113-8656, Japan
E-mail kimura@cim.pe.u-tokyo.ac.jp Tel. +81-3-5841-6455

Abstract: For proper management of total product life cycle, it is fundamentally important to systematize design and engineering information about product systems. For example, maintenance operation could be more efficiently performed, if appropriate parts design information is available at the maintenance site. Such information shall be available as an information infrastructure for various kinds of engineering operations, and it should be easily accessible during the whole product life cycle, such as transportation, marketing, usage, repair/upgrade, take-back and recycling/disposal. Different from the traditional engineering database, life cycle support information has several characteristic requirements, such as flexible extensibility, distributed architecture, multiple viewpoints, long-time archiving, and product usage information, etc. Basic approaches for managing engineering information infrastructure are investigated, and various information contents and associated life cycle applications are discussed.

Key words: Engineering Information Infrastructure, Product Life Cycle Management, Digital Engineering

1. INTRODUCTION

Due to the very severe competition in global market and rapid technological progresses, manufacturing industry is now facing fundamental changes and renovation of its operations and organizations. There are several keywords which characterize such new trends of manufacturing:

- very short time-to market,
- quick and drastic changeability,
- customer-pull production instead of manufacturer-push,
- service production instead of product production, etc.

By thorough adoption of such trends, it becomes possible to achieve highly efficient manufacturing and large reduction of various kinds of losses for time, cost, human and physical resources.

Based on the above consideration, a vision for future manufacturing is summarized as shown in Figure 1. According to very strong social demands and constraints, environmental consideration should direct manufacturing activities into more resource-saving and environmentally benign manners. At the same time manufacturing industry must be competitive to survive in very severe global market, as discussed above. Information technology is clearly an enabler to accommodate both requirements, and to lead to a new manufacturing paradigm: from product manufacturing to function or service manufacturing.

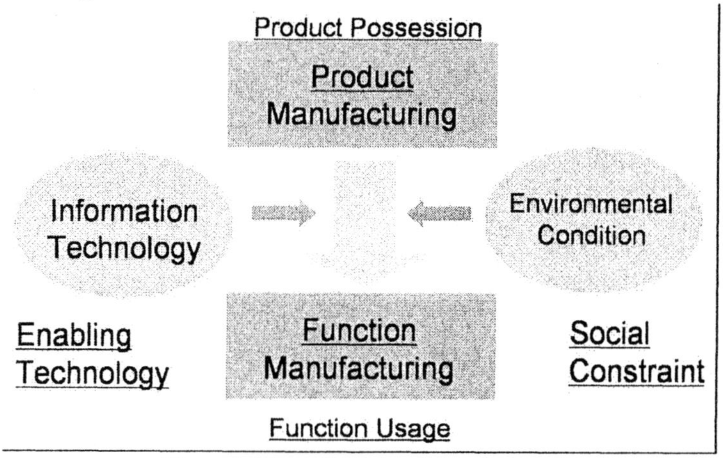

Figure 1. Vision for Future Manufacturing.

In such new manufacturing paradigm, where service providing plays an important role, it is essential to realize an engineering information infrastructure, which is shared by all stakeholders of manufacturing, such as manufactures, users, society, etc. Engineering information infrastructure should contain all aspects of product and production related information, and, as a ubiquitous information environment, can be utilized for optimally efficient usage of products. Therefore, design of information infrastructure systems for manufacturing is one of the most keen issues for future manufacturing.

In this paper, characteristics of future manufacturing are briefly discussed. As a core technology for future manufacturing, life cycle management and modeling are explained. And some examples of practical approaches for engineering information infrastructure are shown.

2. FUTURE VISION FOR MANUFACTURING

There are many issues for realizing competitive and environmentally benign manufacturing for the future:

- how to achieve drastic reduction of lead time to market,
- how to achieve agile changeability under rapid technology progresses and market fluctuations,
- how to achieve efficient negotiation and collaboration activities in global environment,
- how to efficiently generate and reuse intelligent resources,
- how to achieve comprehensive automation based on mature technology,
- how to achieve quick-response production through customer-pull processes,
- how to increase value-addition by mass-customization or tailor-made production,
- how to adapt to and to maintain variety of product usage, etc.

For solving all such issues, engineering information infrastructure for product life cycle is mandatory. Particularly it is important to support information sharing in the early planning stages of manufacturing and after-sales/service stages, as shown in Figure 2.

It is also important to install appropriate mechanisms for total life cycle management of engineering knowledge based on information technology based infrastructure, as shown in Figure 3. For competitiveness of manufacturing, it is essential to create new innovative knowledge, but at the same tine, it is important to systematize such innovative knowledge. Innovative knowledge gradually becomes mature, and it is important to manage such mature knowledge and to reuse it for efficient manufacturing. For such life cycle management of knowledge, engineering information infrastructure plays a very essential role. Here are some critical issues for knowledge management:

- early and lossless capturing of knowledge,
- flexible knowledge sharing,
- transparency for knowledge evolution,
- re-usability of knowledge, etc.

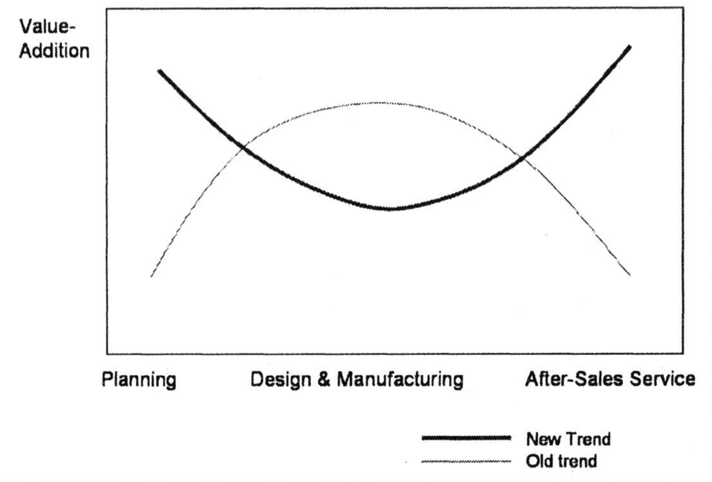

Figure 2. Value-Additive Processes in Manufacturing.

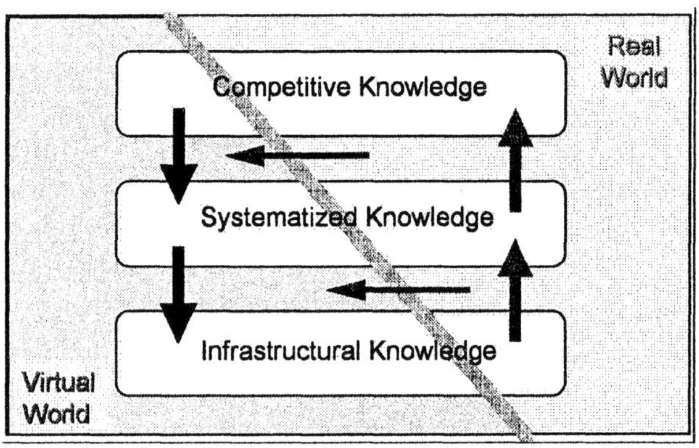

Figure 3. Life Cycle Management of Engineering Knowledge.

Those issues need to be fully investigated for engineering information infrastructure.

Based on such information infrastructure, various kinds of manufacturing activities can be loosely integrated or federated, and new service oriented manufacturing can be organized, as shown in Figure 4. There are different kinds of "Factories" with different focuses on life cycle stages. "Service Factory" provides services to end customers, "Inverse Factory" accepts used products for reuse/recycling, and "Automated Factory" produces mature goods for daily life. "Innovation Factory" offers new knowledge and technology to other Factories. Based on comprehensive information infrastructure, various kinds of new manufacturing activities can be easily visualized.

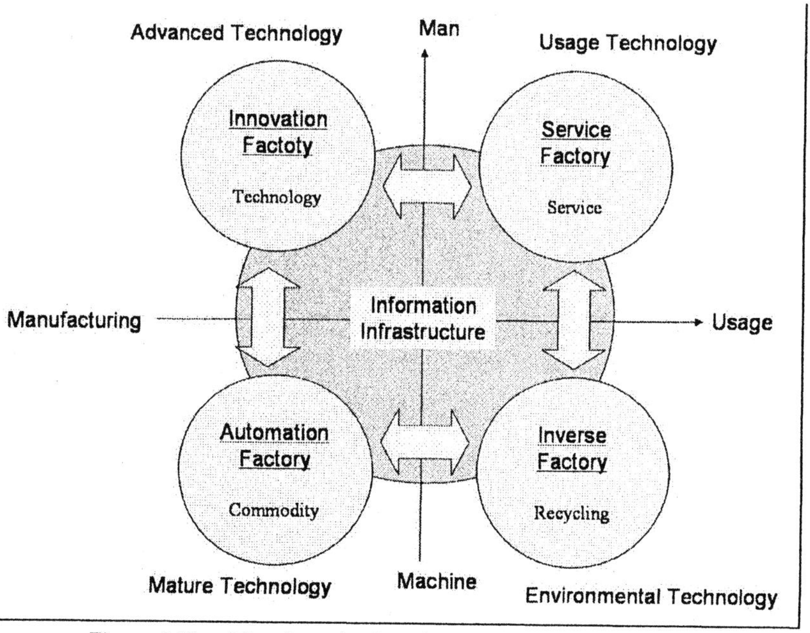

Figure 4. New Manufacturing based on Information Infrastructure.

3. LIFE CYCLE MANAGEMENT AND MODELLING

In total life cycle of products, product usage phase is becoming more and more important. As shown in Figure 5, product usage phase includes many activities, such as repair, upgrade, refurbishment, etc. It can be said as another kind of manufacturing or extension of traditional manufacturing. It is a crucial issue how to rationalize such new manufacturing activity for environmentally conscious manufacturing.

In practice, products are not used as designers plan, and many inefficiencies happen, such as:

- unexpected early disposal of products,
- non-use or idle products in users' hands,
- long-term use of old inefficient products, etc.

All those are primarily due to the lack of communication and common understanding between users and manufacturers. Therefore it is important to set up common information infrastructure, and to integrate product life cycle design and management activities with product design activity, as shown in Figure 6.

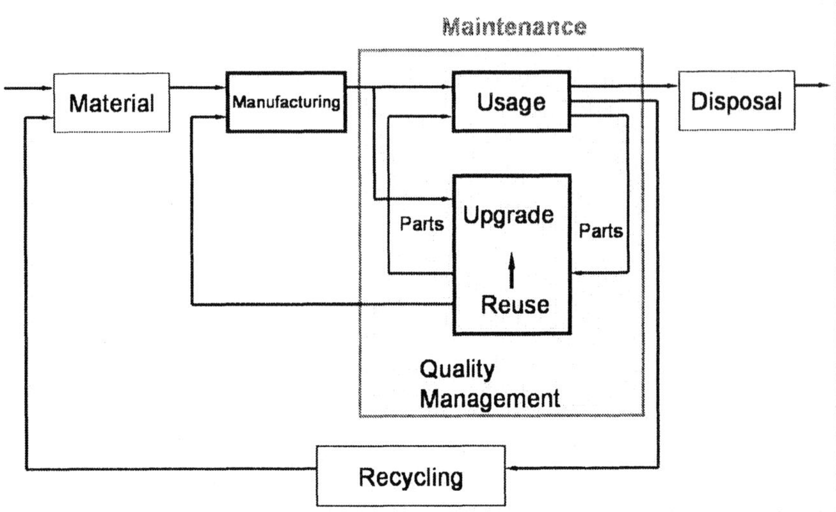

Figure 5. Total Product Life Cycle Management.

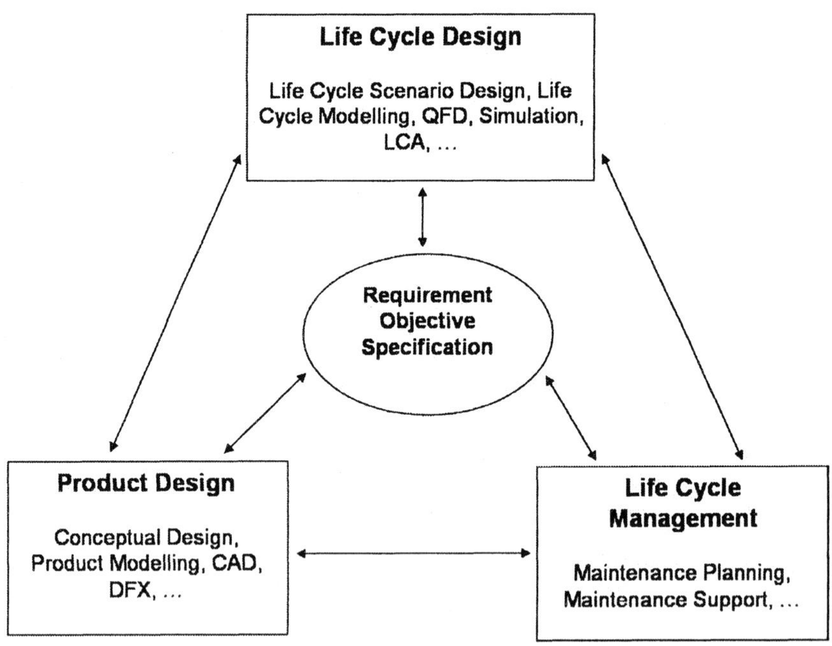

Figure 6. Product Life Cycle Design.

For proper life cycle management, product life cycle modelling is important, which includes modelling of product deterioration during usage[1]. Based on deterioration simulation, life cycle design and management can be rationalized based on computer supported technology. An approach to deteriorated behaviour simulation is shown in Figure 7.

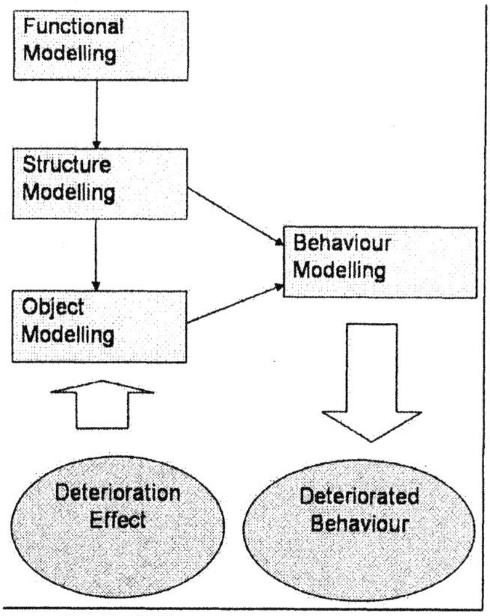

Figure 7. Modelling of Deteriorated Behaviour.

An example of total life cycle modelling for elevator maintenance planning is shown in Figure 8[2]. A core of this model is Failure Model of Elevator and Monitoring Unit for elevator operations.
Much of such model information comes from elevator design activity, and it is clearly important to share engineering information among various stages of product life cycle.

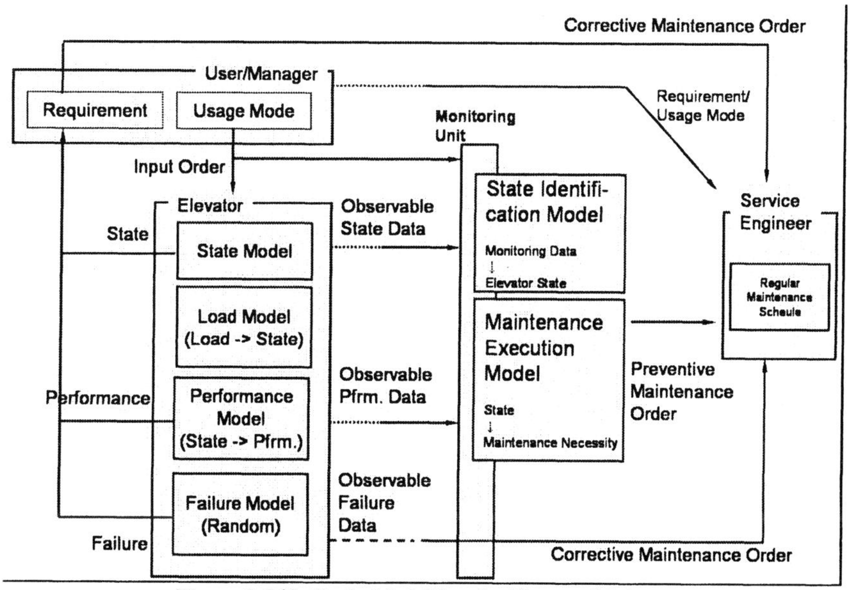

Figure 8. Life Cycle Modelling for Elevator Maintenance.

4. ENGINEEREING INFORMATION INFRASTRUCTURE

After many years of research and development about computer aided technology for product development, such as CAD/CAM/CAE, engineering databases and PDM, it is now feasible to consider the total integrated support of product creation processes by computer. As a basis for such comprehensive support, total life cycle modelling is important, from product planning, through product creation, production preparation, usage support and down to reuse/recycling/disposal. Such whole life cycle modelling is summarized in Figure 9, where some modelling is well developed in relation with product and process engineering, whereas some other modelling is still in primitive stage, such as deterioration modelling, reliability modelling, functional modelling, etc.

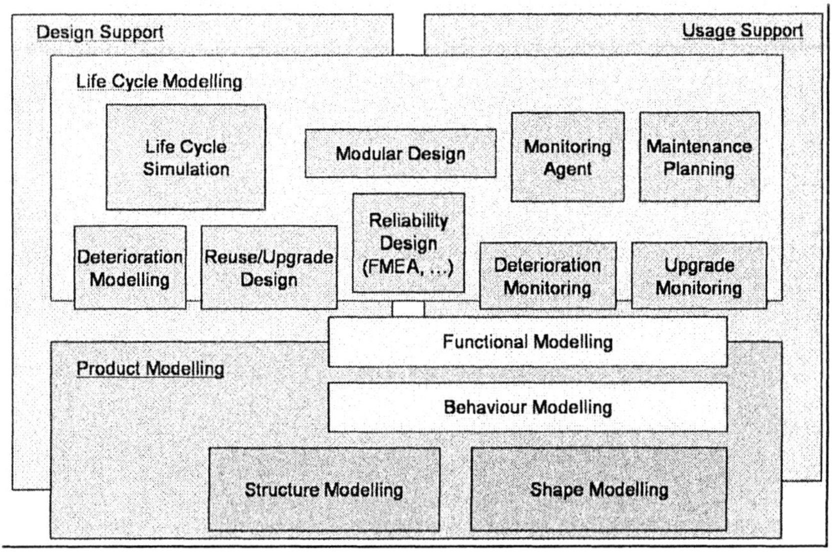

Figure 9. Total Product Life Cycle Modelling.

For construction of engineering information infrastructure for representing total life cycle models, the following points shall be noted:

- Hierarchical systematization of related engineering knowledge is necessary. As a basis of generic information infrastructure, basic engineering concepts shall be systematically organized.
- Modelling framework shall be flexible enough to enable loose and evolutional federation of various kinds of modelling information.

For modelling framework, there are many critical issues remaining:

- how to integrate or federate various kinds of models, such as shape, engineering constraints, product configurations, etc.,

- how to manage modelling processes in dynamically evolutional and collaborative environment,
- how to systematize information infrastructure based on generic primitive models, etc.

The above issues are theoretically very complicated and difficult. However there are many practical development activities for coping with these issues. One recent example is iViP development supported by German government[3]. A similar approach is pursued in Japanese project[4], and is shown in Figure 10. Here a common information bus is constructed, and available information resources can be connected to this bus via a generic wrapper mechanism.

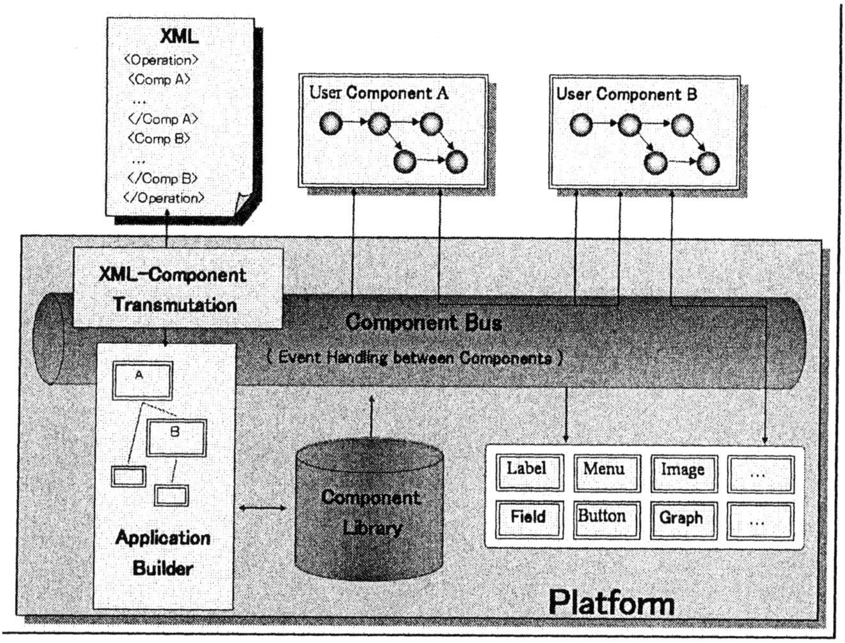

Figure 10. Software Integration Platform for Engineering.

5. CONCLUSION

Future manufacturing could be more competitive and environmentally benign due to change of paradigm from products to services. This paradigm change can be realized by intimate information sharing among all stakeholders of manufacturing based on engineering information infrastructure. Engineering information infrastructure facilitates rationalized life cycle management of products, particularly at the product usage phase.

In recent years, such life cycle management becomes popular, and practical implementation is emerging under the concept of PLM (Product Lifecycle Management). However there are still many open issues for powerful implementation of engineering information infrastructure, for example:

- consistent modelling of engineering semantics throughout the total product life cycle,
- various information modelling standards for federating multiple models,
- light-weighted frameworks for information representation and management.

In the future, engineering information infrastructure will be merged in the social information infrastructure, and will become fundamental industrial backbone for advanced countries.

REFERENCES

1. Fumihiko Kimura: Comprehensive Product Modelling for Designing Life Cycle Scenarios, EcoDesign, IEEE Computer Society, 2001, p.1-6.
2. Fumihiko Kimura and Tomoyuki Hata: Life Cycle Design and Management based on Simulation of Service Quality Variation, CIRP Life Cycle Engineering Seminar, 2003.
3. Frank-L.Krause, Trac Tang and Ulrich Ahle (eds): Project iViP (Integrated Virtual Product Creation) Final Report, The Federal Ministry of Education and Research (BMBF), 2003
4. Norio Matsuki: Digital Meister Project Report, AIST, 2003.

3

ARCHITECTING AN UBIQUITOUS & MODEL DRIVEN INFORMATION INFRASTRUCTURE

J.B.M. Goossenaerts
Eindhoven University of Technology, the Netherlands
Em: J.B.M.Goossenaerts@tm.tue.nl

Abstract: A model driven architecture (MDA) approach is applied to the architecting of a Ubiquitous and Model-driven information Infrastructure (UMI). Our focus is on the stakeholders of the ubiquitous infrastructure, the distinction between the infrastructure, the enterprises and applications accommodated by it, and the dependencies among the conceptual models at different levels. A small example illustrates the proposed concepts and constructions.

Keywords: Model Driven Architecture, Information Infrastructure, Ubiquity

1. INTRODUCTION

There is an increasing understanding of modeling techniques and their support for communication with the stakeholders in (information) systems, prior to systems implementation and deployment. As a result, methodologies and tools come available for the model driven building and deploying of information systems and software applications. The recent OMG-proposed [1] Model Driven Architecture (MDA) puts the model, a specification of the system functionality, on the critical path of software development, prior to the implementation of that functionality on a specific technology platform. *"The MDA approach and related standards allow a same model to be realized on multiple platforms, and allows different applications to be integrated by explicitly relating their models, enabling integration and interoperability and supporting system evolution as platform technologies come and go."*

Accepting a model driven approach, this paper separates three levels at which to apply MDA: the enterprise, the application and the information infrastructure. Most publications on MDA [2] target application development, and publications on information infrastructure tend to focus on the ICT platform and its performance. Complementary to these other contributions, this paper focuses at models and architecting at the information infrastructure level, and at the consequences for enterprise and application development of using infrastructure level models.

2. ANCHORING ARCHITECTURE BY MODELS

Intuitively, the vision of a model driven architecture can be linked to a combination of Boehm's Win-win Spiral model [3] and Kruchten's 4+1 view model [4] of (software) systems architecture. The Win-win spiral is used to ensure that the end-users drive the architecture and development work for the whole duration of the project. The model also introduces milestones to anchor the development process, and to assess and mitigate risks. The 4+1 view model is adopted because projects are situated in an engineering context where a large portion of specifications (expressed as models), software systems and data, and hardware systems are (re-) used and/or have to inter-operate (in a software intensive system), and evolve over time.

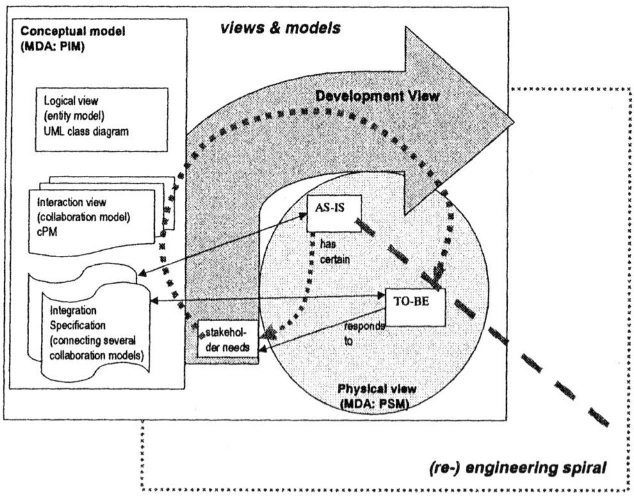

Figure 1. A re-engineering spiral anchored by views and models

The UML offers modeling constructs for each of the 4+1 views. In a modified approach we use a conceptual (pseudo) collaboration model (pCM) combining notational elements from high-level Petri nets (HLPN), IDEF-0, and UML activity diagram. Our notion of collaboration is similar to that of ebXML (http://www.ebxml.org/specs/). The hierarchy of activities is specified using a parent-child connector (|̲), which is frequently used in product structures. A swimming lane layout separates the activities to be performed in the different roles with a controlling stake in the collaboration. The input, output, control and support conventions of the IDEF-0 generic activity model are applied, they connect the activity with (Petri-net-like) places containing an expression (over the entity model) that indicates which entities are involved in the activity. Figure 3 illustrates the collaboration modeling technique. The Integration Specification deals with the integration and aligning of the different collaboration models. All models in the conceptual model block are platform independent models (PIM) in the sense of MDA. The platform specific models (PSM) are part of the physical view: the ICT platforms need them to carry out their share of the work.

Assume now that there is an existing system (AS-IS) that needs to be improved. Then the re-engineering spiral in Figure 1 is model enabled: problem analysis delivers additional stakeholder needs, requirements analysis and design deliver extended or new collaboration models, optionally with refinements in the entity models, and a new integration specification. The latter is an input to the development and implementation to deliver the TO-BE physical realization.

3. UMI, COMMUNITIES AND APPLICATIONS

An information infrastructure consists of the information models, data, and information processing services and tools that are shared by the different autonomous entities that collaborate or interact in a community or society. The trend towards a ubiquitous information infrastructure builds on the connectivity and low-cost high-performance computing and communication facilities provided by computers, the Internet and wireless communications, ranging from Bluetooth to satellite-based. A UMI is defined for and embedded in a society to support all the society's members and communities.

The term *society* is used here with the meaning of "all people, collectively, regarded as constituting a community of related, interdependent individuals". A *community* is "a group of people having interests or work in common, and forming a smaller (social) unit within a larger one." This definition thus covers enterprises, public bodies, sports clubs, schools, hospitals, etc. All members of a society are *persons* with equal rights and, in principle, the ability to use the UMI. Each person may belong to several communities. A community has no member outside society.

Typically, each community will enact processes and install applications to sustain its interests. Maybury for instance, describes Collaborative Virtual Environments for distributed analysis and collaborative planning for intelligence and defense [5]. The DIISM conferences have been dedicated to the design of the information infrastructure systems for manufacturing and engineering enterprises. Virtual communities in relation to Peer-to-Peer collaboration architectures are discussed in [6]. Table 1 lists products and artifacts that typically are involved when the re-engineering spiral is applied at the levels of infrastructure, community and application.

Table -1. Levels of applying the re-engineering spiral

Level	Typical services	Conceptual model	Dev. view	Physical view
infrastructure	authentication & personalization collaboration standards	market & collaboration models (e.g., ebXML, SimpleEconomy)	J2EE/EJB WSDL	SOAP/XML CORBA/CCM DCOM/.NET
community	production & services	enterprise model, process model	BPR and its tools	operational processes
application	purchasing CAD ERP	orders&invoicing eng. product struct. log. product struct. MDA:PIM	ARIS, DEM Rational Telelogic TogetherSoft	ERP systems PDM systems PPC systems MDA:PSM

Whereas the development and physical view components in Table 1 are working systems or accepted standards, most of the infrastructure level state-of-the-art components lack (public) models or trace-ability to stakeholders needs. In fact, we have no comprehensive and stakeholder/end-user-driven set of criteria to evaluate the infrastructure level components for their fitness to serve in an UMI. In a step towards a more rigid foundation, the further sections will highlight some of the issues. Relying on piecemeal ontological commitment[7] the focus is on simple application scenarios for a minimal societal ontology of objects and activities[8]. We do not consider the content of the entity classes[9]. In Section 4, the view and spiral model (Figure 1) is applied to UMI and some basic models are given. In Section 5 we briefly consider infrastructure-enabled application development. At the infrastructure level the focus is at members and their roles in typical collaborations. Applications support specific collaborations, which they may also partially control.

4. ARCHITECTING UMI

The current state of the information infrastructure is that physical view aspects of its architecture are better understood than the conceptual view aspects. Our position is that conceptual models are an integral part of an information infrastructure because of their role in anchoring a model-driven architecting process for the communities and the applications.

At the infrastructure level three kinds of stakeholders are identified: *society*, *member* and *community*. The stratification of the common context for these stakeholder's requirements is addressed in another paper[10]. Some generic win conditions are given here.

The society as a whole pursues compliance to its enacted models and agreed upon policy goals (e.g. fair trade and protection of property in the global society). With goals such as rapid implementation of new "laws" or charters, it could use the subsidiarity principle to organize its institutions and ensure that each problem is addressed at the level at which it is common for all the lower-level stakeholders.

The success of a community depends on the support that its members receive for their relevant actions, conform the processes enacted and the society's law or rules. E.g., the certification of a new type airplane by the relevant authorities, or the carrying out of tax payments and elections. Change, i.e. improvements of the operational processes, must happen smoothly, without disruption of the community's services, and with a minimal burden to its members.

The member's win conditions include a.o. empowerment, legal security, efficient operations, optimal propagation of change, minimal risk of inconsistencies, data protection and privacy [11]. Infrastructure facilities that contribute to enabling these requirements include personalization [12] everywhere and anytime.

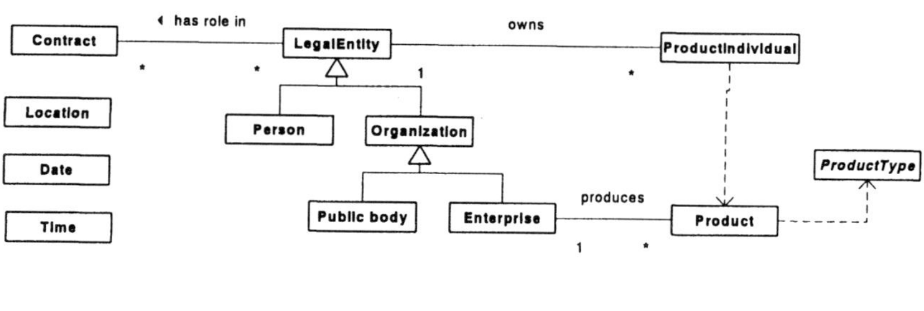

Figure 2. A SimpleEconomy entity-model

A platform independent model of an UMI includes a model of the persons and communities interacting in society. Because quite a few of these interactions are concerned with the production, exchange and consumption of goods or products, it is evident to also include classes for products. Persons can join or leave communities (e.g., organization) (Figure 2). The Sale collaboration illustrates the SimpleEconomy interactions (Figure 3). Collaborations in SimpleEconomy must meets market rules that are part of the integration specification and constrain the choices of the entities involved in combinations of collaborations.

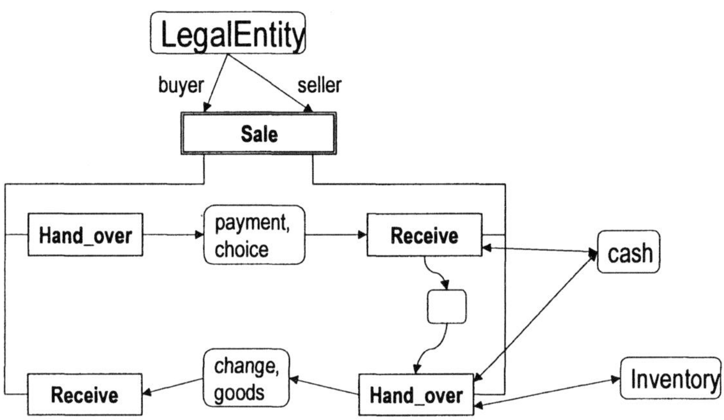

Figure 3. The Sale collaboration in SimpleEconomy

The above models are part of the conceptual view. A model driven infrastructure requires also the elaboration of a physical view. The infrastructure should manage a "proxy", or unique representant, for each instance (entity) in society. In one of many possible implementations, this proxy could be an XML document instance that is conform to the schemas expressing the ontological and collaborative commitments shared in society.

5. UMI ENABLED COMMUNITIES

Given the ontological commitment of the society domain, any community, e.g. a company, will be the result of the execution of community formation steps (Join, Exchange, Leave activities) as well as proprietary formation steps and refined ontological commitments, which are not shared with society as a whole. For instance, a company may decide to source parts from several suppliers, to assemble them, and then exchange them for money.

In its proprietary conceptual model, the company's enhanced ontological commitment is embodied in a refined classification hierarchy often complemented by an enhanced meta-model, e.g. one that gives consideration also to product and facility structure or product family. Company specific resource sub-classes such as Storage and Walkway, and the Product sub-class Part, illustrate the refined classification hierarchy. The company's collaborations then refer to the enhanced ontological commitment.

In the physical view, the refined classification hierarchy and enhanced meta-model give rise to extended document instances as proxies for the entities within the context of the company. To the extent that the information infrastructure is model-driven and has a proper architecture, any community will be able to reuse society models, and to align its proprietary models with its core competences.

6. CONCLUSION AND FUTURE WORK

This paper has clarified the interwove ness of infrastructure and enterprise level conceptual models within a MDA approach. The UMI architecture description was addressed and briefly illustrated for an abstract society using a fairly simple ontology of individuals. One challenge for future work is to scale up the ontology from individuals to objects with a state-of-the-industry complexity. To this end, piecemeal ontological commitment and multi-strata conceptual modeling must be combined.

REFERENCES

1. OMG Architecture Board ORMSC (2001) Model Driven Architecture (MDA), Doc. Nr. ormsc/2001-07-01.
2. de Miguel, M., Jourdan, J., Salicki, S. (2002) Practical Experiences in the Application of MDA. In: Jezequel, J.-M., Hussmann, H., Cook, S. (eds) UML 2002, LNCS 2460, pp. 128-139, Springer Verlag Berlin Heidelberg.
3. Boehm,B., Egyed,A., Kwan,J., Port,D., Shah,A., Madachy, R. (1998) Applying the WinWin Spiral Model: a Case Study, IEEE Computer, July 1998, pp 33-44.
4. Kruchten, P. (1995) Architectural Blueprints - The "4+1" View Model of Software Architecture, IEEE Software, 12 (6)

5. Maybury, M. (2001) Collaborative Virtual Environments for Analysis and Decision Support, Communications of the ACM, 44(12) pp 51-54
6. Lechner, U. (2002) Peer-to-Peer beyond File Sharing. In: Unger,H., Boehme,T.,& Mikler, A. Innovative Internet Computing Systems, LNCS 2346, pp 229-249, Springer Verlag.
7. Borst, P., Akkermans, H., & Top, J.(1997) Engineering Ontologies. Int. J. Human-Computer Studies, 46 (1997) 365-406
8. Goossenaerts, J., Pelletier, C. (2002) Ontological Commitment for Participative Simulation. In: H. Arisawa, Y. Kambayashi, V. Kumar, H.C. Mayr, I. Hunt (eds) Conceptual Modeling for New Information Systems Technologies. LNCS 2465, Springer Verlag, pp 127-140
9. Bickhard, M.H. (2001) Why Children don't have to solve the Frame Problems: cognitive representations are not encodings. Developmental Review, 21 (2001) pp 224-262
10. Abramov, V.A., Goossenaerts, J.B.M., De Wilde, P., Correia, L. (2003) Ontological Stratification in an Ecology of Infohabitants, this volume.
11. Berkers, F., Goossenaerts, J., Hammer, D.K., Wortmann, J.C. (2001) Human Models and Data in the Ubiquitous Information Infrastructure. In: H. Arisawa, Y. Kambayashi, V. Kumar, H.C. Mayr, I. Hunt (eds) Conceptual Modeling for New Information Systems Technologies. LNCS 2465, Springer Verlag, pp 91-104
12. Popescu-Zeletin, R., Abranowski, S., Fikouras, I., Gasbarrone, G., Gebler, M., Henning, S., van Kranenburg, H., Portschy, H., Postmann, E. & Raatikainen, K. (2003) Service architectures for the wireless world, Computer Communications, 26(1), pp 19-25

4

SERVICE MODELING FOR SERVICE ENGINEERING

Yoshiki Shimomura[1] and Tetsuo Tomiyama[2]
[1]Research into Artifacts, Center for Engineering, the Univ. of Tokyo, Japan
[2]Faculty of Mechanical Eng. and Marine Tech., Delft Univ. of Technology, The Netherlands
e-mail: simomura@race.u-tokyo.ac.jp

Abstract: Intensification of service and knowledge contents within product life cycles is considered crucial for dematerialization, in particular, to design optimal product-service systems from the viewpoint of environmentally conscious design and manufacturing in advanced post industrial societies. In addition to the environmental limitations, we are facing social limitations which include limitations of markets to accept increasing numbers of mass-produced artifacts and such environmental and social limitations are restraining economic growth. To attack and remove these problems, we need to reconsider the current mass production paradigm and to make products have more added values largely from knowledge and service contents to compensate volume reduction under the concept of dematerialization. Namely, dematerialization of products needs to enrich service contents. However, service was mainly discussed within marketing and has been mostly neglected within traditional engineering. Therefore, we need new engineering methods to look at services, rather than just functions, called "Service Engineering." To establish service engineering, this paper proposes a modeling technique of service.

Key words: Service Engineering, Service Modeling, Service CAD, Artifactual Engineering

1. INTRODUCTION

Constraints and limitations imposed by natural resource availability, energy supplies, and the ability of the Mother Nature to accept industrially generated waste, have led to the obvious problem of making, using, and disposing of increasing numbers of artifacts. In addition to these environmental limitations, social limitations are also evident, as are limitations of markets to accept increasing numbers of mass-produced artifacts. Consequently, unless we develop revolutionary technologies, environmental and economic limitations will severely restrict economic growth. To fundamentally attack and remove these problems, we need to reconsider the current mass production paradigm and to pursue a new manufacturing paradigm. The new paradigm should reduce the production and consumption volume of artifacts to an adequate, manageable size and bring this volume into balance with natural and social constraints. This new idea is called the Post Mass Production Paradigm (PMPP) and it aims at qualitative satisfaction rather than quantitative sufficiency and decoupling economic growth from material and energy consumption [1-3]. To achieve PMPP, the current practices, methodologies, and tools to design artifacts should be revisited. Products should have more added values largely from knowledge and service contents, rather than just materialistic values, to compensate volume reduction under PMPP [2, 4].

This dematerialization of products requests to enrich service contents. To this end, we need engineering methods to look at services, rather than just materialistic values, called service engineering and to intensify service contents of product life cycles.

This paper focuses on a service modeling method which is needed in the first step of the Service Engineering. Firstly, we clarify why service should be discussed and argue that intensifying service contents is crucial not only for arriving at environmentally conscious design and manufacturing but also for creating more added values in future advanced societies. Secondly, we present essential concepts of service engineering that is required to intensify service contents. Furthermore, we discuss about subjectivity of service. In general, service is related not only to objective elements but also to the receivers' perception of service contents. In this context, we propose a service modeling method which can represent services with such subjective property of service by clarifying actual service cases.

2. SERVICE ENGINEERING

Service is generally perceived as an activity that changes the state of a service receiver [5]. This means that service cannot be stored as opposed to materials and it disappears instantly. Second, to conduct services, we need certain items, including service channel to deliver and amplify services and service contents to be delivered. Usually, artifacts play roles of service channels or service contents.

Figure 1 defines service within a framework of service provider, service receiver, service contents and service channel. A service receiver receives service contents from a service provider through a service channel in order to change own states. Service contents are material, energy, and/or information. The service channel in delivering service contents consumes material, energy, and information. These imply that service cannot be free from environmental impacts. Service sent by the service provider changes the state of the service receiver, which is the most important feature of service as activity.

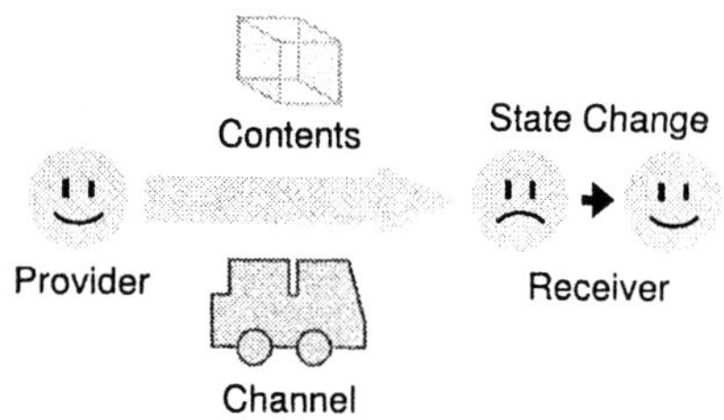

Figure 1. Definition of Service

In this definition, artifacts are devices to deliver, automate, or amplify services used as service channels. Artifacts have their own function, behaviors, and states. The generated total added value of a service is given by the function of the service channel and the quality of the service channel which may includes capacity (amplitude), efficiency, and channel quality (e.g., timeliness, frequency, punctuality, etc.). And service engineering aims at intensifying, improving, and automating this whole framework of service generation, delivery, and consumption. To increase total added value of a service, we can either improve function or quality. Traditionally, engineering design aimed at improving only function. However, this does not suffice in light of service engineering that should look at improving also total added value of a service. For instance, we need to address not only customization of products but also customization of service delivery.

3. SERVICE MODELING

In analogy with product development, service development could be facilitated with a variety of engineering tools and methodologies that can be called service development engineering. For instance, we may consider service modeling inspired by function modeling [6] and this will serve as a basis for establishing service engineering to design services. Namely, such service design assisted with a service modeling tool can help a service designer to design innovative services with more added values and less environmental impacts. According to the definition of service in Figure 1, we can outline some of its elements as follows.

First, service is an activity to change the state of the service receiver. This means that we also need to define activity. Our working definition of activity is that an activity is a series of actions or procedures performed by the involved participants (i.e., service provider and service receiver). Service has the following elements to be modeled. Having defined service elements, we might be able to develop a service modeling tool of which implementation is yet a research issue. It should be equipped with knowledge bases storing knowledge about activities, service environment, and service channel. We are currently planning to implement such a modeling tool.

As we mentioned above, service is defined as an act of a service provider, who is a supplier of service, to cause the service receiver's state change, which is the purpose of service. Service contents are defined as elements of service, which cause the receiver's state change directly in supply of service. A service channel is defined as a device that indirectly contributes to the receiver's state change by delivering, supplying, and amplifying. Service contents and channels are realization methods of service.

Given a purpose to achieve, to synthesize its realization method can stand in a contrast with conventional design of artifacts. Design of artifacts is a process to synthesize physical realization methods for given functions. This correspondence between service design and conventional artifact design suggests that methods to design artifacts (in particular, those for functional design) can be applied to service design.

The state of a service receiver can be represented by parameters. Therefore, these receiver state parameters (RSPs) can represent the receiver's state change that is the result of service. Representing the service receiver's demand with the RSP is the most crucial process in service design. A service content has also parameters. Those parameters that directly influence the service receiver's state change are called content parameters (CoPs). A service channel contributes indirectly to the service receiver's state change as well. A parameter of a service channel is called a channel parameter (ChP).

3.1 Subjectivity of Service

As discussed above, service design has a similarity with artifact design in that functions of artifacts corresponds to service purposes (i.e., the service receiver's state change), and physical mechanisms to service realization methods (such as service contents and service channels). However, even though given the same service as an action, the effect of the service could be different, depending on how the service receiver recognizes the service. Therefore, just like functions are subjective [6], service is also very much subjective and depends on the viewpoints of the service provider and of the service receiver. These two viewpoints must be considered in modeling service.

Because of this subjectivity, the division between the service contents and service channel could be sometimes unclear; *i.e.*, it depends on the viewpoints of the service provider and service receiver, or a service content can sometimes be recognized as a service channel..

3.2 A Model of Service

In building a model of service, it is essential to take into consideration the information about the viewpoints of the service provider and of the service receiver. For this purpose, we here propose a model of service with two assistant models that are called a view model and a scope model (see Figures 2).

A view model of a service represents the viewpoints of the service provider and should include service components and the relationships among them; namely, RSPs, service contents, and service channels in the functional context. Figure 2 illustrates a view model of cooling service containing relations between RSPs that describe the receiver's state changes, and CoPs and ChPs that trigger these state changes. Here, service contents are cooling function while the service channel could be a refrigerator. Functions of a service channel and a service content are represented with a set of function name (FN), function parameter (FP), and function influence (FI). FP is a main target parameter of the function and FI is a main influence of the function to the FP.

In general, functions in the view model are related to other function's FPs or RSPs. We define that a CoP is a FP that is directly related to RSP and a ChP is a FP that is indirectly related to RSP through other FPs. We consider that the essence of service design is to build such relationships among these service elements through functional structure containing RSP, CoP, ChP, and relations among these parameters. Because we also include artifacts in this functional expression of service, we can obtain images of artifacts as service channels or service contents.

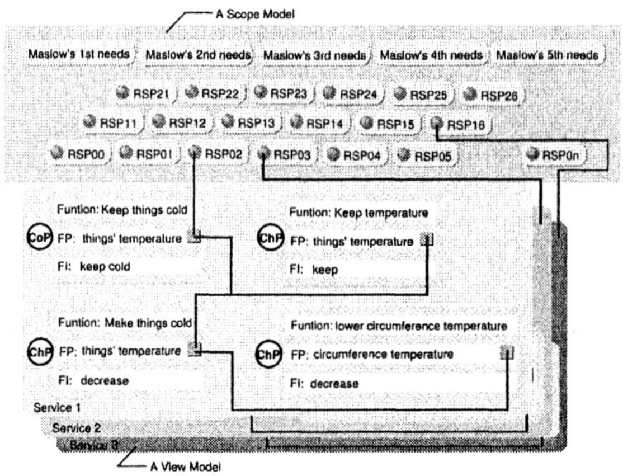

Figure 2. The View and the Scope Model of Cooling Service

Table 1. An Image of the Service Matrix

Service Name	ID No.	Entry Person	RSP	RSP Description	Needs Category	View Model
Manufactuiring / selling a refrigerator	001	xxxxxxxxxx	RSP02		Maslow's 1st Needs	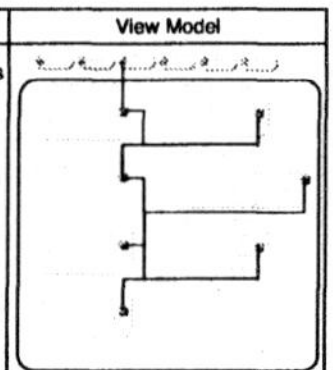
Manufactuiring / selling a refrigerator	002	xxxxxxxxxx	RSP03		Maslow's 1st Needs	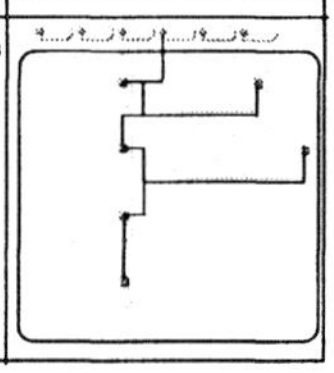

On the other hands, a scope model of a service represents the viewpoints of the service receiver. Because in reality services can form complicated structures with multiple service providers and service receivers, it is necessary to specify the scope of a service for which the service is intended. A scope model expresses a target area of the current service and represents relations between different services through relations between RSPs (see Figure 2).

In Figure 2, we used *Maslow's needs hierarchy* [7] to give rough classification for needs to represent the service receiver's viewpoints. This scope model represents the relationships among multiple RSPs.

By modeling service based on the above-mentioned definitions, we can arrange existing service cases with a matrix form in which the properties of services are used as keys and it is possible to use this matrix as a basic data for service design. Table 1 shows an image of this *service matrix* data. The cases of services in this service matrix can be organized with various kinds of key information combination (e.g., artifacts, changes of the receiver, functional structure as a realization method of a service, CoPs, ChPs, etc..) By arranging actual service cases in this way, we can generate service data that has high reusability. In addition it is possible to develop a new CAD system for service design by applying various design operations on the service cases represented in this data scheme.

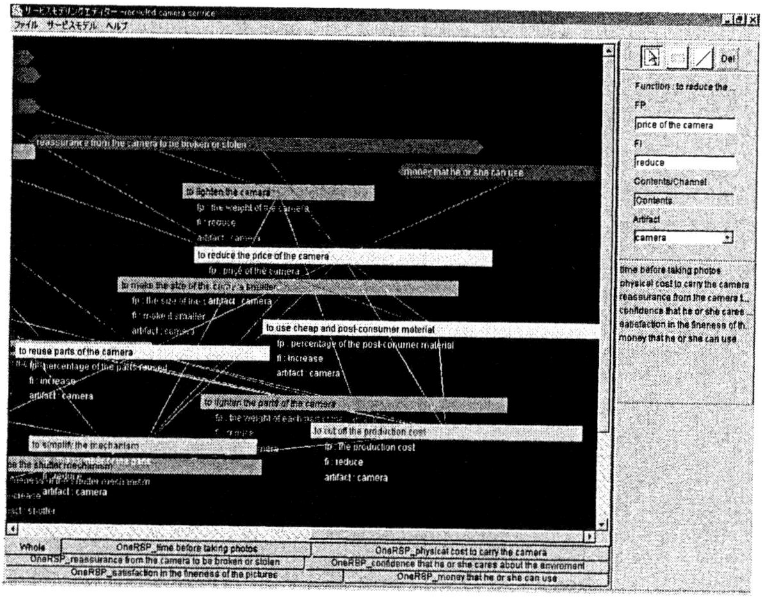

Figure 3. A Screen Shot of the Service Explorer

4. SERVICE EXPLORER

Based on above-mentioned definitions of our service model, we have developed a prototype system of computer aided service modeling tool, which is called Service Explorer. The current version of Service Explorer is developed using Borland JBuilder7 with Java SDK version 1.4.1 and XML version 1.0 on Microsoft Windows2000 environment. Figure 3 shows a screenshot of the prototype system. By using this service explorer, we can represent various services on a computer with the scope and the view model description. In Figure 3, a view model of 'disposable camera service' is expressed in the following manner. A hexangular node mean a RSP of the 'disposable camera service' and the other square nodes represent the functions with its FN, FP and FI in this service.

5. CONCLUSION

Motivated by the concepts of PMPP, this paper proposed a modeling technique of services that is essential to establish service engineering. First, we introduced fundamental concepts of service engineering which aims to intensify service contents (servicification) for creating more added values in future advanced societies and explained the definition of service in the proposed service engineering. Second, we explained the concepts of service modeling by identifying elements of service and then proposed a modeling technique of services introducing a view model and a scope model. A prototype system for service modeling is illustrated.

While many of these ideas are yet to be implemented, this paper is our first attempt to scientifically deal with service. In the future, a service designing tool should be developed on which service design engineering, service production engineering, and service development engineering will be established.

REFERENCES

1. T. Tomiyama, T. Sakao, Y. Umeda, and Y. Baba, "The Post-Mass Production Paradigm, Knowledge Intensive Engineering, and Soft Machines," in Krause, F. L. and H. Jansen (eds.), Life Cycle Modeling for Innovative Products and Processes, Chapman & Hall, London, 1995, pp. 369-380.
2. T. Tomiyama, "A Manufacturing Paradigm Toward the 21st Century," Integrated Computer Aided Engineering, Vol. 4, 1997, pp. 159-178.
3. T. Tomiyama, "The Post Mass Production Paradigm," in Yoshikawa, H., R. Yamamoto, F. Kimura, T. Suga, and Y. Umeda (eds.), Proceedings of the First International Symposium on Environmentally Conscious Design and Inverse Manufacturing (EcoDesign '99), February 1-3, 1999, Tokyo, IEEE Computer Society, 1999, pp. 162-167.
4. T. Tomiyama: "Knowledge Intensive Engineering towards Sustainable Products with High Knowledge and Service Contents," in Horvath, I., A.J. Medland, and J.S.M. Vergeest (eds.), TMCE 2000, Third International Symposium on Tools and Methods of Competitive Engineering, April 18-20, 2000, Delft University Press, Delft, the Netherlands, 2000, pp. 55-67.
5. T. Tomiyama: "Service Engineering to Intensify Service Contents in Product Life Cycles," in Proceedings of the Second International Symposium on Environmentally Conscious Design and Inverse Manufacturing (EcoDesign 2001), December 11-15, 2001, Tokyo, IEEE Computer Society Order Number PR001266, IEEE Computer Society, (2001), pp. 613-618.
6. Y. Umeda, M. Ishii, M. Yoshioka, Y. Shimomura, and T. Tomiyama: "Supporting Conceptual Design based on the Function-Behavior-State Modeler," Artificial Intelligence for Engineering Design, Analysis and Manufacturing, Vol. 10, No. 4, 1996, pp. 275-288.
7. A. H. Maslow: Toward a Psychology of Being, 3rd Ed., New York, Wiley, 1998.

5

THE EXTENDED PRODUCTS PARADIGM, AN INTRODUCTION

Kim Jansson[1] and Klaus-Dieter Thoben[2]
[1]*VTT Industrial Systems, VTT, Finland*
[2]*BIBA, Bremen Institute of Industrial Technology and Applied Work Science, Germany*
e-mail:kim.jansson@vtt.fi

Abstract: This paper explains and illustrates the extended product paradigm. The main dimensions to extend a product to be considered are the product life cycle, the range of product related information and the resources requested to provide extended (i.e. enterprise networks) on the market. The paper is based on work done in the European Information Society Technologies (IST) project Extended Products in Dynamic Enterprises - EXPIDE. The project objective is to bring together research projects, stakeholders (such as suppliers, manufacturers, consultants, service and technology providers, etc.) involved and interested in the provision of extended products, etc. The project EXPIDE aims to identify future value creation potentials as well as future needs related to the extended product paradigm. The Intelligent Manufacturing Systems (IMS) Global Engineering and Manufacturing in Enterprise Networks - GLOBEMEN project contributes to the specification of the Extended Product. The GLOBEMEN project results, especially in the sales and service life cycle phases, illustrate in practice the extended product paradigm. This paper delivers examples from the GLOBEMEN project.

Key words: Extended Product, Value adding, Service

1. INTRODUCTION

The arising new digital society does radically change the way industrial enterprises deliver goods and services. Digital business is one of the forces behind emerging new concepts for products, new approaches for production and in fact a promising way of new value creating processes for the future.

Due to the open markets and an increased competition producers are facing a tremendous pressure. To keep the worldwide leadership in product quality, manufacturers have to further develop their products in terms of new customer value creating concepts. Today for all kind of industries the key challenge is to come up with new integrated product approaches, which must allow new quality products, organisational innovation and the efficient management of information and its transformation into new value adding services, Figure 1. However, new integrated product approaches are needed which fits to the paradigm and the technologies of today's information and knowledge society.

Accordingly, compared to the past, providing a single (tangible) product is not sufficient anymore to be competitive on a global market. Today customers are requesting for services, benefits or even requesting a guaranteed success when buying a product. Accordingly the customer focus has moved on from the ownership of a physical product or systems towards guaranteed benefits based on a provided offering. Therefore the so-called core product needs to be extended or packaged up with different kinds of value adding services, based on new and ICT. These intangible product assets, which are very often information and knowledge intensive, can consist of engineering, software, maintenance, customer support services and many others. The extended product paradigm represents such an integrated approach.

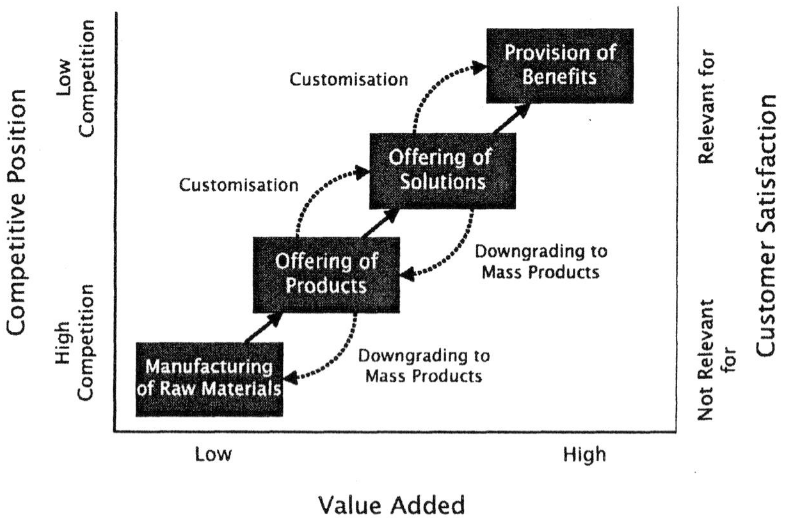

Figure 1. The Progress of Competitive Advantage (elaborated from Pine, 1999)

2. EXTENDED PRODUCT – DEVELOPING THE TERM

According to literature there are various views on how to "extend" a product:

- Extended Product Responsibility is an emerging principle for a new generation of pollution prevention policies that focus on product systems instead of production facilities. Although supportive, out of the scope.
- The terms of Extended Product Life intensified product use and extending product lifetime should emphasise concepts of a closed-cycle industry. The terms point out that the value given to a product in manufacturing should be used as intensely as possible and maintained as long as possible.
- Extended Product Support is a term used by some software provides e.g. help desks open 24 h/day and 7 days/week. This approach could easily be adapted to other types of products.
- This term Extended Product Information is used for pharmaceutical and natural product content description

However, the EXPIDE project has focused on elaborating a specification for the term Extended Product which is related to manufacturing. According to EXPIDE the term includes the following elements:

- A combination of a physical product and associated services / enhancements that improve marketability.
- Intelligent, highly customised, user-friendly tangible product assets such as embedded features like maintenance.
- Intangible product assets, which are information and knowledge intensive
- Customer focus is on a value-added service or guaranteed success and not anymore the physical product.

Production activities are normally caused by the needs of potential customers. Extended products can even help to define or identify new customer needs. Servicing the customer means also supporting the customer in deriving new business ideas. Consequently product suppliers have to extend their offerings in a dramatic way. Suppliers have to offer everything, which might support the customer in achieving the goals of the new business ideas.

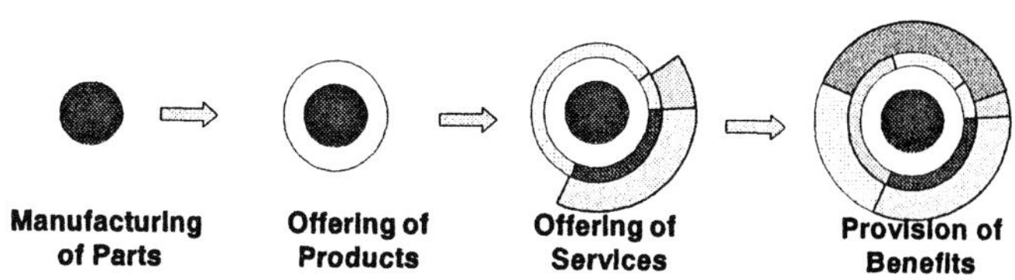

Figure 2. Changing the focus: From manufacturing of parts to provision of benefits

3. EXTENDED PRODUCTS CHARACTERISTICS

Figure 3 shows the three main dimensions to be considered when characterising extended products. The "perfect" extended product contains complex information (incl. embedded systems, knowledge, etc.), is produced by dynamic, complex enterprise networks (incl. concepts like the extended enterprise, the virtual enterprise, etc.) providing services over several phases of the life-cycle. The following chapter briefly discusses the three dimensions illustrated in figure 3.

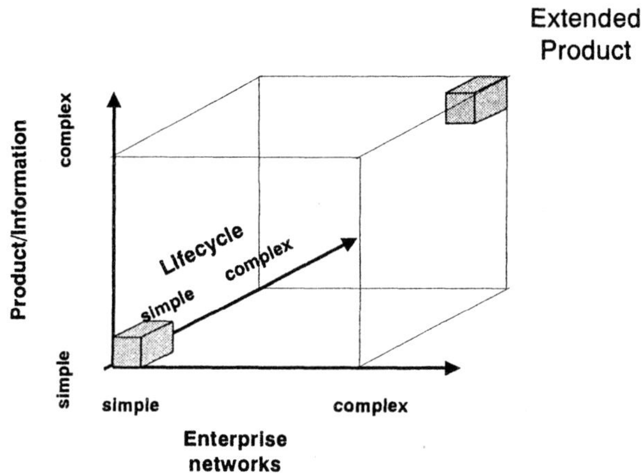

Figure 3. Characteristics of extended products

3.1 Increased Product/Information Complexity

In Figure 4 a layered model has been developed in order to structure the extended product approach. The three rings can be described as follows:

- There is a core product which is closely related to the core function (s) of a product. As an example the core function of a car, which is the ability to move objects from location A to location B is enabled by parts like the engine, wheels, etc
- The second ring describes the "packaging" of the core functions, the features that makes the product attractive to the customer. In the car example this could be a nice design, a nice colour painting, air conditioning etc. From a functional point of view these features are not really necessary. However, these features are the drivers for marketability of the product.

- The third ring summarises all the intangible assets of the product. Intangible assets surrounds the tangible product. In the car example this could be an efficient 24h/7day service network. To increase the competitiveness by offering added value for the customer nowadays OEMs have to provide ever more intangible product assets.

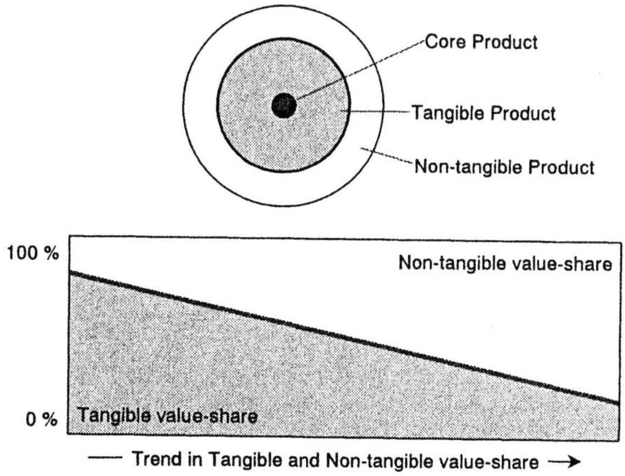

Figure 4. Concept of the three rings

The lower graph in figure 4 illustrates how the intangible value-share volume of the product is expected to increase over time. In addition the information content grows as product gets complex. For example in a modern paper machine the share of automation systems and software will soon correspond to more the half of the costs.

3.2 Life-cycle information of the extended product

Today the lifecycle of a complex product includes phases like: Marketing & Sales, Design, Implement, Operate & Service, and Demolish. However the life cycles of large one-of-a-kind products can be much more complex. The operational phase by far is the longest phase and it may include everything ranging from small modifications and repair work to large modernisation and redesign works. Even the purpose or main functions may be changed. For example in a chemical plant the main product can be changed.

In future a successful management of a product along its entire life-cycle must be based on extensive information and communication management. Thus the product continuously needs to be accompanied with life-cycle history, status and usage information. The cumulated life-cycle information can also be used to improve delivery and the functionality in future related products and their life-cycle phases.

3.3 Network

Extended products tend to be complex and therefore difficult to produce by a single enterprise. Extension to products will often, as mentioned, constitute physical products as well as associated services or accessories. The increasing complexity of products, the consideration of the full life-cycle and the concentration on core competencies (incl. the availability of appropriate ICT) make it difficult to produce extended products by a single company. Thus, depending on the core competencies required to supply associated as well as appropriate services, infrastructures, etc., several business partners are required to collaborate very closely towards a common goal: the delivery of the extended product. For example, the service of a process plant can be provided by a service virtual enterprise -- see case Bühler below.

4. EXTENDED PRODUCTS IN THE IMS GLOBEMEN PROJECT

The IMS Globemen project focuses on issues related to global manufacturing in enterprise networks. Globemen is an international project with partners from several IMS regions: EU, Australia, Japan and Switzerland. The approach of Globemen is to address three main aspects of manufacturing: sales and services, inter-enterprise management and distributed engineering. Based on industrial requirements specifications the work will be co-ordinated and integrated into a Virtual Enterprise Reference Architecture and Methodology, VERAM.

The extended product paradigm is not explicitly in the focus of Globemen, however the industrial case studies conducted within the project give proof of the concept. The Globemen case studies are explained in more detail in other related papers and presentations during the DIISM 2002 conference. The Japanese region case studies have a major role in DIISM 2002. Accordingly this chapter will only highlight the extended product features of three European cases.

4.1 E-service extension

Bühler AG Switzerland is a global developer, manufacturer, and supplier of machines, installations, and systems in the food processing industry. It is increasingly important for producers of one-of-a-kind machinery or plants to offer after-sales services, which provide the best possible support to customers across the globe in the use of their machines and plants. If production problems or breakdowns occur, it is imperative that the producer respond rapidly and offer suitable help measures.

In those cases where the customer is located far from the producer, it will in future no longer always be possible for the producer to offer the assistance of its own company service technician, for costs in time and money are too high. One possible solution consists in providing the customer or local service partners sufficient support that they themselves can execute service tasks under the direction of the producer through the aid of modern ICT. Applications like Video Conferencing combined with Application Sharing, Wearable Computing, and interactive Internet applications (trouble shooting guides and animated manuals, for example) can conceivably be put to use.

These services offered by Bühler is good example of intangible extension existing products. The operational users of the product can highly benefit from additional services offered. (Hartel)

4.2 A knowledge creation environment to supporting the tendering phase of Extended Products

Fortum Engineering (FE) is part of Fortum Group Finland, a diversified group of companies concentrating on the energy industry and focusing on countries around the Baltic Sea. FE acts as an engineering, procurement, and construction provider for power plants.

In such investment projects the investor, constructor and operator are the principal players. They also own and produce the information that is most helpful during the inception of a new project. FE has developed a knowledge creation environment with the objective to increase competitiveness by connecting the tacit knowledge of investors and operators with FE's own experience and offering FE possibilities to develop new features for its power plants. The effective use of modern ICT is a method by which FE expects to create a specific knowledge to support the client during the inception of the investment and, thus, pave the way for successful tenders.

This is a good case example of how knowledge related to a complex product over several of its life-cycle phases can be used in subsequent product deliveries. The service provided to the customer already in the tender phase is the ability to offer improved products based on cumulated experience from previous deliveries. The knowledge can of course in addition be used throughout the delivery project.

4.3 Product models as a mean to provide life cycle support for extended products

The YIT Group Finland provides total service for construction, industry and telecommunication networks. The company offers residential, property, infrastructure and industrial investment and maintenance services. The YIT Group's service chain spans the entire life cycle of the investment, from design and implementation to continuous maintenance and operation.

In construction project the company uses product models throughout the delivery process from architectural design, cost estimation, distributed detail design, scheduling, and procurement to installation. The comprehensive product model is based on international standards. The model is constantly kept up to date and the level of detail is increased by model merging form different partners.

The complete intangible product model is thus an extension to the actual physical product, which can be used for a variety of maintenance functions by the customer or the service branches in the YIT Group.

5. CONCLUSION

The evident and ongoing trend in the one-of-a-kind business environment is to develop value adding services to product. The objectives of suppliers are to offer these services over the product life-cycle. The idea is not only to provide benefits for the customer, but also to learn and acquire knowledge also for the own company and business partners. The paper has presented the concept of extended product and the associated layer model. The three dimensions network, information and life-cycle that make up and extended product are explained. The paper also explains how the extended product can be seen in three industrial cases. The practical cases show that the extended product concept is a useful approach for creating value adding e-business services.

Further approaches, methods and tools are needed to make to most of the concept e.g. to design the features of extended products, to design enabling infrastructures for the provision of extended products and to compose enterprise networks for the provision of extended products.

ACKNOWLEDGEMENTS

This work has been partly funded by the European Commission through IMS Project GLOBEMEN (IST-1999-60002) and EXPIDE (IST-1999-

29105). The authors wish to acknowledge the Commission for their support. We also wish to acknowledge our gratitude and appreciation to all project partners for their contribution to the development work.

REFERENCES

1. Klaus-Dieter Thoben, Pierre Kirisci, Sebastion Kicin, Jens Eschenbächer and Paul Higins. Holistic approach for structuring the various facets of e-business in enterprise networks. Proceedings of the 8th International Conference on Concurrent Enterprising. Rome 2002.
2. Ingo Hartel, Stephan Billinger, Gerhard Burger and Yoichi Kamio, Virtual organization of after-sales service in the one-of-a-kind industry, PRO-VE'02, 3rd IFIP Working Conference on Infrastructures for Virtual Enterprises, May 2002, Sesimbra, Portugal
3. Kim Jansson, Martin Ollus and Iris Karvonen. Methods and IT Models to Support Information Management in the Sales and Service Life-Cycle Phases of a One-of-a-kind Product. Automation 2001, Automation days, Helsinki 2001
4. Bernd E. Hirsch, Klaus.-Dieter Thoben and Jens Eschenbaecher. Using e-business to provide Extended Products. Automation 2001, Automation days, Helsinki 2001
5. Bernd E. Hirsch, Jens Schumacher, Jens Eschenbächer, Kim Jansson, Martin Ollus, Iris Karvonen Extended Products: Observatory of Current Research and Development Trends. Proceedings of E-work and E-commerce conference, voume 1, Venice 2001
6. Joseph Pine II and James Gilmore, The Experience Economy – Work is Theatre & Every Business a Stage; Harvard Business School Press, 1999
7. The EXPIDE project, www.expided.org
8. The GLOBEMEN project, globemen.vtt.fi

6

PROCESS PLANT INFORMATION INTEGRATION IN THREE DIMENSIONS

Iiro Salkari, Kim Jansson, and Iris Karvonen
VTT Industrial Systems, VTT, Finland
e-mail: iiro.salkari@vtt.fi

Abstract: The paper discusses the management of product information of complex one-of-a-kind products. Information and knowledge can be seen as important resources for design, construction, operation and modernization of process plants. The research problem is approached through three cases that form the basis for a generic conceptual model for information integration over process plants lifecycle. The model for managing and integrating information includes three dimensions: between interest groups, inside plant lifecycle and between plants or plant projects. This paper is based on a project called Novel Information Integration Techniques and Models (NIITM).

Key words: Process Information, Information Integration, Plant Projects

1. INTRODUCTION

The process industry has increasing challenges concerning environmental friendliness, profitability and efficiency. Competition has changed from local and regional to global. In this kind of business, increasing collaboration between interest groups from different phases of the lifecycle of the process plant is needed. Operating in a network becomes the driving force behind successful companies. Common feature to this networked collaboration is creation of information and its use and reuse. Systematic information management practices and procedures enable this.

The key issue is effective utilization and integration of the information that has accumulated over the lifecycle of the process plant and during previous plant projects. Typically decisions that are made in one stage of the lifecycle have far-reaching effects on the forthcoming stages and eventual flaws in these decisions are expensive and difficult to fix later [1]. Thus, by exploiting this information, the interest groups can intensify their operations and thereby meet the future expectations and gain competitive edge in respect of competitors. Effective utilization of the accumulated information requires communicating the information bidirectionally over the lifecycle, between projects and interest groups.

2. PRESENTATION OF THE CASES

The first case (Alma Software) is an IT-system provider that develops a software tool for automation design, process electrification design, field instrumentation planning and managing related information during operation of a plant. The system also provides tools for managing plant's maintenance. The main objective of the system is better usability of once produced information during plant lifecycle by integrating information management in design, delivery, operation and maintenance. The case provided knowledge about typical design tools and IT-systems that are used during plant projects and in operating plants. The case introduced an interesting approach to plant information management in general.

In the second case (Kemira Chemicals Kokkola Plants), the aim was to develop plant design information management during maintenance of a chemistry plant and thereby intensify the entire maintenance process. Also information management during the plant delivery was dealt with since information that is produced then, forms the basis for planning activities during the maintenance. The chemistry plant consisted of two units: one that manufactures bulk chemicals and other that produces small amounts of chemicals according to customer order. The studies during the case were mainly carried out in the first one. The case provided the operator view to the plant information management process during operation and maintenance.

The third case took place at Power IT, which is a part of the PVO Group. The PVO Group together with its subsidiaries operates, designs and takes overall responsibility of plant deliveries. Power IT's role is to provide information systems and tools that support these other functions. Typically, the company deals with several other companies in its business and has noticed the commonly known development areas in plant design information management. In our case study, the objective was to clarify the information management process from design to the delivery of the plant. This included defining different interest groups, their roles and requirements for information, information contents, and IT-systems. The case offered views concerning the delivery process and related information management between the (main) contractor(s) and the operator. Table 1 summarizes the major development areas and actions of the cases.

Table 1. Major development areas and actions in the cases

Case	Development areas	Development actions
1	New ideas on how to develop the product further	Based on other cases and the analysis of Alma tool, a "future requirement specification" was composed. The idea is that the document supports further development of the tool: it gives some guidelines and ideas on what areas should be developed.
2	Better information management during maintenance	Short term actions: o Several new working practices that promote cooperation and bringing the information commonly available o Intensified use of existing IT Long term actions: o Improvement ideas for IT o Improved communication during bigger projects
3	Better management of design information during a plant delivery project	Modeling the overall plant delivery project process, developing the process further and agreeing on common working practices and tools during the project. This resulted in an aim to continue with a new project during which an IT-system that supports the new practices and tools will be developed and taken into use.

3. ANALYSIS OF THE CASES AND GENERIC FRAMEWORK FOR INFORMATION MANAGEMENT

The following chapters analyze the cases and at the same time build a generic framework, which is depicted in figure 1. The objectives of the generic framework are to model the different information flows and clarify the relations between information integration dimensions. It provides one way for planning information management of plant projects and plants.

3.1 Interest Groups and Boundaries between them

The generic framework presents four types of interest groups: (machine) supplier(s), engineering office, operator and maintenance service provider. The boundaries between these interest groups may be internal or external and the strength of the boundaries may vary. This means that certain interest groups may belong to the same company (for example Kemira Kokkola Plants and Kemira Engineering in the second case) or they may belong to the same network or virtual enterprise (VE). Hardwick and Bolton [2] define VE as follows: "virtually (temporarily) united to exploit a worldwide business opportunity, a consortium of companies manufacture products together non could build alone." Since plant delivery projects require special knowledge from several areas, VE can be seen as a potential organizational way to carry out a plant project.

In the third case the organization during a plant delivery often nears VE depending of the definition of VE: the companies were not strategic partners but they cooperated on regular bases. If the interest groups belong to the same company or cooperate on a regular basis, the boundaries between them are not so strong as if they were separate firms without history of cooperation. Operating in a network results in fading of the internal and external boundaries of the networked organizations [3].

Thereby, three different kinds of boundaries can be derived from the cases: 1.) internal between different functions of the same company, 2.) internal of a virtual collaboration network (for example a VE) and 3.) external. When aiming at improved management of plant information between the interest groups, the different organizational boundaries, as well as, the nature of these boundaries should be recognized. In this way, the possible problem points in working over the boundaries can be found and the problems mitigated.

The interest groups that the framework presents are generic in this sense. The number of interest groups may vary in actual projects. In the framework additional interest groups – for example two cooperating engineering offices – would fall under the heading "engineering office". This approach suggests that these two interest groups should be treated equally in respect of the plant information. Collaborating interest groups (who have for example formed a VE) can also be taken care of by the framework: The boundaries of a VE can be set in the model so that different interest groups can form VEs. The boundaries between the VE and other companies and inside the VE can be treated similarly to internal or external boundaries. The boundaries inside the VE are internal, the boundaries between the VE and other firms are external.

3.2 Generic and Specific Plant Information and History Sets

In the framework, the content of plant information is in the roughest level divided into generic and product or project specific information (figure 1). The idea is that the generic information is something that provides the basis for project or product specific planning. The idea of generic information sets comes up in the ALMA Software case in the form of standards that support engineering and from a project information management model presented by Karvonen [4]. Also in the third case, the need to increasingly manage the experiences from previous projects is recognized. The generic information sets in the framework provide one way to do this.

If we look the difference between the generic and specific information from a supplier viewpoint, for example, the generic information forms the basis for constructing and designing a specific product. Specific information is the information that is related to this specific product. The supplier tries to keep the specific product information up-to-date over the lifecycle of the product. If something novel, and applicable also in other projects, came up during the lifetime of a specific product, the supplier may move this information to his generic information set. Thus the information can systematically be utilized in forthcoming projects. In this way, the specific information can be made common within the company.

3.3 Dimensions of Managing Accumulating Information

The following dimensions of information flows were found when studying the cases: down- and upstream between interest groups; between ongoing projects and between operating plants; from generic to specific product knowledge and vice versa. The following generic information flows depict more precisely the dimensions (letters A-E refer to the framework that is presented in figure 1).

A) Accumulation of specific information: The design of a new product and plant begins with studying the generic information that exists from the product or related products. This generic information is based on experiences from the previous projects and from the operation of plants that are taken into use. When the customer specifications are added to the generic information, the project and thus product specific information begins to accumulate.

B) Accumulation of generic information: The generic information accumulates when experiences from the previous projects or from already completed related projects or operating plants is gained and refined into such a format that it can be used to develop the generic information further.

C) Accumulation of information downstream between different interest groups: Accumulation of information downstream means that the information is communicated from earlier interest groups to later ones in a usable format. Communicating the information in this direction enables efficient utilization of the product immediately after the delivery and ensures that the product is used correctly.

D) Accumulation of information upstream between different interest groups: When the plant and its equipment are operated, plant information inevitably accumulates to the operator. Therefore, the operator soon has the best knowledge of the product he uses. This applies when information is systematically gathered and maintained during operation. Operation is typically the longest and thus, from information management viewpoint, the most challenging period in the plant lifecycle. The plant information that has accumulated during the lifecycle of a plant should be communicated upstream to the previous interest groups who then should manage to make this information into common and generic product information for themselves. When information is communicated upstream it helps the interest groups in the early lifecycle phases to develop their products according to the needs and experiences that the users of the products have. Also information about the modifications that are made during later stages of the product lifecycle help the manufacturers and designers to avoid repeating the same mistakes in their forthcoming products and projects.

E) Accumulation of information between projects and plants: Information can be communicated between ongoing projects and between operating plants directly without making it generic or it can be communicated by making the specific information into generic. The difference between these is that in the former one, the information remains specific to few cases whereas in the latter one the information adds one new layer to the generic information history set and is thus usable for all forthcoming projects.

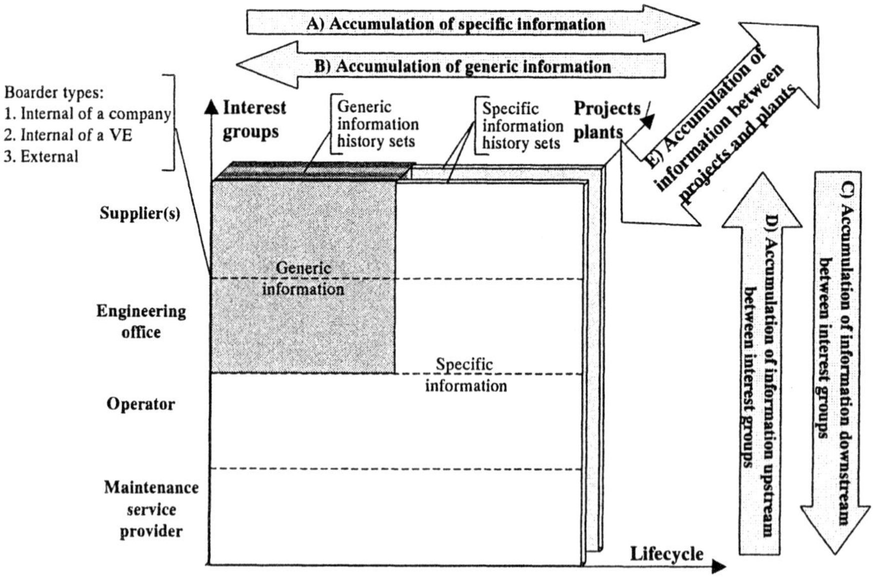

Figure 1. Generic framework for information integration over process plant lifecycle

3.4 Information Flows and Contents during the Lifecycle

Opening the framework up to a more detailed level provides a view to the actual generic information flows that were derived from the cases. The second level of the framework is depicted in figure 2.

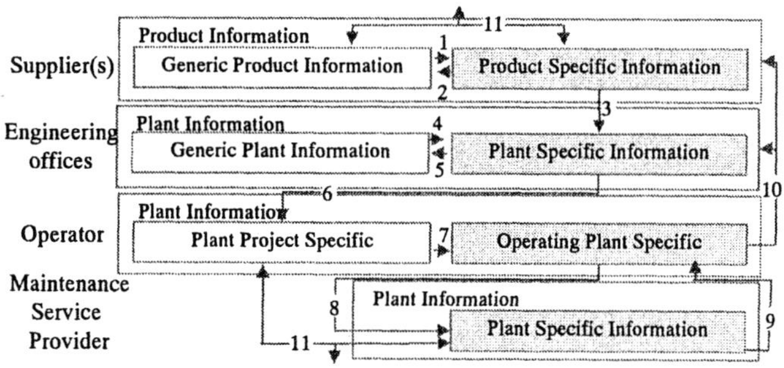

Figure 2. Generic information flows over the lifecycle of a process plant

The information flows that are numbered in the framework's second level with numbers 1 to 11 are clarified in table 2. In the table, the information content is shortly described as well as the type and dimension of the information.

Table 2. Generic contents of the information flows of the framework

No	Type of the information	Dimension of the flow	Content
1	As designed	A	Generic information gained during previous similar or related projects. Based on this generic information new specific projects can be planned
2	Accumulated as designed	B	Such specific information of some product (machine) that can be reutilized in other similar or related projects
3	As designed	C	Information about some machine that is included in a plant on reasonable level of detail and in usable form: accuracy of information is such that it can be used as a part of the plant information
4	As designed	A	Generic information gained during previous similar or related projects. Based on this generic information new specific projects can be planned
5	Accumulated as designed	B (E via generic)	Such specific information of some product (plants) that can be reutilized in other similar or related projects
6	As built	C	Information from engineering office about a specific plant including several machines. Management of information from those who have realized the plant project to the operator(s)
7	As built	A	Operator takes the new plant into use and experiences start

No	Type of the information	Dimension of the flow	Content
			to accumulate
8	As operated	C	Information from operation to maintenance
9	As maintained	D	Information about maintenance to operation
10	As operated and as maintained	D	Information from operation and maintenance to previously involved interest groups
11	As operated and as maintained	E	Information (experiences & solutions) from one plant, which is utilizable in other plants (information flow goes either via the generic information set or directly between plants)

4. CONCLUSIONS

The framework models the information management processes, identifies on generic level the typical interest groups, information flows and the contents of these flows. It presents the idea of generic and specific information together with history sets. It also stresses the different dimensions of information management. Thus the framework supports planning information management over the lifecycle of many different kinds of plants and plant projects.

The framework can also be used as a preliminary framework for designing information management systems:it gives guidelines for designing the processes that the plant design information system should support.

5. SUGGESTIONS FOR FURTHER RESEARCH

Structuring experience based information so that it can be communicated efficiently and systematically. Study how different interest groups' strategies affect on the plant information management.

REFERENCES

1. Tommila Teemu, Viitamäki Paavo, 1991. Vaatimusmäärittely prosessiautomaatiossa – Lähestymistapoja esi- ja perussuunnitteluun, Espoo, VTT – tiedotteita 1292, p. 98.
2. Hardwick Martin, Bolton Richard, 1997. The Industrial Virtual Enterprise. Communications of the ACM, September, Vol. 40, No. 9, pp. 59-60.
3. Ollus Martin, 1998. Verkostotalouden lähtökohdat in Yritysverkostot – kilpailua tiedolla, nopeudella ja joustavuudella, Ollus M., Ranta J., Ylä-Anttila P. (eds.),Vantaa,Sitra,pp. 1-7.
4. Karvonen Iris, 2000. Management of one-of-a-kind manufacturing projects in a distributed environment, Espoo, VTT Research Notes 2044, 54 p.

7

USING CONTEXTS IN MANAGING PRODUCT KNOWLEDGE

John J. Mills[1], J.B.M. Goossenaerts[2]
[1]*Department of Mechanical and Aerospace Engineering, The University of Texas at Arlington, Arlington, TX, USA; Em: jmills@mae.uta.edu*
[2]*Department of Technology Management, Eindhoven University of Technology, the Netherlands; Em: J.B.M.Goossenaerts@tm.tue.nl*

Abstract: The appropriate use of context is a powerful tool for managing complexity asexemplified by the realization of a modern product in a globally distributed environment. We show that computational contexts can be created from the properties of entities such as the company, the project, the user, his organizational role, and the product itself. Our paper briefly presents the architecture of a product knowledge environment that is based on computational contexts. It then discusses several possible uses of computational contexts in the product realization process including management of product knowledge, managing the product realization project and supporting users while they perform the wide variety of tasks necessary to design, develop and produce a modern product.

Key words: Product Knowledge, Context

1. INTRODUCTION

Several authors have pointed out that industry has to move away from thinking about managing product data to managing product knowledge. There are many viewpoints as what product knowledge and indeed what knowledge is. We have proposed elsewhere that knowledge comes into existence when someone identifies in sets of data, in information or in existing codified knowledge, one or more patterns or relationships which exist across multiple contexts. A pattern or relationship which exists only in a single context is only information. Further, we pragmatically define codified *Product Knowledge* as the combination of the set of all *data sets* which depict everything we need to know about a product, the context independent *relationships* within and among these data sets, and the *contexts* in which they are created and used. A data set is just a set of data which, when properly interpreted, depicts some aspect of the product (e.g. requirements, geometry, production plans, inspection programs). A data set can be just a file. The word "codified" is placed in front of Product Knowledge to indicate that in this work we do not attempt to deal with tacit knowledge which is an other, separate and equally important area of research.

Our approach is based on the idea of a *computational context* which seeks to mimic how humans use context. We show that computational contexts can be created from the properties of entities such as the company, the project, the user, his organizational role, and the product itself. The architecture of a product knowledge system is proposed. It is partially validated by addressing the following three issues:

- How to detect context changes?
- What areas of the process can the context help support?
- What advantages does using context have over other approaches?

2. DEFINITIONS

2.1 Knowledge and Product Knowledge

Our view of product knowledge and indeed knowledge itself is based on the concept that the progression from data to information to knowledge is through the recognition and understanding of patterns[1]. Patterns in data make information. Information becomes knowledge when some pattern is recognized and understood in someone's mind in information in more than one context. Thus, Newton created new knowledge (his laws of motion) when he observed patterns in the behavior of different objects in different contexts.

Applying these definitions to product data, patterns in data imply relationships within and among sets of data. A CAD file is just data until it is associated in some way with a pattern that brings meaning to it. STEP recognizes this with the idea of a data model which describes the specific pattern (i.e. the relationships within) the data within the file. Thus files of data in general (e.g. CAD, Spreadsheets, data bases, project plans) have no meaning until other information is associated with it: for instance, until it is associated with a data model. Product Data Management Systems (PDM's) manage the complex relationships among sets of data describing all parts of the product. Workflow management systems (WFM's) manage the complex relationships that occur in the flow of product data along the product realization process and that occur among the various tasks that are required in that process at a higher level than that of a CAD system. However, it is not clear what can be called knowledge and what information because contexts are not identified. Moreover, the approach taken in both appears to be ad-hoc with no fundamental basis. Both of these management systems manage what we call "data sets". A data set can be a CAD file, a set of requirements, an NC program, etc. A data set has no meaning until other information is associated with it. The source of this other information is discussed below after we have introduced the concept of "context".

We can identify some product knowledge immediately. One fact that is true, independent of context, is the suggestion by various authors [2] and generalized by us that a product consists of a set of "product units" arranged in some topology. This is a pattern that exists for all kinds of products. Elsewhere we have suggested that each "product unit" can be represented in a computer (or on paper or some other medium) by abstract entity which we call its "Product Unit Representation" (PUR). The PUR consists of two kinds of "Aspects"[3]: a single "Core Aspect" which contains properties which are intrinsic to that product unit and a set of Aspects which contain all the context dependent information about the product unit. The Core Aspect contains several facts that are true about the product unit, independent of context. A product unit has a color, a weight, a material it is made of and a cost. The value of all of these is independent of context because they are intrinsic to that product or component. An Aspect consists of a data set with an associated list of context dependent properties assigned to that Data set at the time of its creation or modification. A product unit has many Aspects.

2.2 Contexts

PDM's and WFM's essentially seek to manage the complex relationships among data sets involved in product realization *across multiple contexts*. Users sitting in their offices perform tasks that eventually and in totality realize a product. Each works in their own "context": a situation which constrains how they perform that task, indeed, what task they perform.

Contexts are used daily by everybody as a method of managing the complexity of their daily lives. We think nothing of switching from being father to husband to driver to navigator to executive to negotiator. Yet contexts have been woefully neglected in research until recently. Recent workshops have started to reverse this trend, and research results are becoming available in a wide variety of fields. Despite this research, there is still no universally accepted definition of a "context"[4]. Brézillon and Pomerol have recently reviewed the state of the art in context research[5]. In product realization, context has been mentioned a number of times [3,6,7].

Our current view of context is based partially on observations of how humans use contexts, albeit in an instinctive way. Humans seem to be able to build complex contexts automatically and instinctively without really knowing what they do. First the context (a pattern) is recognized and understood. Then the set or properties required to deal with that context are automatically assembled. Humans create the context from entities that surround them, extracting from those entities, properties that are relevant to themselves. The properties of the entities involved in building the context are learned by experience and constitute knowledge when those patterns of entity combinations are used in multiple context. Properties of entities can be other entities, procedures, attributes, rules, etc.. We use the word "entity" in its fundamental sense[8].

3. CONTEXTS IN A PRODUCT KNOWLEDGE SYSTEM

3.1 Computational Context

In a computational environment supporting humans performing tasks, we suggest the concept of a computational context defined as follows: *a computational context surrounding an entity of interest is a set of properties (with values), that are (a) provided by a set of entities in the same symbolic or physical space as the entity of interest, (b) relevant to the entity of interest in that situation of interest during some time interval and (c) added to the properties of that entity only within that context* [5].

The entity of interest is the thing which is surrounded by the context. It may be a person, a task, a file or a set of data. In product realization, there are two generic entities of interest: The set of data which describes some aspect of the product and the human performing the task. Computational contexts surrounding a set of data has been discussed elsewhere [9].

The set of possible entities that can form contexts for data set creation is quite large, including : the company, the operating system of the computer of the user, the industry sector to which the company belongs, the person performing the task, the role they are playing in the project, the product, the application used to create the data set, and so on. Each of these brings different properties that are relevant to the data set. The product may define the process to be used (e.g. novel or parametric).

The properties of entities forming the context can by anything including other entities, attributes, rule sets, constraints. An example is a project context which has properties such as ID, Name, Description, Responsible person, Goals and other entities such as the company, which has as one of its properties a country context. This nesting of context properties is important because it allows complex relationships to be established among entities of interest. The idea is that entities only become contexts when they are relevant to a particular entity of interest.

3.2 System Architecture

The system we are proposing to create computational contexts for people performing tasks and the aspects they create and use is illustrated in Figure 1. The Entity Data Base (EDB) contains specific entities with particular values for their properties. The Context Ontology (CO) contains rules for organizing and combining the entities appropriate to the specific context. The rules include a set which use the idea of context levels (syntax, semantics and pragmatics) as discussed in an earlier paper. The CO contains the rules which allow the PKE to govern when general entities such as user, project, company etc., become relevant to a particular entity of interest and allow the PKE to build a specific Computational Context for it. The Aspect Repository (AR) contains this specific computational context information for that particulare entity of interest, say a data set like a CAD model. The name of its file is combined with the computation context information to form the Aspect for that CAD model and this information is then stored in the Ar for future reference. The AR stores no data sets. These can be stored anywhere on the WEB. The computation context information contains the name of the file and its location so that it may readily be accessed at a later date. For enetities of ineterst such as users, the computational context would contain such information as appropriate processes for the product realization, tool lists, other similar products to facilitate their tasks. The Product Knowledge Environment (PKE) provides functions to users. This includes the ability to (a) create user computational contexts with suggested Aspects, procedures, protocols, rules, standards etc., (b) create computation contexts for data sets which then become Aspects stored in the AR and (c) allow users to search for specific data sets (i.e. Aspects) using the properties of the Aspects as indices. This latter allows users to create multiple directory structures for data sets based on the values of the properties in the Aspect. For instance the Aspects can be organized by the PKE by product structure, organization unit, project organization by team structure, all of which are properties of the computation context formed from the entities in the EDB.

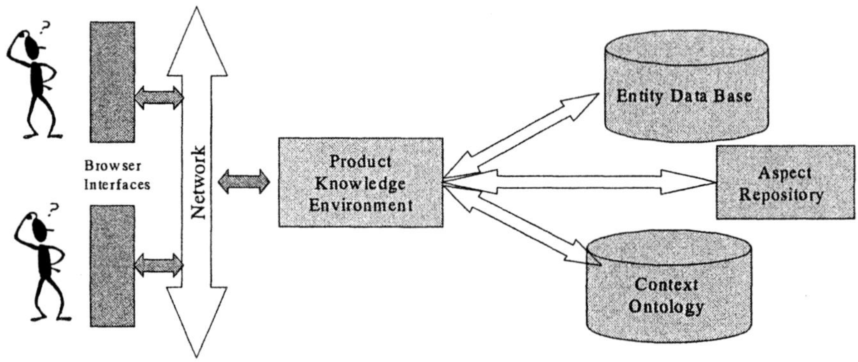

Figure 1. Architecture of a Product Knowledge Environment

3.3 Development

In order to partially develop the architecture we are addressing the following several issues:

- What advantages does using context have over other approaches?
- What areas of the process can the context help support?
- How detect context changes?
- What are the critical rules for the CO?

The main advantage of the context-based approach to supporting the product realization process is that it builts on what people to every minute of every day: detecting and using contexts to manage the complexity of their lives.

For the second issue, we believe that context-based system described above can help manage the data, information and knowledge more efficiently since the aspect concept described elsewhere allows one to organize data sets along multiple dimensions: process, project, company, industry, product structure, and can also be used to automatically provide required data sets for a particular task. With an appropriate set of relevant properties a context can supply a user with (a) sub processes and procedures that change with the users experience, training and task, (b) appropriate Aspects for a task, (c) standards that change with the industry segment and country, and (d) policies and procedures from the company and project. While not all of these will be needed for a particular task, the PKE can provide easy access to them.

Our current thinking about detecting contexts is that it will depend on the Entity of Interest. For a data set and its accompanying Aspect, we envisage a set of key properties that will be compared: The context into which it is being brought vs the context in which it was created. We would start the comparison at the most deeply nested entity and move outwards. Since most context changes for Aspects will be in the syntactic level, this will allow for fast detection, while still allowing an Aspect to be moved globally.

When a user changes context usually this will be some action taken which can be detected through keystrokes or mouse clicks. One example would be a user changing tasks by selecting a new task. Another might be when a user has multiple tasks running in different windows. The move to a different window would indicate a change in context.

4. DISCUSSION

We have briefly presented our idea on knowledge, product knowledge context and a suggested system which makes use of these ideas. As an example of how this system architecture might be used consider a CAD model. This entity of interest is created in a context created from a particular combination of particular entities such as the user, his computer environment including the application they use to create it, their task, the project phase, the project, the company, and the country in which they work. The PKE has identified this specific computation context and created in according to the rules it finds in the Context Ontology. When the user saves the CAD model, under a specific file name in a specific directory on a particular computer, the PKE collects that information as part of the syntactic context level and combines it with the rest of the computational context information (the semantic and pragmatic levels) and stores it as an Aspect in the Aspect Repository. Note that the file - a data set - can be stored anywhere on the World Wide Web. Part of the information in the Aspect is the list of tasks that need this file and the people responsible for these tasks. The PKE would then inform these people that the file is available. In this example, the PKE acts like a Work Flow Manager as well as a Product Data Manager but instead of being based on ad-hoc assignments of files to cabinets and people, it is based on natural contexts. When one of these people enters the PKE and chooses one of these tasks, the PKE knows from the Aspect where it is, how it is stored and in what format, and can immediately transfer the file to that computer and make it available to the user. It is foreseeable that the

PKE could also provide automatic translation of the file format to that of the application the new user is running.

This system has the potential for being the next generation of product realization knowledge managers. Our vision is that entities in the EDB can include high level tasks which specific procedures for dealing with certain types of product units (novel units, redesigned units), experience level of users (novice, 40 years experience) which are used by the PKE and the CO to create specific computational contexts with suggested approaches, standards, protocols, and Aspects appropriate to the users training and background. Space precludes mentioning all the possible uses that the proposed system may have in managing product realization knowledge.

ACKNOWLEDGEMENTS

J. J. Mills would like to acknowledge the financial assistance and support that he obtained from the Information Technology Group of the Technology Management Faculty of Eindhoven University of Technology and the Institute of Mechanical systems of the Swiss Federal Institute of Technology, Zurich, during the Sabbatical leave he spend with those institutions. During this time many of the ideas in this paper were created and clarified during informative and insightful discussions with their faculty and students.

REFERENCES

1. Mills, J.J. and J.B.M. Goossenaerts (2001) Towards Information and Knowledge in Product Realization Infrastructures. In: J.P.T. Mo and L. Nemes (eds.) (2001) *Global Engineering, Manufacturing and Enterprise Networks*, Kluwer Academic Publishers, Boston pp 245-254
2. Henderson, M. R. and L. E. Taylor (1993) A Meta-Model for Mechanical Products Based upon the Mechanical Design Process, Res. in Eng. Design, Vol. 5, pp 140-160
3. Talukdar, S. N. and S.J. Fenves (1991) Towards a Framework for Concurrent Design. In: D. Sriram, R. Logcher, S. Fukuda (eds.) *Computer-Aided Cooperative Product Development, MIT-JSME Workshop*, MIT, Cambridge, USA, November 20/21, 1989, Proceedings. LNCS 492 Springer-Verlag
4. Bouquet, P. L. Serani and P. Brézillon (1999) Preface to *2nd International and Interdisciplinary Conference on Modeling and Using Context (CONTEXT'99)*, LNAI 1688, Springer.
5. Brézillon, P. and J.-Ch. Pomerol (1996) Context in Artificial Intelligence: II. Key elements of contexts. Available at URL: http://citeseer.nj.nec.com/502533.html
6. Kannapan, S. and K.M. Marshek (1992) Engineering Design Methodologies: A New Perspective. In A.N. Kusiak (ed.) *Intelligent Design and Manufacturing*, John-Wiley & Sons, pp. 3-38.

7. Hale, M. A. (2000) Presentation at the *Gordon Conference on Theoretical Foundations of Design and Manufacturing*, Plymouth NH, June, 2000.
8. Sowa, J. (2000) Knowledge Representation: Logical, Philosophical and Computational Foundations, F. Brooks/Cole Thompson Learning, New York.
9. Mills, J.J., J. Goossenaerts and H.J. Pels (2001) The Role of Context in the Product Realisation Process, *Proc. of the International CIRP Design Seminar*, held in Stockholm, Sweden, June 6-8, 2001.

8

OBJECT-ORIENTED DESIGN PATTERN APPROACH TO SEAMLESS MODELING, SIMULATION AND IMPLEMENTATION OF DISTRIBUTED CONTROL SYSTEMS

Satoshi Kanai[1], Takeshi Kishinami[1], Toyoaki Tomura[2], Kiyoshi Uehiro[3], Kazuhiro Ibuka[3], and Susumu Yamamoto[3]

[1] Dept. of Systems Engineeering, Hokkaido Univ., Japan
[2] Asahikawa National Colledge of Technology, Japan
[3] Motorola Japan Ltd., Japan
e-mail: kanai@coin.eng.hokudai.ac.jp

Abstract: Distributed control systems (DCS) come into wide use in automation areas. In this paper, an object-oriented design pattern approach for modeling, simulation and implementation of the DCS is proposed. The proposed design patterns enable the uniform modeling of the static structures and dynamic behaviors of the DCS, the transformation of the models into simulation program, and the generation of the embedded codes. The Java-based modeler and simulator, and code generator were developed based on these patterns. Applications to the building automation and factory automation systems proved its effectiveness.

Key words: object-oriented modeling, design pattern, distributed control system, simulation, UML, LonWorks, FieldBus.

1. INTRODUCTION

Distributed Control Systems (DCSs) using open networks such as Fieldbus, CAN and LonWorks, are rapidly replacing traditional centralized control systems in a factory, process and building automation areas [1]. As shown in Figure 1, the DCS generally consists of many devices and an open network interconnecting them. A device consists of one control node and several device components (sensors, actuators), while a composite device consists of a group of devices. The DCS can make the system more scalable, and its building and

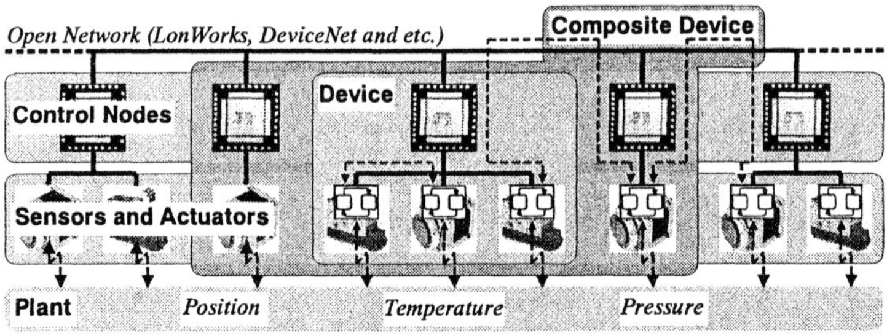

Figure 1. The structure of the DCS

wiring cost much less. On the other hand, the communication traffic among the devices tends to concentrate on a network. This causes an unreliable control performance. The system integrators have to test and avoid such situations after building the DCS. However, these works are time-consuming and costly.

To solve the problem, the system integrators need the computer-aided tools where they can simulate performances of the designed DCS, and can generate the embedded control codes installed into devices. The systematic method is needed to develop such tools. The method has to satisfy several requirements: 1) static structures of various devices and device components can be modeled, 2) dynamic behaviors in the devices and communication among the devices can be modeled, 3) the models can be easily used in the executable simulation program, and 4) the models can be automatically transformed to the embedded control codes installed in each device.

The object-oriented methods have been used for modeling various manufacturing systems. So far, SEMI/CIM-Framework[2], OSE[3], SEMI/OBEM[4] and GEM[5] specified the reference models of manufacturing resources described by UML or Coad&Yordon methods. However, their modeling scopes do not fit to the above requirements of the DCS.

The purpose of this research is to propose an object-oriented and design pattern–based method for seamless modeling, simulation and implementation of the DCS. Five design patterns are newly proposed. The DCS modeler and simulator, and the embedded code generator are also developed based on these patterns. A case study for controlling material handling system proves the effectiveness of our method and tools.

2. OVERVIEW OF PROPOSED DESIGN PATTERNS

Design pattern is a reusable structure for collaboration and interaction among classes or objects applied to capturing problems in a general domain [6]. However, existing design patterns are too generalized to express the DCS simulation models. Therefore, in this paper, five design patterns are proposed specializing in modeling, simulating and implementing the DCS. Figure 2 shows

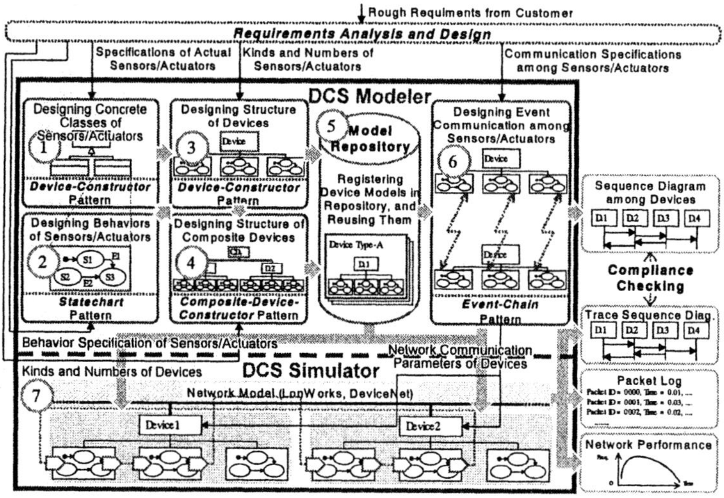

Figure 2. The design pattern-based modeling and simulating tools for the DCS

our modeling and simulating tools of the DCS and related design patterns.

Device-Constructor pattern describes the instantiation mechanisms for structuring the device models composed of many kinds of sensors, actuators, and local controllers. Composite-Device-Constructor pattern describes ones for structuring the composite device models composed of many kinds of devices. Statechart pattern defines the state-transition mechanism for realizing the dynamic behavior of each device and its device components. Event-Chain pattern defines the event dispatching mechanism among the sensors and actuators inside the device and inter-device on the network. Statechart-compiler pattern defines the mechanism for transforming the behavior of the DCS simulation model to the low-level codes embedded in the device. The first four patterns are designed for the DCS simulation, while the Statechart-compiler pattern is for the implementation. The details are described in section 3 and 4.

3. DESIGN PATTERNS FOR MODELING AND SIMULATION

3.1 Device and Composite Device Constructor Patterns

The Device-Constructor pattern and Composite-Device-Constructor pattern are proposed to model the static structure of the DCS[7]. Figure 3 shows the UML class diagram of the patterns. The Composite-Device-Constructor pattern

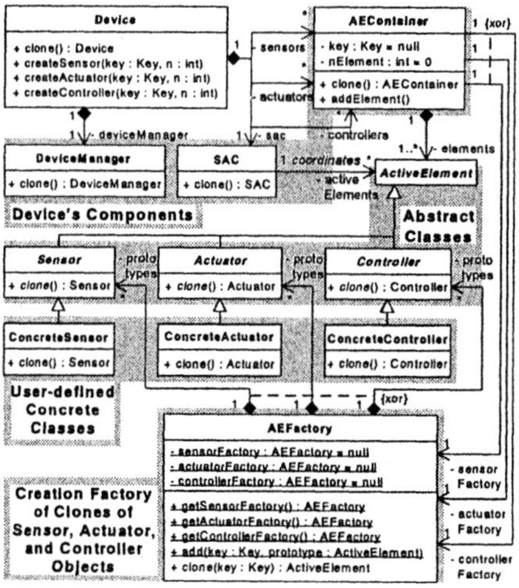

Figure 3. Device Constructor and Composite-Device-Constructor Patterns

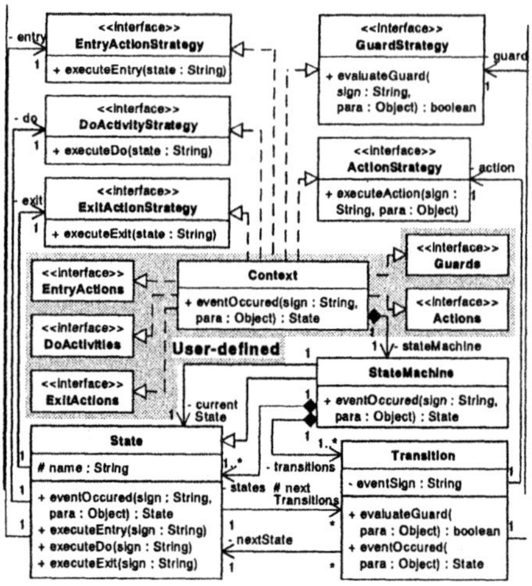

Figure 4. Statechart Pattern

is a pattern that extends three classes (*CompositeDevice*, *Device Container*, and *Device Factory*) of the Device Constructor pattern to the composite device. In this pattern, *AEContainer* represents the container class for all of the objects of a particular concrete class. *AEFactory* represents the factory class for creating the cloned objects of the concrete classes. The object of the *AEFactory* class contains the objects of each concrete class with a key string. *ConcreteSensor*, *Concrete Actuator*, and *Concrete Controller* represent the concrete classes of the three abstract classes *Sensor*, *Actuator* and *Controller*.

The features of this pattern are as follows. Firstly, a particular kind of device component can be modelled as a concrete subclass, so that we can directly represent the system structure that various types of sensors, actuators are connected to the control node. Secondly, the connection of a control node with its constituent device components can be expressed explicitly. Thirdly, various types of devices can be flexibly modelled only by changing the kind or the number of device components in the predefined device models. These features are similar to the ones of the Composite Device-Constructor pattern.

3.2 Statechart Pattern

The dynamic behavior of each device component in the DCS can be specified as the Statechart. The Statechart provides finite state machines with the notions of sub-states, entry and exit actions, do-activities and guards [8]. To accurately model these notions of the Statechart and realize the executable Java code, the Statechart patterns are newly proposed.

Figure 4 shows the class diagram of the Statechart pattern. *Context* has the dynamic behavior described by the Statechart. *StateMachine* describes a finite state machine composed of sets of states and transitions. *State* corresponds to either the state itself or its sub state machine. *EntryActions*, *DoActivities*, and *ExitActions* describe the actions and the activities in a state as interfaces. *Guards* and *Actions* represent guards and actions of a transition as interfaces. *EntryActionStrategy*, *DoActivityStrategy*, and *ExitAction Strategy* represent the concrete implementation methods to execute actions and activities. *GuardStrategy* and *ActionStrategy* also represent the ones to evaluate transition conditions and to execute actions according to occurred events.

In the proposed Statechart pattern, the one-to-one simple mapping from the design pattern to the Java code can be explicitly defined. Moreover, because the general structure of the Statechart itself and the implementation of actions, activities and transitions depending on the context can be completely separated in the pattern. So the system designer can easily identify and modify the methods of actions, activities and guards only in the Context's class

3.3 Event-chain Pattern

To complete event-driven DCS simulation models, the event links between device components transmitted both at the inter-device and intra-device levels have to be modeled and dispatched as shown in Figure 5. For modeling these event links, we propose the Event-chain Pattern.

Figure 6 shows the class diagram of an Event-Chain pattern newly proposed. *IOVariable* represents the class of input/output variables of the device components, device, and composite device. Any subclass of the *IOVariable* class can be defined freely according to the specification of network. *Link* represents the class of links between the input and output variables, including peer-to-peer, multicast, and broadcast communications. *EventDispatcher*

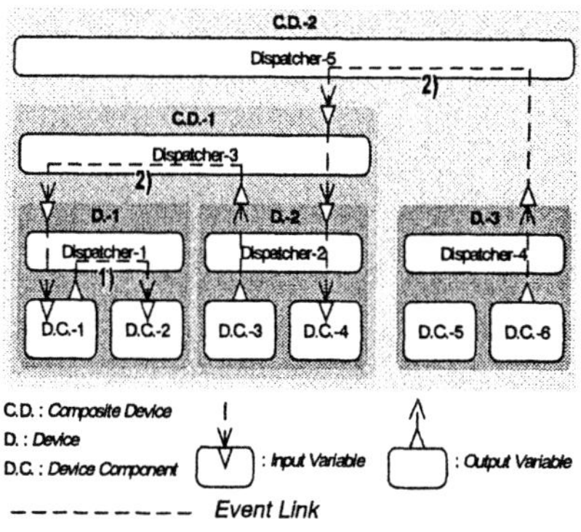

Figure 5. Event links and dispatchers

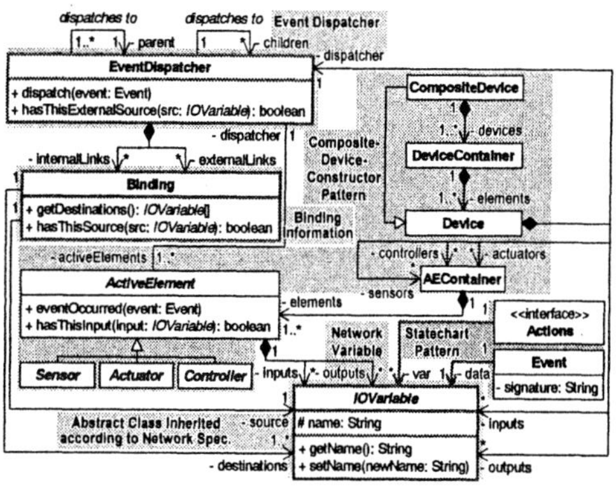

Figure 6. Event-chain Pattern

represents the class of event dispatchers that determine the destination of the generated events of the device.

In the Event-Chain pattern, the *Device* and *CompositeDevice* objects only have to relate to a uniform *EventDispatcher*, and they do not have to care if the event is going into or out from the device. The existing *Device* and *CompositeDevice* objects including pre-defined links are also reusable in the modelling of other *CompositeDevice* objects. The change of *IOVariable* objects' value is interpreted as an event or an action in the Statechart pattern of concrete device classes. Using the procedure, the DCS simulation model can automatically dispatch all the events among the suitable device components.

4. STATECHART-COMPILER PATTERN

The behavior of the control code executed in the device can be modeled as Statechart, but in case of LonWorks-based DCS, the code is eventually implemented as Neuron C. The Neuron C [9] is a subset of ANSI C, so that the Statechart behavior must be implemented without "class" concept. The Statechart-compiler pattern is proposed to bridge this gap. This design pattern can transform the device behavior modeled as Statechart to the Neuron-C code.

Figure 7 shows the process to apply the Statechart-Complier pattern. Firstly, the Statechart is converted to the textual formal description whose syntax is specified by the extended BNF[10]. Secondly, the formal description is parsed to obtain a syntax tree. Each node of the tree corresponds to the object defined in

Figure 7. The process of control code implementation using Statechart-Compiler pattern

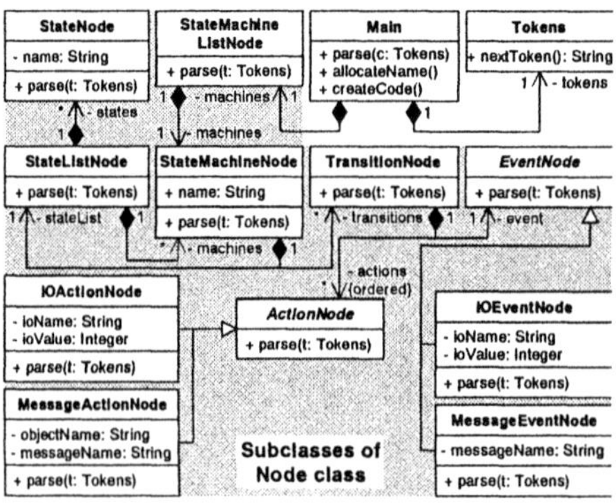

Figure 8. Statechart-Compiler pattern

Table 1. A mapping rule from Statechart-Compiler pattern to Neuron C

Event and Action			Neuron C code

Event	I/O event	(*Start of event chain*)	when(**io_changes**(*io_name*) to *value*) {
			if(**eventOccurred**(&*machine*, &*event*)) {
			/* Its action(s) */ }
			}

	Message event	From different device (*Start of event chain*)	when(**nv_update_occurs**(*message_name*)) {
			If(message_name == ST_ON) {
			If(**eventOccurred**(&*machine*, &*event*)) {
			/* Its action(s) */ }
			}
			}

		From same device	If(**eventOccurred**(&*machine*, &*event*)) { /* Its action(s) */ }
Action	I/O action		**io_out**(*io_name*, *value*);
	Message action	To different device	*message_name* = *value*;
		To same device	*is equal to message event from same device.*

the Statechart-complier pattern. Finally, the Neuron C code can be automatically generated by the mapping rules from objects of the syntax tree to the Neuron C statements. Table 1 is a part of these rules.

Figure 8 shows a class diagram of the Statechart-Compiler pattern. *Main* executes a method of generating Neuron C code after reading the formal description. A *StateMachineNode* represents one state machine described in the formal description. A *StateNode* represents a state in the statechart, and *TransitionNode* does a inter-state transition. *EventNode* and *ActionNode* represent an event or an action defined in the transition. *IOEventNode* represents an incoming I/O signal event from a sensor or an actuator, while *IOActionNode* represents an outgoing signal event to them. *MessageEventNode* and *MessageActionNode* represent message events received from or sent to other state machines.

The syntax tree does not depend on any programming language. Therefore, this pattern is applicable to implementing the control codes by the languages besides Neuron C. The control code of the other language can be built by re-defining the language-specific mapping rules from the syntax tree to the statements of that language. The automatic generation of the embedded control code for DCS can be realized by this systematic procedure.

5. A CASE STUDY OF DCS DEVELOPMENT

We have implemented the Java-based DCS modeler and simulator software, and Neuron-C code generator based on our patterns and procedures. In the simulator, a designer can predict network traffic and a packet log shown in Figure 9. By applying our design patterns, the modeler and simulator development could be completed only for two months. These simulation tools were introduced to several system integrators of building automation, and their effectiveness on the DCS development has been verified[7].

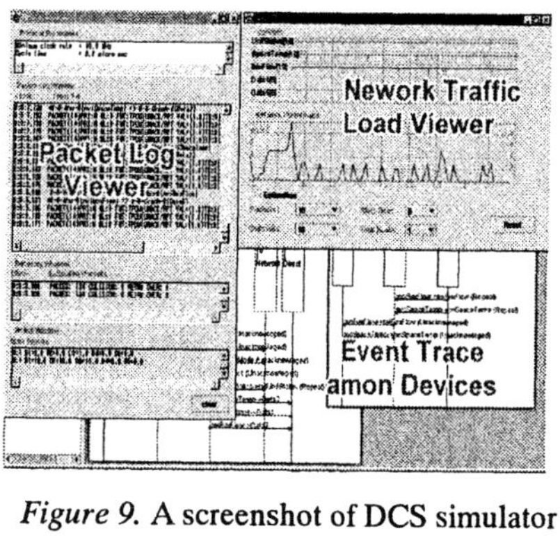

Figure 9. A screenshot of DCS simulator

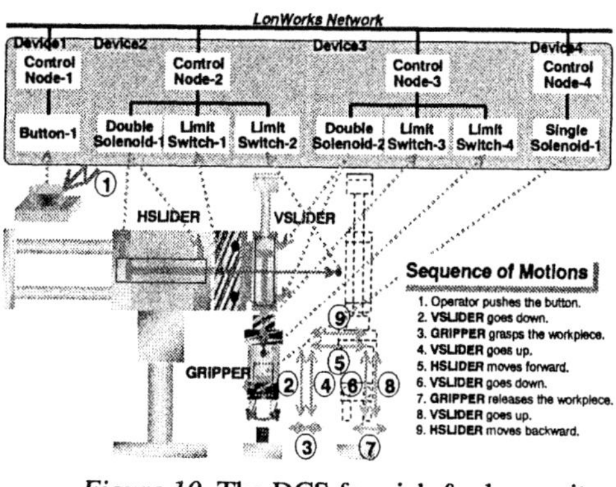

Figure 10. The DCS for pick & place unit

Moreover, as an application to the factory automation, we consider the DCS for controlling a pick-and-place unit driven by the pneumatic circuit. Figure 10 shows the structure of the DCS and the motion sequence of the unit. The DCS consists of LonWorks and four devices each of which consists of one control node, buttons, limit switches and valve actuators. Firstly, the task motions are described as a sequence diagram. Secondly, the Statechart of an object's behavior participating in a sequence diagram were identified. They were used with the Statechart and Event-Chain patterns. Using a Statechart-compiler pattern, four executable Neuron C codes have been generated and were installed into each of the four control nodes. As a result of the process, the motion sequence of the actual pick & place unit could be correctly worked by the DCS whose code was implemented from the simulation model.

6. CONCLUTIONS

In this paper, we proposed five object-oriented design patterns specializing in seamless modeling, simulation and implementation of the DCS. The DCS modeler, simulator, and embedded code generator could be efficiently developed based on the patterns. The result of using the modeler and simulator in system integrators, and the one of building the DCS for the material handling system controlled LonWorks proved the effectiveness of our approach.

REFERENCES

1. Dietrich, D., Neumann, P., and Schweinzer, H. (1999), *Fieldbus Technology – Systems Integration, Networking, and Engineering*, Springer-Verlag.
2. SEMATECH (1995), CIM Application Framework, Specification 1.2, 6.
3. OSE (1996), OSEC Architecture Version 2.0.
4. SEMI (1998): SEMI E98-0302 Provisional Standard for the Object-based Equipment Model (OBEM), SEMI.
5. SEMI(1998) : SEMI E30-0298 Generic Model for Communications and Control of SEMI Equipment (GEM), SEMI.
6. E. Gamma, R. Helm, R. Johnson, and J. Vlissides (1995), *Design Patterns: Elements of Reusable Object-Oriented Software*, Addison-Wesley.
7. Tomura, T., Kanai, S., Kishinami, T., Uehiro, K., and Yamamoto,S. (2001), Developing Simulation Models of Open Distributed Control System by Using Object-Oriented Structual and Behavioral Patterns", *Proceedings of Fourth IEEE Int. Symp.Object-oriented Real-time Distributed Compiting*, 428-437.
8. OMG (1999), Unified Modeling Language Specification Version 1.3.
9. Loy, D., Dietrich, D., and Schweinzer, H. (2001), *Open Control Networks: –Lon Works/EIA 709 Technology*, Kluwer Academic Publishers.
10. Aho, A. V., Sethi, R., and Ullman, J. D. (1985), *Compilers: Principles, Techniques, and Tools*, Addison Wesley.

9

AN INTEROPERABILITY FRAMEWORK AND CAPABILITY PROFILING FOR MANUFACTURING SOFTWARE

M. Matsuda[1], E. Arai[2], N. Nakano[3], H. Wakai[4], H. Takeda[5], M. Takata[6] and H. Sasaki [7 ¶]
[1]*Kanagawa Institute of Technology, Japan*
[2]*Osaka University, Japan*
[3]*Mitsubishi Electric Mechatronics Software Co., Japan*
[4]*Komatsu Ltd., Japan*
[5]*National Institute of Informatics, Japan*
[6]*The University of Electro-Communications, Japan*
[7]*Yokogawa Electric Co., Japan*
matsuda@ic.kanagawa-it.ac.jp

Abstract: ISO/TC184/SC5/WG4 is working on ISO16100: Manufacturing software capability profiling for interoperability. This paper reports on a manufacturing software interoperability framework and a capability profiling methodology which were proposed and developed through this international standardization activity. Within the context of manufacturing application, a manufacturing software unit is considered to be capable of performing a specific set of function defined by a manufacturing software system architecture. A manufacturing software interoperability framework consists of a set of elements and rules for describing the capability of software units to support the requirements of a manufacturing application. The capability profiling methodology makes use of the domain-specific attributes and methods associated with each specific software unit to describe capability profiles in terms of unit name, manufacturing functions, and other needed class properties. In this methodology, manufacturing software requirements are expressed in terns of software unit capability profiles.

Key words: Manufacturing software, Interoperability, Capability profiling, International standardization

¶ All authors are members of Japanese mirror committee of ISO/TC 184/SC 5/WG 4.

1. INTRODUCTION

Few years ago, an ISO/TC184/SC5 Study Group (Title: Manufacturing software capability profiling, Convenor: Dr. U. Graefe) identified the problem that a unified characterisation might be expected to address in terms of the two view points of user requirements and contribution to interoperability. User requirements on the manufacturing application were identified assembling a new functionality, selecting appropriate software, substituting one software component with another, migrating to another platform, managing software inventory, certifying software to a capability profile, distributing software to the mass market, and registering new software. Interoperability issues in manufacturing software were identified as the ability to describe software in unambiguous terms to enable a common understanding, the characterising of the business benefits delivered by software components, the ability to find enabling candidate software components automatically using search engines, expressing the dependencies of one software component on other application or operating system components, and the management of the implications of change. As a conclusion, the study group suggested launching a standardization work item on these view points [1].

At present, ISO/TC184/SC5/WG4 (Title: Manufacturing software and its environment, Convenor: Dr. M. Matsuda) is developing a 16100 series International Standard which is titled "Manufacturing software capability profiling for interoperability". This project addresses concerns of users and suppliers of manufacturing software with regard to user requirements and interoperability of software in the area of industrial automation. This paper discusses a manufacturing software interoperability framework and a capability profiling methodology which were proposed and developed in this international standardization activity.

2. MANUFACTURING SOFTWARE INTEROPERABILITY FRAMEWORK

2.1 Manufacturing software unit interoperability

The interoperability framework for manufacturing software is based upon a more general interoperability framework for manufacturing applications. An integrated manufacturing application is modeled as a combination of a set of manufacturing resources and a set of information units whose data structure, semantics, and behaviour can be shared and exchanged among the manufacturing resources. The set of integrated manufacturing resources forms a manufacturing system architecture that fulfils a set of manufacturing application requirements. These manufacturing resources, including the manufacturing software units, provide the functions associated with the manufacturing processes, as shown in Figure 1. The combined capabilities of the various software units, in an appropriate operating environment, provides the required functionality to control and monitor the manufacturing processes according to the production plan and the allocated resources [2].

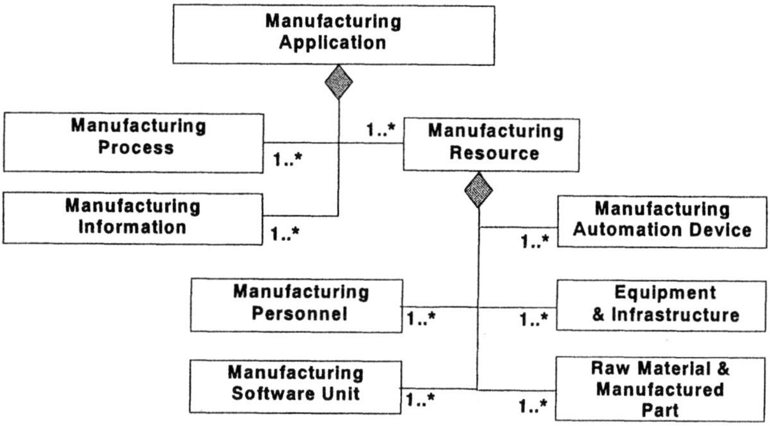

Figure 1. Class diagram of a manufacturing application [2].

A manufacturing software unit is a class of software resource, consisting of one or more manufacturing software components, performing a definite function or role within a manufacturing activity while supporting a common information exchange mechanism with other units. The software interoperability of a set of manufacturing activities is described in terms of the interoperability of the set of software units associated with each manufacturing activity [2].

2.2 Conceptual framework for manufacturing software unit interoperability

A manufacturing software interoperability framework consists of a set of elements and rules for describing the capability of software units to support the requirements of a manufacturing application. The capability to support the requirements cover the ability of the software unit to execute and to exchange data with other software units operating in the same manufacturing system or in different manufacturing systems used in the manufacturing application. A manufacturing software interoperability framework is based on syntax and semantics shared between manufacturing software units, functional relationships between the manufacturing software units, services, interfaces, and protocols offered by the manufacturing software units, and ability to provide manufacturing software unit capability profiling [2].

Figure 2 shows the conceptual interoperability framework. In Figure 2, the interoperability of software units can be described in terms of their capabilities that are associated with the aspects of functionality, interface and structure. The profiling of a software unit involves the generation of a concise statement of manufacturing capabilities enabled by the software unit in terms of the functions performed, the interfaces provided, and the protocols supported as required by the target manufacturing capability. The software units capability profile definition is registered in an appropriate capability profile database after passing the conformance test. The profile database has a set of taxonomies for use in describing the capability profiles. When developing new manufacturing software or reusing a software unit, the profile database is referred to and searched [3].

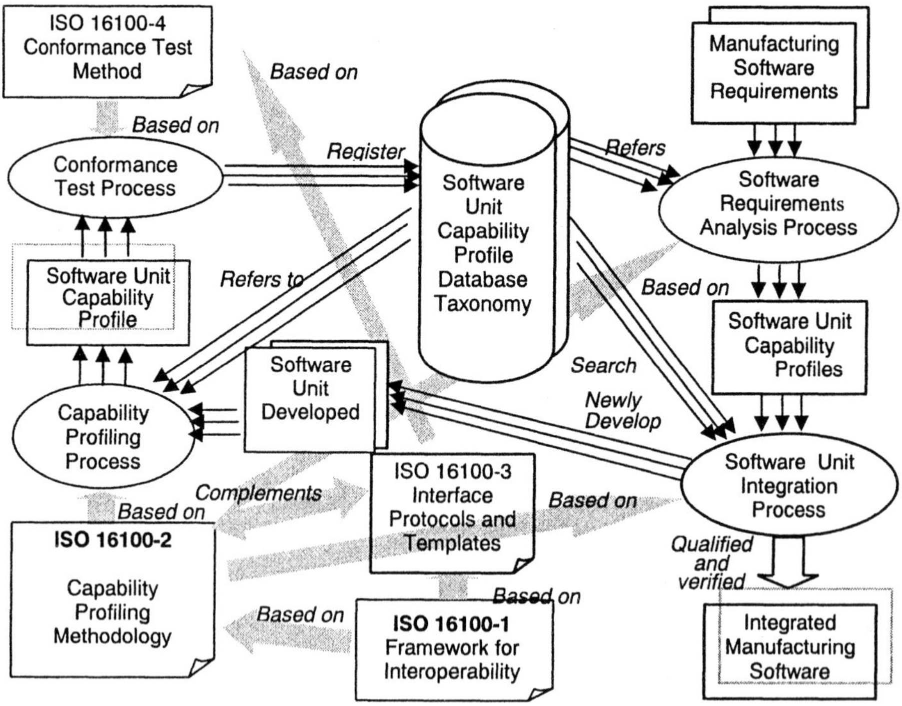

Figure 2. Conceptual framework for manufacturing software unit interoperability.

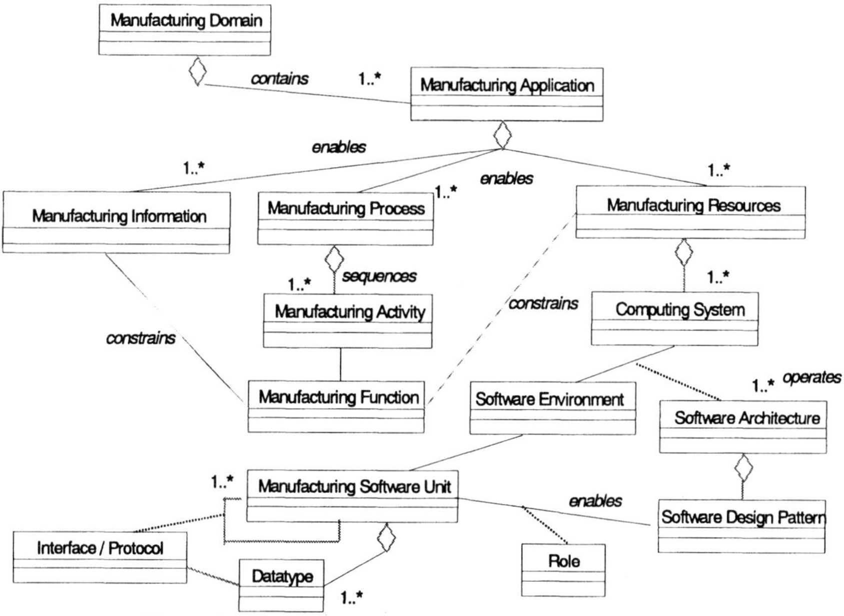

Figure 3. Class diagram of a software unit and its surroundings and associations within a manufacturing application [2].

3. ELEMENTS IN A MANUFACTURING SOFTWARE INTEROPERABILITY FRAMEWORK

3.1 Capability classes

The capability of a manufacturing software unit is expressed in terms of capability classes. These classes is derived from the manufacturing activities noted in Figure 3. These classes also denote the manufacturing function, resource, and information handled by the manufacturing software unit according to the requirements of the manufacturing process [3].

A manufacturing process has a structure that is both nested and hierarchical. At each level, the manufacturing software requirements can be modelled as a set of capability classes organized in a similar structure as shown in Figure 4.

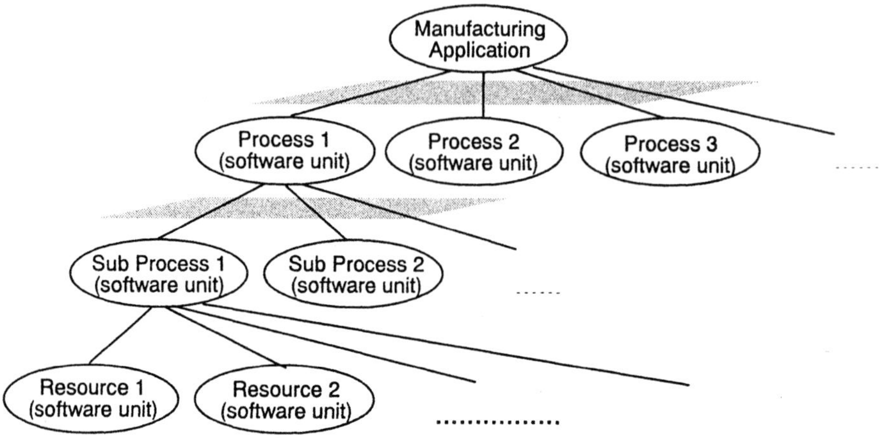

Figure 4. Hierarchical structure in manufacturing application.

3.2 Capability templates

A software unit that enables or supports an activity with an associated capability class is concisely described in a capability template. The structure of a software capability template follows the structure of a manufacturing capability class as shown in Figure 4. In a hierarchical structure, a capability template is associated with each capability defined at each level of the structure. In a nested structure, a similar association exists between each capability class and a template at each level of the structure [3].

3.3 Capability profiles and software unit profile database

Capability profiles are capability templates with, at a minimum, a profiled software unit name instantiated. Other items are fulfilled according to the specification level [3].

A set of taxonomies, a set of capability classes, a set of capability templates, and a set of capability profiles are stored in software unit profile databases, and are distinguished by their differing dictionary names. The databases may be structured as a free combination of the above four elements to provide the necessary services. A taxonomy, a capability class, a capability template, and a capability profile are unique when entered in an adequate corresponding dictionary.

Matching capability profiles are used in the analysis of software unit in the capability profiling process, the decomposition of requirements in the manufacturing software requirements analysis process, and the database search for each profile in the software unit selection and verification process. Matching is attempted between software unit descriptions, manufacturing software requirements, or required software unit capability profiles in these processes and that of capability profiles in the database [3].

4. DETAILED PROCESSES IN THE MANUFACTURING SOFTWARE INTEROPERABILITY FRAMEWORK

4.1 Capability profiling process

The capability profiling process shown in Figure 2 is detailed in Figure 5. A software unit to be profiled is analyzed in terms of the supported paths within the capability class structure. The supported paths then are used in the search for a matching template from the database. When a matching template is found, the fields of the template is filled to make a profile. When no matching template is found, a new template is formed using the set of capability classes.

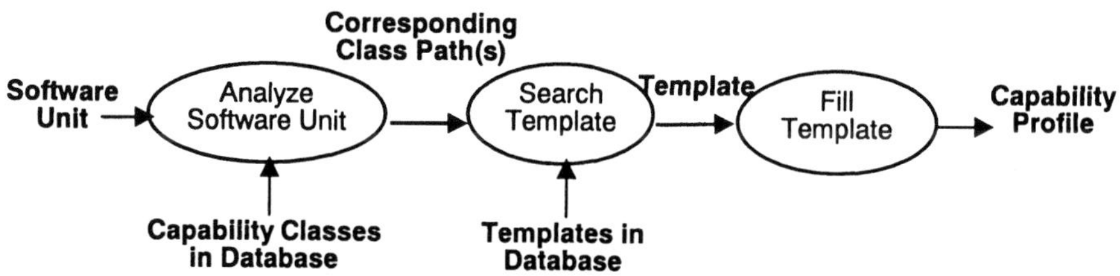

Figure 5. Capability profiling process [3].

4.2 Software requirements analysis process

The software requirements analysis process shown in Figure 2 is detailed in Figure 6. Capability profiles for each manufacturing software unit are derived from manufacturing software requirements in the software requirements analysis process. As a first step, manufacturing software requirements are decomposed into several primitive requirements which are fulfilled by capability classes that are selected from the database. When a template that corresponds to the class exists, the template is filled with specific requirements in order to generate a required capability profile. When such a template does not exist, a new template is created based on rules for template creation.

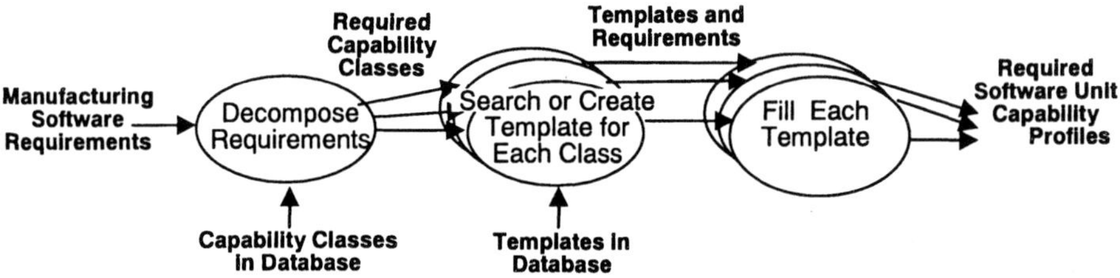

Figure 6. Manufacturing software requirements analysis process [3]

4.3 Software unit selection and verification process

The software unit selection and verification process shown in Figure 2 is detailed in Figure 7. For each required capability profile, a search of matching capability profiles that represent available software units are performed. Matching is performed according to the following rules. When a match exists, the software unit is added to a list of candidates. When a match does not exist, one of the following occurs: a) a new software unit is developed to meet the required profile, b) the required profile is decomposed into a combination of several profiles, or c) requirements are reconsidered against existing profiles. The profile for the new software unit is registered to the database according to the profiling process in 4.1. The selected software units is verified against the manufacturing software requirements according to interoperability criteria.

5. CONCEPTUAL STRUCTURE OF CAPABILITY CLASSES AND TEMPLATES

5.1 Capability class structure

Software capability classes are defined in the structure shown in Figure 8. The contents of a software unit capability class include, but may not be limited to type of manufacturing domain, type of manufacturing activity as differentiated by the process it is part of, the resources involved in conducting the activity, and the information types exchanged during the activity, type of computing system as differentiated by the operating environment, the software architecture, and the design pattern used, type of services, protocol, and data types used in running the software unit, supplier

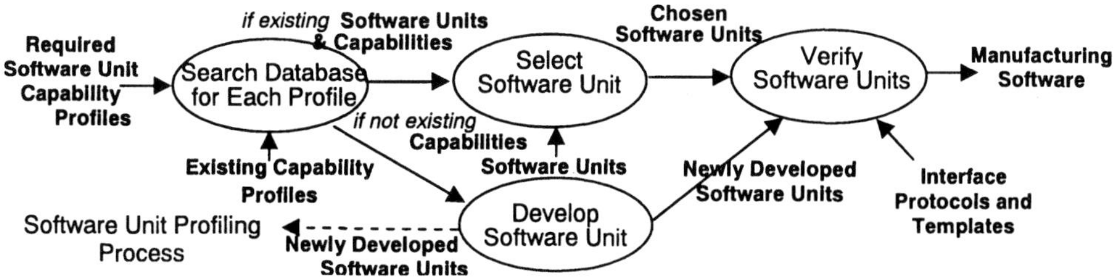

Figure 7. Software unit selection and verification process [3].

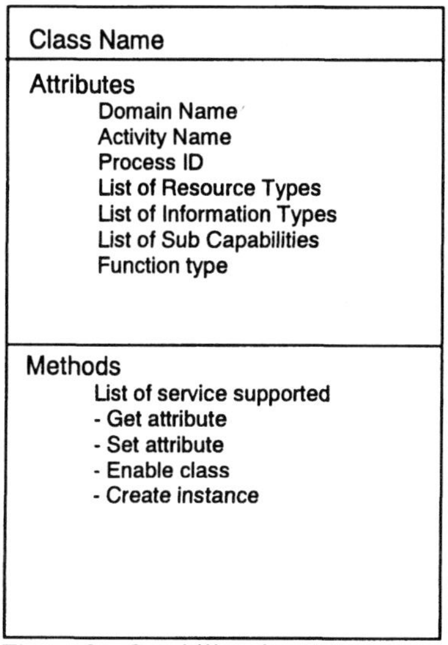

Figure 8. Capability class structure [3].

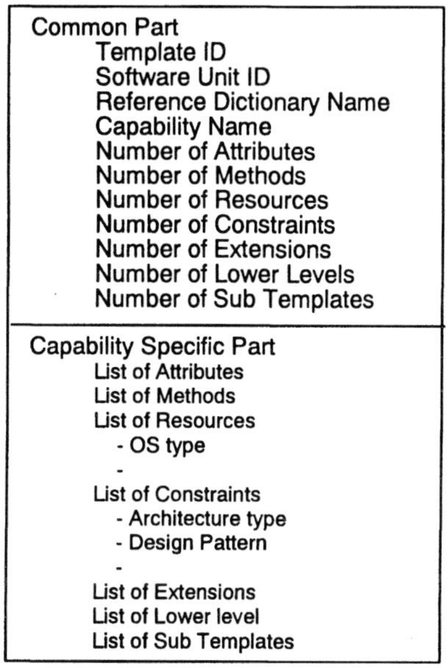

Figure 9. Capability template structure [3].

name, software version, and change history, performance benchmarks, reliability indices, service and support policy, and pricing terms and conditions of use.

5.2 Capability templates structure

A conceptual structure of a capability template is shown in Figure 9. The structure consists of a part that is common to all templates and another part that is specific to capability class. When a capability class is specified in a template and such a class has been instantiated, then the instantiated class represents an object. Two capability templates are identical if their respective attributes and operations are identical. When the attributes of one template form a subset of the attributes of another and the operations of one template form a subset of the operations of another, then the two capability templates are considered to be compatible and have a match.

6. CONCLUSIONS

The ISO 16100 series enable manufacturing software integration by providing the following: a) standard interface specifications that allow information exchange among software units in industrial automation systems developed by different vendors, b) software capability profiling, using a standardized method to enable users to select software units that meet their functional requirements, and c) conformance tests that ensure the integrity of the software integration. The ISO 16100 series consist of four parts. Part 1 specifies a framework for interoperability of a set of manufacturing software products. Part 2 specifies a methodology for constructing profiles of manufacturing software capabilities. Part 3 will specify the interface protocol and capability templates. Part 4 will specify the concepts and rules for the conformity assessment of the other parts of ISO 16100 [2, 3].

ISO 16100 Part 1 has been published. Part 2 is under preparation for publishing. This paper described the concepts and methodology which were newly proposed in the development procedure for Parts 1 and 2. Now, Part 3 is under development.

ACKNOWLEDGEMENTS

The authors thank ISO/TC 184/SC 5/ WG 4 members and its Japanese mirror committee members for fruitful discussions. Especially, the authors are grateful to Mr. E. delaHostoria, Mr. G. Winchester, Mr. B. Sunder and Dr. K. Kosanke of ISO/TC 184/SC 5/ WG 4 member for their useful effort to complete international standards. The authors are also grateful to Dr. U. Graefe, the previous convenor of ISO/TC 184/SC 5/ WG 4 for his helpful assistance with the writing of this paper in English.

REFERENCES

1. ISO/TC 184/SC 5/Study Group (1999). Study Group Report to SC5 Manufacturing Software Capability Profiling, *SC5 study Group N-18*
2. ISO/TC 184/SC 5/WG 4 (2002). Industrial automation systems and integration - manufacturing capability profiling for interoperability - Part 1: Framework, *ISO/IEC 16100-1*
3. ISO/TC 184/SC 5/WG 4 (2002). Industrial automation systems and integration - manufacturing capability profiling for interoperability - Part 2: Profiling methodology, *ISO/TC 184 DIS 16100-2*

10

IT-SUPPORTED MODELING, ANALYSIS AND DESIGN OF SUPPLY CHAINS

Jörg Nienhaus, Robert Alard and Andreas Sennheiser
Swiss Federal Institute of Technology (ETH) Zuerich, Switzerland

Abstract: A common language is a prerequisite for analyzing and optimizing supply chains. Based on experiences with three case studies, this paper identifies the aspects of a supply chain that have to be mapped to take informed decisions on its operations. Current, integrated modeling approaches for supply chains, like the SCOR and the GSCM model, will be analyzed and an advanced approach will be defined. The resulting approach takes advantage of IT-support.

Key words: Supply Chain Management, Modeling, SCOR, GSCM

1. INTRODUCTION

Before supply chains can be optimized, they need to be understood. A hurdle, most companies come across in mapping supply chains, is the lack of a common language, since logistics managers of a couple of companies are involved. Common modeling techniques, like flow charts, fall short when it comes to transcorporate issues, because they only cover a limited number of the aspects a supply chain has. Therefore modeling a supply chain takes an integrated modeling approach that combines different modeling techniques on multiple levels of detail. Those currently are limited in number. Most known are the SCOR ("Supply Chain Operations Reference") and the GSC ("Global Supply Chain") model.

In chapter 2 the relevant aspects of a supply chain that have to be mapped for taking well-founded decisions will be identified. Chapter 3 will give an insight into integrated modeling approaches for supply chains, for what purposes they can be used and where they fall short. While chapter 4 deals with amendments to those models, chapter 5 shows, how the resulting advanced approach to supply chain modeling can be supported by information technology.

2. ASPECTS OF A SUPPLY CHAIN

The supply network a company is involved in is of a very complex nature, because companies have up to 10'000 suppliers of direct material. That is why for analyzing it, restricting attention to supply chains is necessary. A supply chain is the product-specific part of a supply network that is defined by the product's bill of material and by the suppliers that contribute to the product. In Figure 1 gray boxes mark the supply chain of product 14. It differs from the pure bill of material, since suppliers are assigned to its components. In cases of single or multiple sourcing (as opposed to sole sourcing, see [3], p. 47) this leads to multiple occurrences of a product. Product 07, which can be supplied by site A or B, is an example for such a situation.

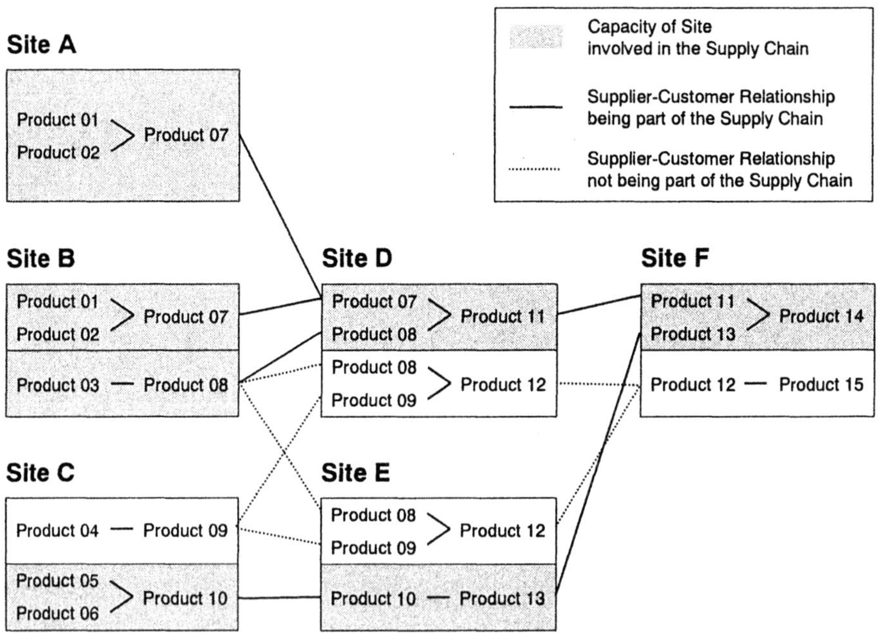

Figure 1. The supply chain – a product-specific part of a company's supply network

The question "What do supply chain operations look like?" is equivalent to "Who does what with which things in which order and according to which rules in the supply chain?" In modeling, four views allow answering this question: The organization, function, object and process view (see Figure 2). Most modeling techniques map two of these views, e.g. a flow chart has process as primary and function as secondary view, while in state diagrams object is the primary and process is the secondary view (see [2] and [6] for more information).

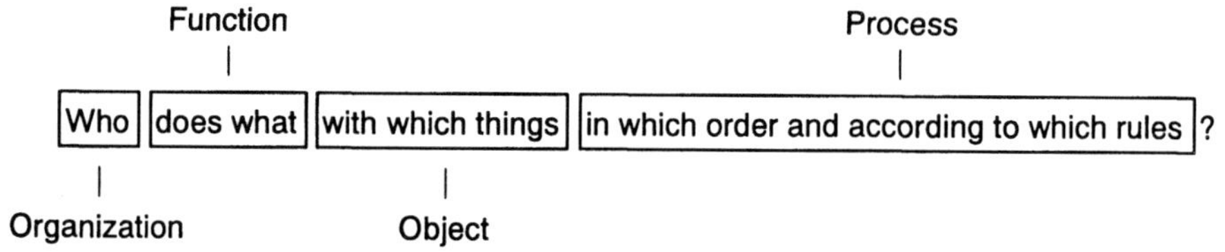

Figure 2. Aspects of supply chains - four modeling views

In the following we will look at which aspects of supply chains should be modeled in the four views and how this should be done to enable logistics managers to take well-founded decisions on their supply chain operations. Process and function view are strongly connected to each other, because order of functions and rules cannot be mapped without mapping functions themselves. Therefore, they are grouped in one sub-chapter.

The aspects are derived from experiences with modeling three supply chains, which are different in respect of production concept within the international research project ProdChain (IMS 99006, IST 2000-61205).

2.1 Aspects of a Supply Chain in the Organization View

Organizations are of interest in supply chains from two points of view:

- Companies own information, sites, capacity and products.
- Companies, their subsidiaries and the subsidiaries' departments are responsible for certain functions within a process. This implies, that organizations being responsible for functions – like functions them selves – have to be mapped on different levels of detail.

2.2 Aspects of a Supply Chain in the Object View

Organizations in a supply chain deal with objects: They buy products, produce and store them in their production and warehouse sites and sell them to customers (see Table 1). Size and weight of the product limit the means of transportation. The transportation costs can be calculated as a function of these product attributes, the mean of transportation and the distance between two sites given by their locations. Because the price of a product varies from supplier to supplier, its value should not be included in the model as an attribute of the product but rather as an attribute of the relationship "Supplier of product to a site", which is listed in Table 2.

Table 1. Aspects of a supply chain - objects

Object	Attributes
Product	Size, Weight
Site: Production	Location, Capacity/Contracted Capacity
Site: Warehouse	Location, Capacity

Distinguishing between production and warehouse sites is necessary in order to map the stocking level of products within the supply chain. This is reflected in the relationship "Stock" in Table 2. Similar to transportation costs, storage costs for products on stock can be calculated as a function – here mainly of product size and value.

Three types of flows take place in a supply chain: Material, money and information. While material and financial flow are directly related to a product and can be described in a structured way, information is as manifold that there probably is no structured approach to describe its flow. Though financial flow is mainly relevant on the company level, it should be mapped on the more detailed level of sites keeping the relation of material to money flow. Aggregation to the company level is possible by the relationship "Ownership of site".

Table 2. Aspects of a supply chain – relationships between objects

Relationship	between objects	Attributes
Bill of material	Product - Product	Number of components used for assembled product
Supplier of product to a site	Company - (Product - Site)	Price of product, delivery reliability
Stock	Product – Site	Number of products, storage costs
Material flow	Product - (Site - Site)	Time, mean of transportation, transportation costs
Financial flow	Product - (Site - Site)	Time, duties/tariffs to deduct
Information flow	Site – Site	Time, type of information, frequency
Ownership of site	Site – Company	
Ownership of product	Product - Company	

2.3 Aspects of a Supply Chain in the Process/Function View

The basic functions of a company are planning (balancing aggregated demand and supply to develop a course of action that best meets the requirements), sourcing (procuring goods and services), production (transformation of products to a finished state), product delivery (provide finished goods or services) and managing product returns. Each of these processes has interfaces to other partners in the supply chain – be it suppliers or customers. For understanding the supply chain, the processes within these five high-level groups (Plan, Source, Make, Deliver, Return) need to be mapped to identify the rules according to which the supply chain works. The SCOR model (see [4] and chapter 3) provides a good basis to cover aspects of supply chains in the process and function view.

3. CURRENT APPROACHES FOR MODELING SUPPLY CHAINS

As mentioned earlier, only integrated modeling approaches are able to cover the aspects of a supply chain. Such modeling approaches combine pure modeling techniques on different levels of detail and allow users to navigate between models as well as between levels of detail. Objectives of modeling usually are to analyze and improve a system. Integrated modeling approaches offer different ways to reach improvement (see Figure 3):

- Some modeling approaches stop with visualizing the current state. ICON SCC, for example, provides users with a full view of stock levels throughout the supply chain. But taking decisions based on this information is in the hand of users.
- Other approaches provide users with reference models. Firstly, those can be used to configure the current state from a "common" state of a system. Secondly they are benchmarks and can give hints to what a good system should look like.
- Simulation models give numeric feedback on the current and possible future states of a system under different scenarios. Users decide, which possible future state is the one to choose.
- Finally, optimization models calculate the best future state of a system by an algorithm (see [5], part two).

For humans modeling, analyzing and designing a supply chain only visualization and simulation models are in scope. In contrast to optimization models, those are little demanding in terms of modeling effort needed and rich in visual feedback. From reference modeling approaches, only process reference models cover a number of the aspects of supply chains mentioned.

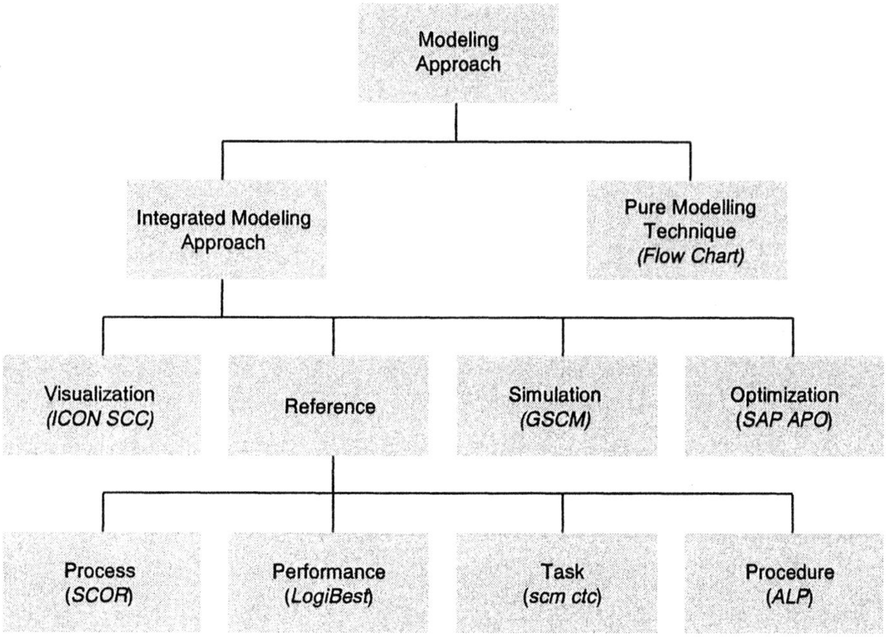

Figure 3. Hierarchy of modeling approaches for supply chains

The SCOR model has its focus on the process and function view. In its current version 5.0 the model (see [4]) describes 26 core processes down to a detailed level of elementary functions. Organization view, performance indicators and best practices are arranged around this main view. Its drawback is, that it is not capable to map product-related and financial information. GSCM, on the other hand, mainly covers the object view, having weaknesses in mapping organizational issues and the rules according to which a supply chain operates. For more information on GSCM, please refer to [1].

4. AN ADVANCED APPROACH TO SUPPLY CHAIN MODELING

Table 3 shows five charts that in combination answer the question "Who" (O) "does what" (F2, F3) "with which things" (B, G) "in which order and according to which rules?" (F2, F3). To reduce complexity, aspects covered by the chart are partly optional (marked in gray in Table 3), so that the end user can filter information.

Table 3. Combination of five modeling techniques covering the aspects of a supply chain

Chart	Organization View	Object View	Process/Function View
Organizational Chart (O)	– Company – Subsidiary – Department	– Site: Production – Site: Warehouse – Product: Stock Level – Flows – Costs	
Bill of Material (B)	– Company	– Site: Production – Site: Warehouse – Product:Bill of Material – Stock Level – Flows – Costs	
Geographic Overview (G)	– Company – Subsidiary	– Site: Production – Site: Warehouse – Product: Bill of Material with suppliers (see Figure 1) – Product: Stock Level – Flows – Costs	Processes from the five SCOR level 1 categories (Plan, Source, Make, Deliver, Return)
Flow Chart,Level 2 (F2)	– Company – Subsidiary – Department	– Flows	Processes from the 26 SCOR level 2 categories
Flow Chart,Level 3 (F3)	– Company – Subsidiary – Department	– Flows	Processes from SCOR level 3

This approach integrates the strong process and function view of the SCOR model and enriches it with product-related and financial information like financial flow, stock levels, storage and transportation costs.

The charts are connected to each other in two ways. On the one hand, they may include the same aspect (e.g. organizational chart and flow chart, which both have "company" as entity) allowing navigation on the same level of detail while keeping focus on the shared aspect. On the other hand, a chart may allow drilling down to a more detailed view of the same chart (e.g. from geographic overview to flow chart, level 2).

5. HOW INFORMATION TECHNOLOGY CAN SUPPORT SUPPLY CHAIN MODELING

Enriching the SCOR model with product-related and financial information has an obvious drawback: While the structural data necessesary for mapping a supply chain according to the SCOR model can be collected from logistics managers of the companies involved in about 9 hours, product-related data and financial information is of high volume and only available from a company's ERP system. Thus support of software with a link to an ERP system is crucial for applying the approach presented in chapter 4.

Most of the structural data gained from logistics managers in workshops can be obtained from an ERP system as well: Starting from the bill of material, suppliers of components can be identified. The result is the bill of material view (B). Geocoding suppliers then yields the geographic overview (G). Of course, an ERP system only can provide information on a company, its suppliers and its customers. Thus, mapping supply chains on more than these three levels means coordinating multiple ERP systems.

6. CONCLUSION

It has been presented what the aspects of a supply chain are, that logistics managers want to have mapped for taking informed decisions. These currently are covered by integrated modeling approaches only to a limited extend. The advanced approach to supply chain modeling enriches the SCOR model with product-related and financial information, for which it needs to link to an ERP system.

REFERENCES

1. Global Supply Chain Association. *Global Supply Chain Model*, URL: http://www.gsca.org/ (Accessed September 27, 2002).
2. Scheer, A. W. (1994). Business process engineering: reference models for industrial enterprises, Springer-Verlag, Berlin et. al, Germany
3. Schönsleben, P. (2000). Integral Logistics Management: Planning & Control of Comprehensive Business Processes, CRC/St. Lucie Press with APICS, Boca Raton, FL, USA
4. SCOR: Supply Chain Operations Reference Model, Version 5.0, August 2001, Supply Chain Council, Pittsburgh, PA, USA
5. Shapiro, J. (2001). Modeling the Supply Chain. Duxbury, Pacific Grove, CA, USA
6. Specker A. (2001). Modellierung von Informationssystemen: ein methodischer Leitfaden zur Projektabwicklung, vdf-Hochschulverlag, Zürich, Switzerland

11

MULTI-STRATA MODELING IN MCM AND CLM FOR COLLABORATIVE ENGINEERING

Kiyoshi Itoh[1], Ryo Kawabata[1] Akiko Hasegawa[1], and Satoshi Kumagai[2]
[1]*Dept. of Mechanical Eng., Sophia Univ., Japan*
[2]*Yamatake Corporation, Japan*
e-mail: itohkiyo@sophia.ac.jp

Abstract: In order to analyze collaboration of people, the author enhances the modeling techniques with specific constructs for the multiple strata in Multi-Context Map (MCM) and Collaborative Linkage Map (CLM). The first is the collaboration stratum where a collection of collaboration tasks is enumerated as top view for systems analysis. The second is the workflow stratum where, for each of collaboration tasks, workflow of request is clarified, and collaborators, tokens, material and information are enumerated. The third is the state transition stratum where state transitions for collaborators, material and information are clarified.

Key words: Multi-Strata Modeling, Collaborative Engineering

1. INTRODUCTION

Collaboration tasks are often observed in various fields. Collaboration task is an activity which involves multiple participants with a shared goal including individuals, personnel, organizations and facilities with different roles or perspectives in the process. Collaboration task is not a completely concurrent but appropriately coordinated process, because all of the participants in the process do not always produce, assemble, or modify material and/or information in a simultaneous and parallel manner.

To grasp collaboration tasks effectively, it is very important to analyze what workflows are, what multiple participants, i.e., collaborators, are doing, which position they are in, and how they communicate with each other, and how material and/or information are dealt with. When monolithic view or single strata modeling is adopted, the interrelationships between various entities and actions such as participants, workflows, materials, information resources, events, positions, communications are often mixed into a single description. Multi-strata modeling view is very effective for analyzing systems with collaboration tasks by distinguishing, stratification, categorizing, and scope or concept localization of these entities, actions and interrelationship, and by controlling the advancement of systems analysis in stepwise fashion from top view.

A system can be regarded as a collection of collaboration tasks. A collaboration task can be analyzed and modeled by focusing on two sides, general workflow (e.g., [5]) and state transitions. A workflow can be regarded as a collection of contexts, each of which switches one after another among the participants according to the differences of their perspectives. State transitions can represent the behavior of components participating in workflow. The authors devised Multi-Context Map (MCM) [4] for the description of workflow and Collaboration Linkage Map (CLM) [1] for the description of state transitions. MCM specifies the feasibility of the collaboration as well as contextual switches among collaborators. CLM specifies state transitions of collaborators and state transitions of material and information that are transferred among collaborators. Analyzing or modeling process in MCM and CLM enables to advance systems analysis in stepwise fashion from top view by multiple strata structure with collaboration tasks, workflows and state transitions.

2. CONCEPT OF CONTEXT AND WORKFLOW

Where an activity is performed in collaboration, there exist two types of collaborators, i.e., clients and servers with their own perspectives. Perspectives are views, opinions, or positions for the activity. Clients and servers represent requestors and performer, respectively. They are individuals, personnel, organizations and facilities or combinations. The authors call an activity by clients and servers with perspectives "context". Clients and servers are collaborators of the context. The authors' method is a method to systematically analyze collaboration tasks among perspectives that perform collaboration in the systems, and information and materials dealt in collaboration.

Context Map (CM) as a minimum element is shown in Figure 1. Two perspectives sandwich the context. The perspectives on the left and right of the context are called the "left-hand perspective by client" and the "right-hand perspective by server", respectively. This situation represents where perspectives collaborate with each other, like exchanging information, asking someone to do the job, or so on.

The context deals with three types of entities, - token, material and information – as inputs and outputs. Token represents fact or event, which is mutually recognized and acknowledged by two perspectives. Material represents an entity that is physical and has mass or weight. Information represents an entity that is logical. In the context, resources are used by left-hand perspective by client and the right-hand perspective by server.

CMs are combined as a network to build "Multi-Context Map (MCM)". A MCM can represent a workflow of a collaboration task. In the MCM as shown in Figure 2, the client for the first CM gives a request to the server, and then, the server, as the client for the second CM, gives a request to the server of the second CM. Workflow of request in collaboration task looks like relay of baton.

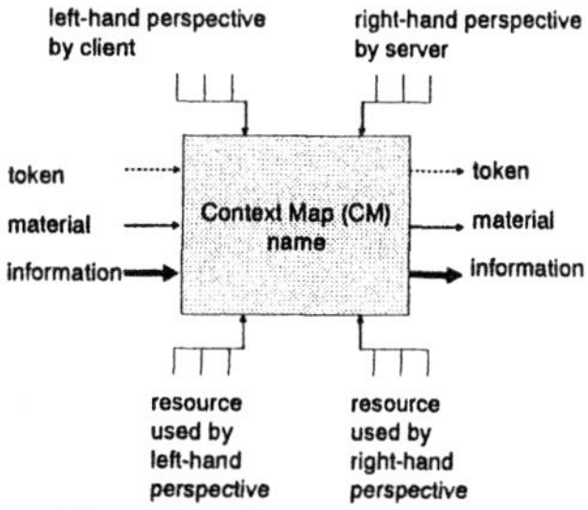

Figure 1. Context Map (CM): General

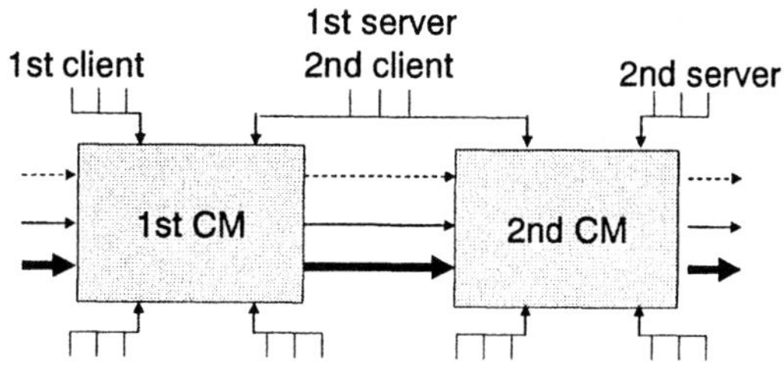

Figure 2. Multi-Context Map (MCM)

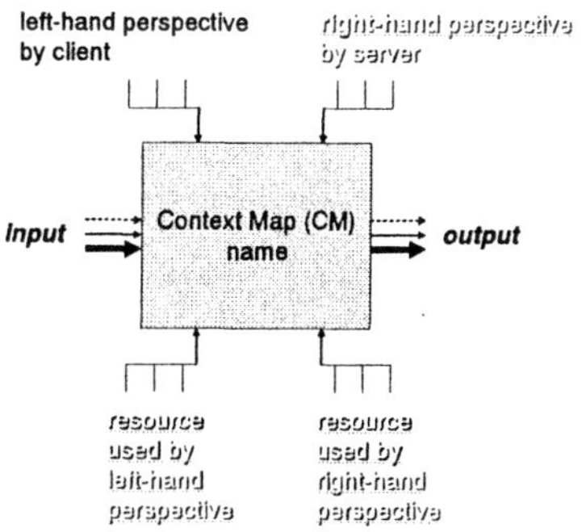

Figure 3. Context Map (CM): Top View

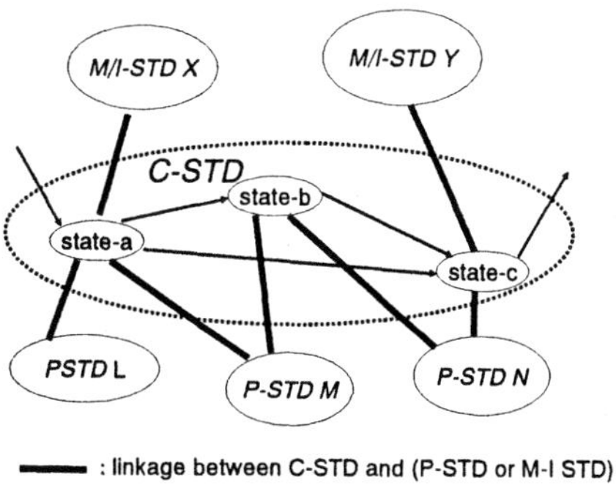

Figure 4. Collaborative Linkage Map

The top views of system can be analyzed from the viewpoint of collaborations. A system is considered to be a collection of collaboration tasks. At the first stage of systems analysis, the existence of collaboration tasks is recognized, and corresponding candidate contexts in top view are enumerated. Such a context is described in "Candidate Context Map (CCM)" as shown in Figure 3. The left-hand perspective by client should be identified, but the right-hand perspective by server may not always be. The differences of token, material and information may not always be identified as inputs and outputs.

3. CONCEPT OF STATE TRANSITION OF COLLABORATION COMPONENT

The workflow described in MCM can be mapped into a state transition diagram (STD). This STD is called "Collaboration STD (C-STD)". In a workflow, material, information, client and server have various behaviors in active or passive fashion. The behaviors of all of those can be mapped into STDs. STD for material and information is called "M/I-STD". STD for client and server is called "Personnel STD (P-STD)". Personnel mean individuals, organizations and facilities. Integration of C-STD, M/I-STD and P-STD is called "Collaborative Linkage Map (CLM)" as shown in Figure 4.

Both P-STD and M-STD represent situations by three types of states. "Enaction state" represents the situation that P-STD is acting on the M-STD and connects both STDs with Collaboration-State. "Commission-state" is defined when one P-STD has any kind of relations with other P-STDs and the information exchanged in such situation is defined as "Communication-Inventory". "Dormant state" is the state neither of them and does not have any relationships to other state diagrams.

4. MULTI-STRATA IN MCM AND CLM MODELING

Modeling in MCM and CLM for collaborative engineering has three-strata as shown in Figure 5. The first is the "collaboration stratum" where a collection of collaboration tasks is enumerated as top view for systems analysis. The second is the "workflow stratum" where, for each of collaboration tasks, a corresponding workflow is clarified, and collaborators, tokens, material and information are enumerated. The third is the "state transition stratum" where state transitions for collaborators, material and information are clarified.

4.1 COLLABORATION STRATUM

The existence of collaboration tasks is recognized. For each of collaboration tasks, a corresponding candidate context in top view is enumerated as Candidate Context Map (CCM). For CCM, the left-hand perspective by client should be identified, but the right-hand perspective by server may not always be. The differences of token, material and information may not always be identified in CCM.

4.2 WORKFLOW STRATUM

For each of collaboration tasks by CCM, a corresponding workflow is constructed by MCM (i.e., network by one or more CMs). The left-hand perspective of CCM becomes the left-hand perspective of the first CM in MCM. Its right-hand perspective should be clarified and becomes the left-hand perspective of the second CM. The situation follows like this to the downstream. The structure of MCM is not only straightforward but also a network one where branch, parallel and loop patterns are permitted. For each CM, perspectives should be determined. For the connecting CMs, token, material and information should be determined.

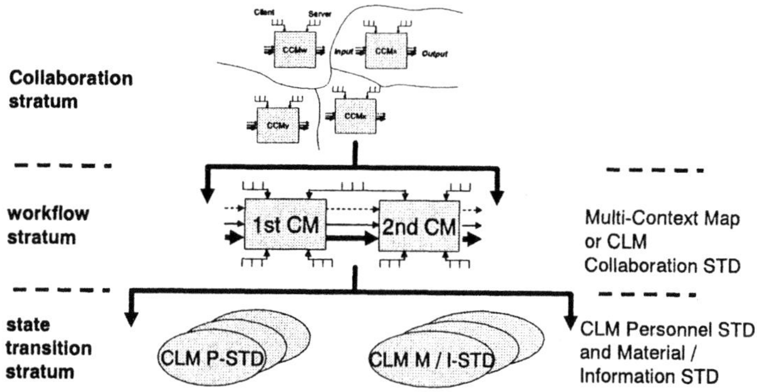

Figure 5. Multi-Strata in MCM and CLM

Junctions are provided for various network structures to show the behavior of the workflow clearly. Branch junction defines the situation that one thing goes only to the one of the outputs and it depends on the condition where to go. Duplication junction copies the input to all of the outputs. Decomposition junction decomposes its input, which means one thing breaks, into two or more outputs. Synchronization junction waits all the supposed multiple inputs come before producing the output. Serialization junction joins the multiple inputs to the output like a flow in the river.

4.3 STATE TRANSITION STRATUM

The workflow described in MCM can be mapped into a Collaboration STD (C-STD). While a request on workflow is processed, material, information, client and server have state transitions actively or passively. The behaviors of all of those can be mapped into P-STDs and M/I STDs.

5. EXAMPLE OF MULTI-STRATA

Figure 6 shows Collaboration Stratum for Academic Affair. “Allocate”, “Enter Registration”, “Lecture”, “Make Exercise for Report”, “Submit Report” and “Certificate” are enumerated as CCMs. Figure 7 shows Workflow Stratum for “Certificate” in Academic Affair. “Determine Certificate”, “Send Certificate” and “Send Failure Notification” are enumerated as CMs. All parameters for these CMs are specified. These CMs are connected into MCM as workflow. Figure 8 shows State Transition Stratum for “Certificate” in Academic Affair. C-STD corresponds to the MCM. P-STDs are specified for “Lecturer”, “Clerk for Certification” and “Clerk for Sending”. M/I STDs are specified for “List of Marks”, “Member List”, “Certificate” and “Failure Notification”.

6. CONCLUDING REMARKS

In order to analyze collaboration of people, the author enhances the modeling techniques with specific constructs for the multiple strata in MCM and CLM. Constructs in MCM and CLM can be organized in different strata with distinct connotations. Each stratum defines a certain group of those constructs with particular semantic content (e.g. collaboration, workflow, and state transitions of resources). In collaborative engineering, it is defined as strata to provide different views of a system The relationships among stratum are not represented appropriately in homogeneous hierarchical structure of diagrams.

MCM and CLM Editors are developed to support to analyze collaboration task [2]. Extraction process of E-R model from MCM and CLM is developed [3].

This research has been partially supported by IMS/RPD & HUMACS program (1999-2001), Open Research Center Project (2002) and Grants-in-Aid for Scientific Research (2002) of Ministry of Education, Culture, Sports, Science and Technology of Japan.

REFERENCES

1. Hasegawa, A., Kumagai, S., Itoh, K.: Collaboration Task Analysis by Identifying Multi-Context and Collaborative Linkage, CERA Journal, Vol.8, No.1, pp.61-71, 2000.
2. Itoh, K., Kawabata, R., Hasegawa, A., Kamimura, M., Kumagai, S.: CHAPTER 1: An Approach for Application of Domain Oriented Systems Development Using Triplex Schema of Key Concepts, in "Itoh, K., Hirota, T., Kumagai, S., ed.: Domain Oriented Systems Development: Perspectives and Practices", Taylor & Francis, 2002, to appear.
3. Kamimura, M., Kato, N., Kawabata, R., Kumagai, S., Itoh, K.: Extracting E-R models from Collaboration Analysis Methods, MCM and CLM, in "Arisawa, H., et al., ed.: Conceptual Modeling for New Information System Technologies", vol. 2465, Lecture Notes in Computer Science, Springer, 2002, to appear.
4. Kumagai, S., Hasegawa, A., Kawabata, R., Itoh, K.: Building Workflow for Collaboration Task using Multi-Context Map, Journal of Integrated Design and Process Science, Vol.4, No.3, pp. 49-61, 2000.
5. Workflow Management Coalition (WFMC): The Workflow Management Coalition Specification, 1996.

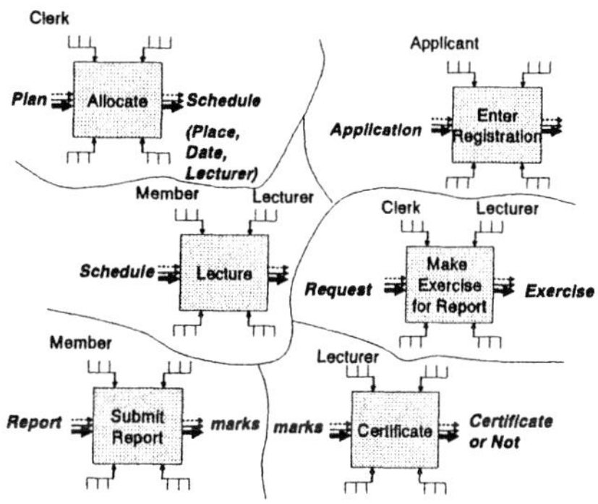

Figure 6. Collaboration Stratum for Academic Affair

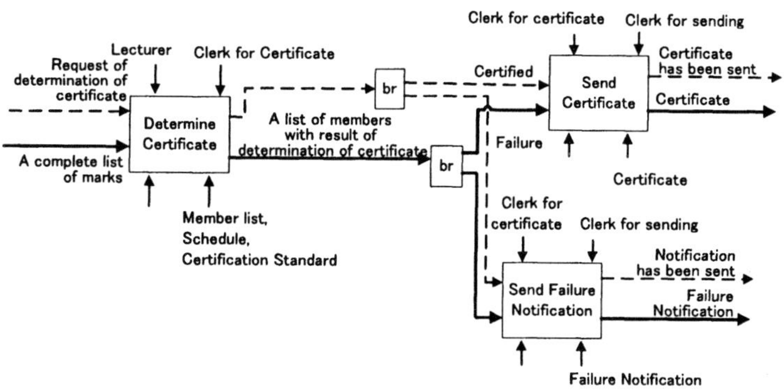

Figure 7. Workflow Stratum for Certificate in Academic Affair

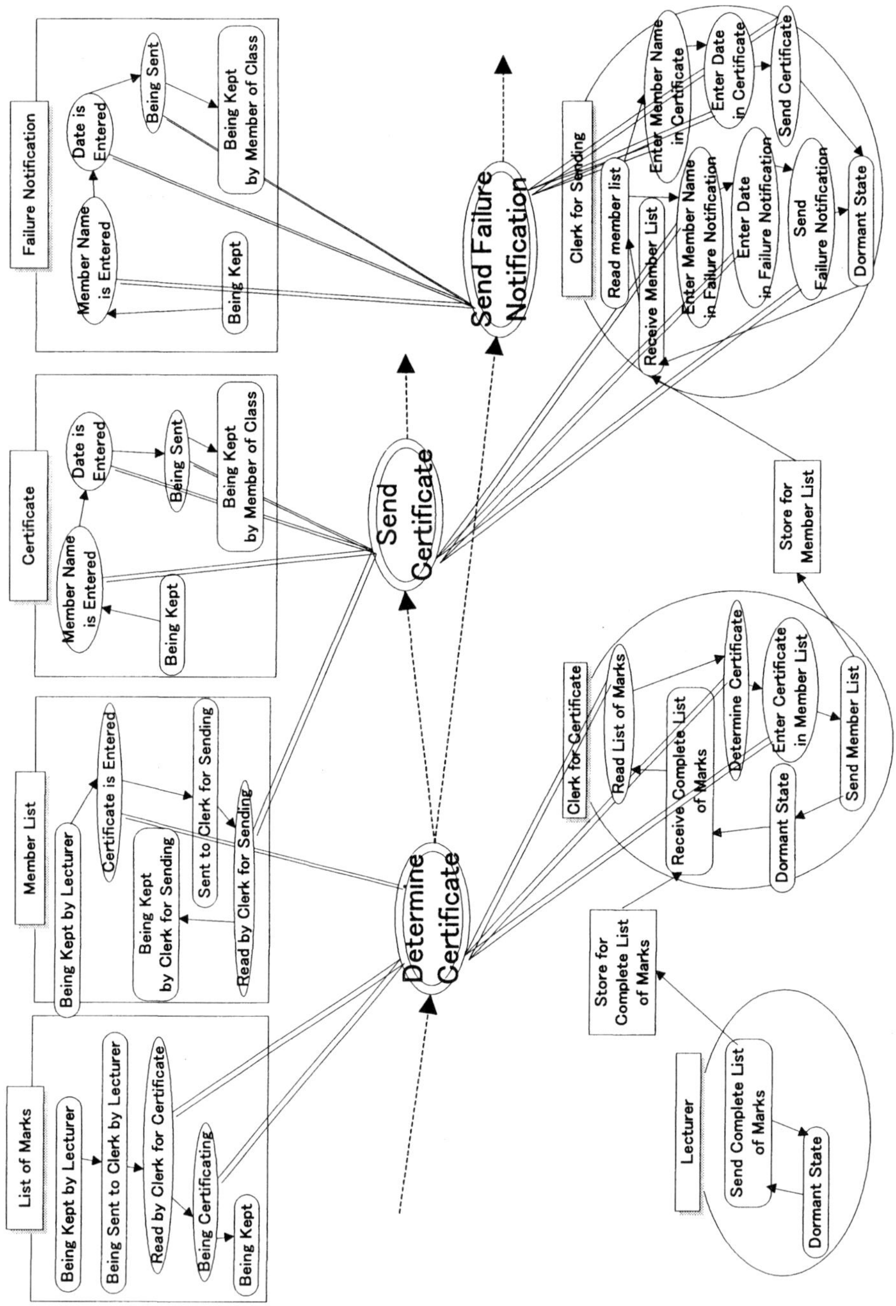

Figure 8. State Transition Stratum for Certificate in Academic Affair

12

ONTOLOGICAL STRATIFICATION IN AN ECOLOGY OF INFOHABITANTS

V.A. Abramov[1], J.B.M. Goossenaerts[1], P. De Wilde[2], and L. Correia[3]
[1]*Eindhoven University of Technology, the Netherlands*
[2]*Imperial College of Science, Technology and Medicine, London, UK*
[3]*New University of Lisbon, Portugal*
e-mail: J.B.M.Goossenaerts@tm.tue.nl

Abstract: This paper reports progress from the EEII research project where ontological stratification is applied in the study of openness. We explain a stratification approach to reduce the overall complexity of conceptual models, and to enhance their modularity. A distinction is made between ontological and epistemological stratification. The application of the stratification approach to agent system design is explained and illustrated. A preliminary characterization of the relevant strata is given. The wider relevance of this result for information infrastructure design is addressed: ontological stratification will be key to the model management and semantic interoperability in a ubiquitous and model driven information infrastructure.

Keywords: Ontological stratification, agent systems, semantic interoperability

1. INTRODUCTION

Agent system research and development has become an active field. There are many examples of using agent systems to solve tasks in so various areas as electronic commerce [1,2], control [3], communication and networks [4], resource management [5], search and information retrieval [6], etc. On the other hand, the novelty and at the same time immaturity of the existing results hinders introduction of agent systems into real life applications. One of the major reasons for that is the fact that most of the research efforts are concentrated on *designing* agents, components of agents, or agent systems. There is little time left for analysis of the properties of the designed systems.

Questions about such system properties as openness, robustness, stability, adaptability, scalability, should be answered before an agent system is deployed in industrial applications and infrastructures. However, only few research results exist that help to answer these questions qualitatively [7] and even less – quantitatively [3]. There is still a large gap between the number of developed agent systems and number of analysis methods and tools for their evaluation.

In order to address this gap the EEII project (Evolution and Ecology of Interacting Infohabitants, EU IST-1999-10304) is developing an ecological approach to modeling and analysis of such agent system properties as *scalability*, *openness*, *stability*, and *adaptability*. The methods applied within this approach include analytical estimations, discrete simulations of agent communities [8], and theoretical considerations regarding system design methods [9]. One of the methods developed is a macro-level ecological analysis method oriented on investigation of agent system openness. Essential part of this method is the practical application of stratification approach influenced by other work on piecemeal ontological commitment in a participative simulation environment [10] and by other multi-strata models [11]. A two-sided stratification approach, with ontological and epistemic stratification, is described in the next section. Then we describe the application of the approach to the agent system design. Finally we draw conclusions about the stratification approach and conjecture about its possible role within a ubiquitous and model driven information infrastructure [12].

2. APPLYING STRATIFICATION

2.1 Ontological Stratification

The concept of ontological stratification has been initially developed in the field of Formal Ontology [13]. In essence, *ontological stratification* is the partitioning of (conceptualization of) an information system's world into several disjoint spaces (or strata) according to the identity criteria used to conceptualize the entities in these spaces. The following set of spaces is proposed by Guarino (with examples of corresponding entities): *static* (a situation, a configuration), *mereological* (an amount of matter), *physical*: (a) *topological* (a piece of matter) and (b) *morphological* (a cubic block), *functional* (an artifact), *biological* (a human body), *intentional* (a person or a robot), *social* (a company).

It is easy to see that the same object may belong to one or more spaces, e.g., an apple may be considered as (i) an amount of matter when we are concerned with transportation of apples, (ii) a food when we want to satisfy our hunger, (iii) a seed when we think about planting an apple tree, or (iv) a product when an apple goes to a market. In all of these cases, different aspects of the same apple are considered, and different laws and rules are applied: transportation of an apple is mostly concerned with the laws of physics, using it as a food – with the functional considerations, using it as a seed – with the biological laws, selling it on market – with the laws of economics in modern society.

2.2 Epistemic Stratification

In modern society, when acquiring disciplinary knowledge, abstraction is used to separate all the abovementioned domains. The partitioning and single-stratum in-depth study of topics accelerates knowledge creation within a particular domain (linked to a stratum) without necessary involvement of knowledge from all the other domains.

Investigation of ontological commitment for participative simulation has revealed a distinction between the being and becoming of entities as studied in ontology proper, and the concept-enabled analyses in various knowledge domains [14,15]. The conceptual models that underlie these analyses are usually also referred to as part of ontology, but conformance with the use of terms in philosophy would position the study of these models in epistemology, the theory of knowledge [16]. Hence the use of the term *epistemic stratification* to denote the diversity of specialized knowledge models that are used by specialists of several domains of expertise, e.g. ergonomy, logistics, physics, sociology.

2.3 Stratification and Information Systems

Complex information systems are often based on models that describe the application domain, its objects, processes, tasks as well as available knowledge related to the task solving. Ontological and knowledge models used in such systems are often related to several spaces or strata in the Formal Ontology sense. Therefore it is reasonable to apply the above stratification approach to these models in order to reduce their overall complexity and increase modularity. This would facilitate modification, extension and re-use of information system models.

In our work, we deal with ontology and knowledge models related to several domains of expertise, and apply the stratification approach to both of these models.

3. STRATIFICATION IN AGENT SYSTEM DESIGN

In our current research work, we are developing an agent system that is to be used within a complex domain. The system has to deal with different objects such as *a physical object*, *a network node*, *a software agent*, *a person*, etc. The system is distributed and networked. It should be used by many users to support their individual task-solving activities. Typical tasks are participation in distributed trade, in supply networks, etc.

The main components of the system are *user* (someone, typically a human, who uses other components of the system to solve its tasks), *agent* (someone, typically a software agent, who assists its user in solving tasks), and *object* (object of task).

These system components naturally belong to several different spaces. By distinguishing and specifying their involvement in specific spaces we can get a more clear description of the overall system. Within each space, more clear and compact specifications are possible of the tasks involving entities present in the space. For example, a (sub)-task of physical transportation of goods is mostly described in terms of entities of physical space. In contrast, a (sub)-task of demand-request matching involves only entities of conceptual space.

Finally and most important for our current research, the ontological stratification allows us to clearly distinguish, model and evaluate different types of agent system openness. For example, we can separately model and study information openness, conceptual openness, intentional openness, etc.

3.1 The stratified agent system model

We consider the agent system consisting of the following high-level types of components: *user*, *agent* and *object*. Components are complex and composite; we consider their *sub-components* or *aspects* to be part of the several distinct spaces. Each sub-component is located in one of the spaces (e.g., think of location in the physical space, network space, etc.) and is related to other entities of the same space in different ways.

Clearly, a product and a client are entities that should belong to the social space, because these entities only emerge in the context of social interactions. On the other hand, an apple is just a physical entity. An apple can as well be a food, or a product, but these entities belong to other spaces – functional and social correspondingly. There is a choice: either to consider such entities as an apple, a food, and a product as a single object or to admit xistence of three different (though dependent) entities. Together with Guarino, we accept the latter choice. As a result, the components of our agent system consist of interdependent sub-components in several spaces. In

Figure 1, this way of ontological stratification is illustrated with concepts of the generic agent system model.

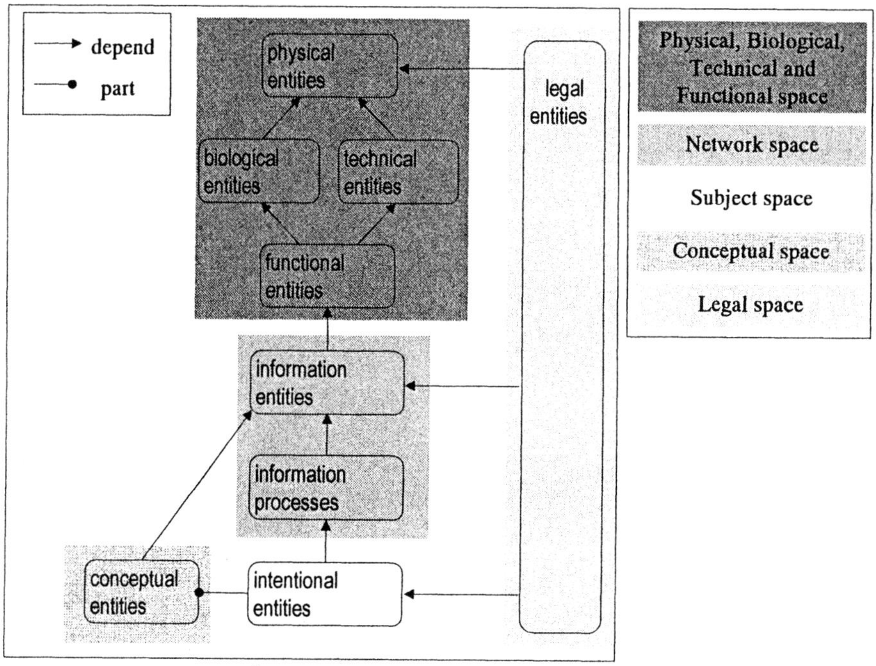

Figure 1. Stratification of generic agent system ontology

In the design of our test agent system, only part of the depicted spaces is used as described here. The *subject space* concerns the component's aspect of having certain goals, and being able to act intentionally. Clearly, *users* and *agents* have this aspect and therefore have presence in this space. The *legal space* contains legal aspects of the system components and legal relations among them. For example, user can *own* some *property*. *Users* and *objects* have presence in this space. The *conceptual space* contains all the *concepts* of an applied domain that are used by all the active components of the system. The *network space* consists of a number of interconnected nodes of an information network. Everything in this space has a network address, i.e., every sub-component of the agent system's components resides at some node. *Users*, *agents* and (information) *objects* have presence in this space. The physical space is organized by a number of geographically distributed sites that are interconnected by transportation links. Each site is associated with a unique geographical address. Every component that has presence in this space has an address, i.e., at any point of time it must be at a unique site. *Users, agents* and *objects* will have their presence in PS. Table 1 summarizes the presence of system components in described spaces:

Table 1. Presence of components in different spaces

Space	User	Agent	Object
Subject space	subject	subject	
Legal space	legal body		legal body
Conceptual space	concept(s)	concept(s)	
Network space	information body	information body	information body (inf. objects only)
Physical space	physical body	physical body (variable for mobile agents, invariable otherwise)	physical body

3.2 An application scenario: the trading task

An agent system simulating a distributed market place has been developed using the stratified ontology. In the system, there are a number of *users* who produce and consume some *products*. The users are distributed in the physical space. Users are interested in buying some products that other users have. Also, they want to sell their own products to earn money. A typical user task is to buy or sell specific amount of certain product. There are many types of products being traded by the users. These types are organized by individual users into partial classifications. All these classifications are individual, not precise and may partially overlap each other. Under these circumstances, the task of the agent system and each of the agents is to satisfy interests of the users.

In order to solve this task, one or more static and mobile agents may be employed by users. Interacting in the network space, the agents find trade partners, products, negotiate deals, sign and implement contracts. The trading process involves making agreement on the object of trade (*ontology negotiation* – each agent has its own applied ontology describing the types of products it can deal with) and price agreement (*price negotiation*). The ontology negotiation is an inherently error-prone process since agents do not share the standard ontology.

The crucial part of the experiment with the system is the study of the agent openness. The agents are provided with certain capabilities that can make them open when these capabilities are utilized. In particular, an agent can share parts of its experience with other agents (information openness). Also, an agent can receive from other agents and adopt concepts about new objects in the environment (conceptual openness). It can recognise those objects when familiar with the concepts. Finally, an agent can choose to share its task-solving intentions with (partner) agents (intentional openness). There are several more types of openness that have been identified in our research. At the moment, the agent system simulations provide a way to study the three types of openness mentioned above. When doing the simulations, we measure individual and population-level performance of agents and relate it to the level and type of agent openness. In Figure 2 the screenshots of the running simulation are presented.

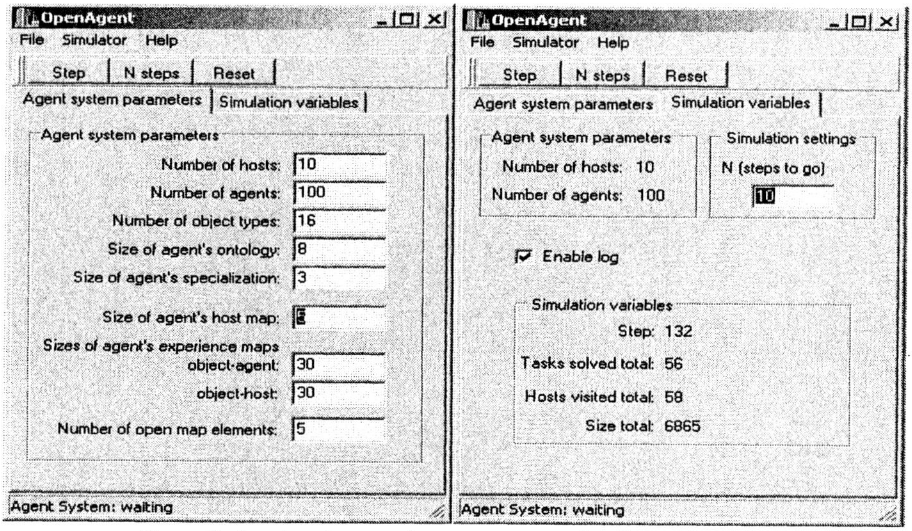

Figure 2. Agent system running an openness-related experiment

During the development of the agent system, a clear advantage has been identified of using a stratified conceptual model of the system. The fact that the key components of the system have been described as sets of sub-components and that all spaces have been separately specified, has increased the system modularity. It has become easily possible to modify the structure of each of the spaces (for example, connectivity model of the network space, or the structure of the conceptual space) without affecting the other spaces. It has been also possible to change the laws governing the entities behaviour in each of the spaces independently.

Separation of conceptual, information and intentional entities into individual spaces has made clear the distinction between the respective types of agent openness. It also made possible independent experiments with each of these openness types. At the current stage of our work, we conduct these experiments to measure dependence of agents' performance on their openness.

4. CONCLUSIONS AND FUTURE WORK

Ontological stratification of the agent system conceptual model has provided us with a better ground for experiments with the system for two main reasons: it provided more structured and modular description of the system thus enabling easier design modifications, and it gave us a way to clearly distinguish types of entities relevant to the openness-related experiments that we conduct with the system.

Ontological stratification in general promises to enable constructing better ontologies for the information systems and their infrastructure. Epistemic stratification recognizes that specialized knowledge models are required for different domains of expertise and allows for easier joint work of specialists from these domains on a single project.

Research continues on the application of the stratification approach in the architecting of a model driven information infrastructure, aiming for results on semantic interoperability and efficient model management.

REFERENCES

1. Chavez, A., and P. Maes (1996) Kasbah: An agent marketplace for buying and selling goods. *Proc. 1st Int. Conf. on the Practical Application of Intelligent Agents and Multi-Agent Technology*, pp 75-90
2. Sycara, K., K. Decker, & D. Zeng (1998) Intelligent Agents in Portfolio Management. In: N. Jennings & M. Woolridge (eds) Agent Technology: Foundations, Applications, and Markets, Springer, pp. 267-283.
3. Fregene, K., D. Kennedy, and D. Wang (2001) HICA: A Framework for Distributed Multiagent Control, *Proc. Int. Conf. on Intelligent Systems and Control*, Tampa, USA. pp 187-192
4. Schoonderwoerd, R., O. Holland, & J. Bruten (1997) Ant-like agents for load balancing in telecommunications networks. In: W. L. Johnson & B. Hayes-Roth (eds) *Proc. 1st Int. Conf. on Autonomous Agents*, ACM Press, pp 209-216
5. Chavez, A., A. Moukas, & P. Maes (1997) Challenger: A multiagent system for distributed resource allocation. In *Proc. Int. Conf. on Autonomous Agents*, Marina Del Ray,California.
6. Knoblock, C.A., Y. Arens, & C. Hsu (1994) Cooperating agents for information retrieval. *Proc. 2nd Intl. Conf. on Cooperative Information Systems*. Univ. of Toronto Press.
7. Lerman, K. (2001) Design and Mathematical Analysis of Agent-based Systems, Lecture Notes in Artificial Intelligence (LNAI 1871), Springer Verlag, pp 222-234
8. Marwala T., P. De Wilde, V. Abramov, N. Szirbik, J. Goossenaerts, L. Correia, P. Mariano, R. Ribeiro (2001) Scalability and optimisation of a committee of agents using genetic algorithm. In *Proc. Int. Symp. on Soft Computing and Intelligent Systems for Industry*, Paisely, Scotland.
9. Abramov, V.A., N. Szirbik, J.B.M. Goossenaerts, T. Marwala, P. De Wilde, L. Correia, P. Mariano and R. Ribeiro, Ontological Basis for Open Distributed Multi-Agent System, *Symposium on Adaptive Agents and Multi-Agent Systems, AISB'01*, York, UK, March, 2001.
10. Goossenaerts, J., Pelletier, C. (2002) Ontological Commitment for Participative Simulation. In: H. Arisawa, Y. Kambayashi, V. Kumar, H.C. Mayr, I. Hunt (eds) *Conceptual Modeling for New Information Systems Technologies*. Lecture Notes in Computer Science 2465, Springer Verlag, pp. 127-140

11. Hasegawa, A., Kumagai, S., Itoh, K. (2000) Collaboration Task Analysis by Identifying Multi-Context and Collaborative Linkage, Concurrent Engineering Research and Applications Journal, 8(1), pp 100-108
12. Guarino, N. (1997) Some organizing principles for a unified top-level ontology. Internal report 02/97 V3.0, LADSEB-CNR, Italy.
13. Goossenaerts, J.B.M. (2003) Architecting a Ubiquitous & Model Driven Information Infrastructure (this volume).
14. Heisig, P., Callot, M., Goossenaerts, J., Kosanke, K., Krogstie, J., Stojanovic, N. (2002) Anchoring Knowledge in Business Process Models to support the Interoperability of Virtual Organizations. In: Kosanke, K., Jochem R., Ortiz, A., Nell, J.G. (eds) *Enterprise Inter- and Intra Organization Integration: Building International Consensus*, Kluwer Academic Publishers.
15. Goossenaerts, J.B.M., Pelletier, C. (2002) Ontology and Enterprise Modelling. In: Eijnatten van, F.M. (ed.) *Participative Simulation Environment for Integral Manufacturing Enterprise Renewal*. TNO Arbeid, Amsterdam, The Netherlands, pp.41-52.
16. Audi, R. (1998) Epistemology – a contemporary introduction to the theory of knowledge, Routledge, London and New York.

13

LOGICS OF BECOMING IN SCHEDULING

Logical Movement behind Temporality

Junichi Yagi[1], Eiji Arai[2], and Keiichi Shirase[3]
[1]*Shimizu Corporation,* [2]*Osaka University,* [3]*Kobe University*
e-mail: jun@sit.shimz.co.jp

Abstract: Scheduling is for optimal allocation of time resource for activities. This paper reconsiders the concept of time as a dynamic framework for series of actions. Time resource is viewed as elastic bodies to incorporate fluctuation of becoming of events and their dynamic behaviour. Lastly, logic of becoming will be presented as foundation of scheduling.

Key words: internal time, external time, Fourier decomposition, action operator, dynamic scheduling, becoming, reflectors, origin of temporality

1 INTRODUCTION

It is the time resource that is dealt with directly under any scheduling scheme. The task is how can be achieved the optimal allocation of a given time resource over a set of assigned activities. The temporal order of the assigned activities is usually described in a Network Chart schematically. Then there are various existing methods how to allocate the activities over time, such as Gantt chart, Multi-Activity Chart [3] called MAC, or else. Most of the existing methods use a parametric time, linear axis of time, on which the activities are orderly allocated in accordance with the order of events. They provide, however, a planner with only a means of static description of time resource distribution at his planning stage, or mere projection of his temporary thoughts in succession. Two basic problems are faced in scheduling, namely; (1) (planning stage) the provided data by which one allocates time resource to a particular construction activity is usually far from precise unlike machine operation in manufacturing. In another word, the beginning and ending nodes of each time interval allocated for a particular activity are both fluctuating to a notable extent. Most of the existing scheduling methods [2][4][6] can not incorporate uncertainty of this sort sufficiently enough or not at all. (2) (operating stage) The need to reallocate the time resource as a construction project proceeds often rises. Although the final goal of the project is clearly defined, no existing scheduling method provides a dynamic adjustment mechanism in reference to the project's goal.

This paper will develop a mathematical model of the dynamic scheduling to deal with both of these fundamental problems directly. The eventual fluctuation is shown given rise originally by the characteristic nature of *becoming* in the temporality. It will be also shown possible that an elastic model can be developed to incorporate the afore-mentioned temporality *per se*.

2 NON-LOCALLY DISTRIBUTED TEMPORALITY

The parametric time is used for the ordinary scheduling scheme in two ways; (1) to display an optimal allocation of the pre-assigned activities on the parametric time in sequence at the planning stage before the project starts, or that of the activities yet to be completed in the middle of the project, (2) to cast its advance of the progressive completion over the parametric time. In both cases, the parametric time is just an indicative tool like a one- dimensional screen over which the project shows its allocation in sequence and casts its progress.

However, a real project cannot be represented solely by the parametric time, for it consists of series of decisions for all sorts of "this or that", this event or that event, this way or that way, this sequence or that sequence, this divergence or that divergence and so forth. At every moment of decision all the potentials are activated [7]. Two streams of activities in the form of potentials then flow into the current activity at present, one from the past and the other from the future. They convolute at present, bounce around their possible ways of settlement, come to a final settlement as to their optimal mixture, and, in the end, transform the integrated potentials into an event at present.

There exist therefore two different temporal orders; the implicit and explicit temporality [1], one for decisional act and the other for its consequential representation. They categorically differ in such a way that *potentials* are dealt with in the implicit order of time, but dealt with *events* in the explicit order of time. However they are distinct, they are dynamically integrated within the dynamics of the current action, two sides of the same action. The combinatory temporal order, Internal Time by External Time is depicted at Fig.1. Each internal time is non-locally distributed over the pre-assigned temporal whole, Ω with density function$f_j(t)$ *such that* $\sum_j f_j(t) = 1$

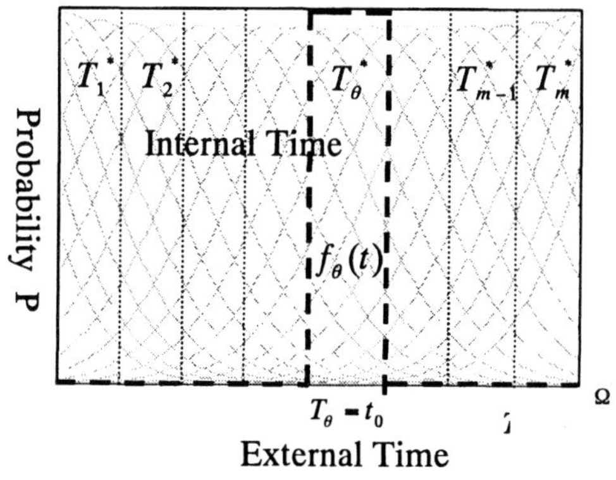

Figure 1. Internal Time and External Time

, $\forall t \in \Omega$ The event produced by the current activity is dichotomous whether it happens (1) or it does not happen (0). This eventual dichotomy however becomes itself explicitly through superposition of all the internal potentials implicated at present by the current action, $\sigma_{op} = \sum_j f_j(t)$.

3 HOLOGRAPHIC MODEL

3.1 Action Operator

A holographic model can be developed based on the combinatory temporality to incorporate a dynamic aspect of scheduling and a fluctuation of work periods.

The set of density functions $\{f_\theta(\tau)\}_{\theta \in \Omega}$ represent collectively the collection of the internal times, each of which is non-locally distributed over the pre-assigned temporal whole Ω. For the sake of convenience, the finite interval of Ω can be chosen as that which ranges from –π to +π without loss of generality. The underlined densities are assumed to satisfy one of the following conditions; (1) symmetric condition and (2) integrally symmetric condition. The latter is more generic condition in the form of $\{f_\theta : \int_{-\pi}^{\pi} f_\theta(\tau) d\theta = \int_{-\pi}^{\pi} f_\tau(\theta) d\theta\}$

By performing the Fourier harmonic decomposition for each density in the collection of internal times, the collection can be rearranged on a unit sphere surface to model the decision process in scheduling. At each moment, an operator of action integrates the whole surface of the sphere from specific angle and produces an temporal event (1 – an event happens, 0 — an event does not happen).

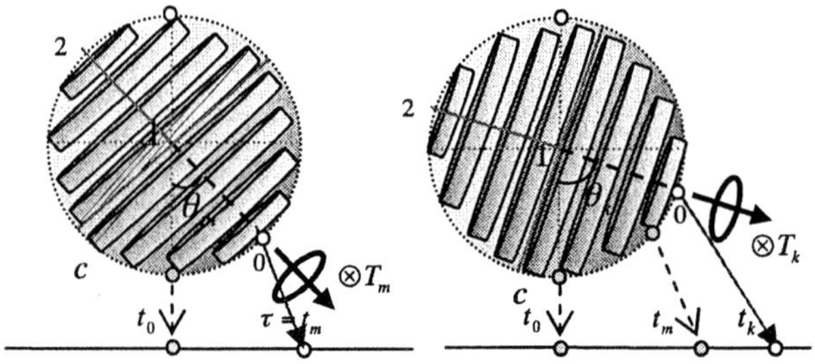

Figure 2. Holographic Operator Acts on the Sphere

The coefficient $\hat{c}_0^*(s)$ on each slice of the sphere are obtained by rearrangement of the Fourier harmonics of the collection of internal time densities. Therefore, the sphere gives a geometric representation for the collection of the internal times, each of which is the probability of "the particular time event happens". The action operator σ_{op} is defined as follows.

$$\sigma_{op} \cdot \otimes T_\tau = 1 + \frac{1}{2\pi}\int_0^{4\pi} \hat{c}_\tau^*(s)ds = 1 \qquad (1)$$

, where the operator acts on the total information of the collection of internal times distributed over the sphere surface at temporal angle τ .

3.2 Temporal Strain

The consecutive actions by the operator on the sphere produce the external temporal events out of the constellations of the internal time potentials. For the future is not realized event, it produces 0 and for the past was realized, it produces 1. Conversely, the internal time potentials are constellated so as to produce either 0 or 1 when integrated, provided that a scheduling is set up well balanced without any strains.

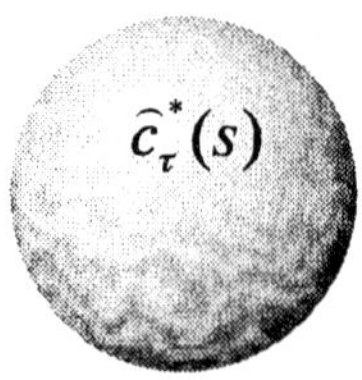

Figure 3. Off-Balanced State of Constellation of Internal Time Potentials

When there are strains in scheduling, overburden or waste of use of time resource, it produces more than 1 or less than 1 respectively. The off-balanced state of the internal time constellation is shown at Fig.3.

4 DYNAMICS

4.1 Restitution Force

Virtual force can be induced out of the potential space described above of the internal times as restitution force. The restitution force is a force required to restore a scheduling with positive or negative strains to an ideal scheduling with no strain. Activities are allocated in sequence on a critical path with some period as depicted at Fig.4. Each activity is the aggregate of the densities of the internal times within the period. All of the potential curves, obtained by aggregation, of the activities cross at the probability of 0.5 in a scheduling with no strain. Whereas, the intersections of the off-balanced sequence of the activities are away from the balancing point, where the restitution force acts on by the Hooke's law.

$$F_i = -k_i x_i \tag{2}$$

The larger the standard deviation of the underlined density of an internal time is, the smaller the modulus coefficient is. The modulus is the inverse of the standard deviation, for the stronger non-locality of the density distribution allows the internal time more elastic. Suppose the density is normally distributed, that is the case for most of application, the relation of probability density and restitution force becomes explicit as shown at Eq. 3.

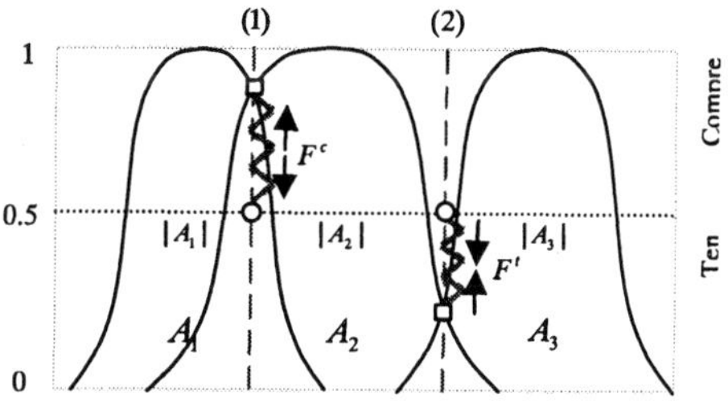

Figure 4. Restitution Force and Internal Time

$$F(x) = -\sigma^{-1}x$$
$$p(x) = \frac{1}{\sqrt{2\pi}\sigma}\exp[-\frac{1}{2}(\sigma^{-1}x)^2] = \frac{1}{\sqrt{2\pi}\sigma}\exp[-\frac{1}{2}F^2(x)] \tag{3}$$

4.2 Coupled Elasticity

Each consecutive work activity on a critical path generally is distributed as normal distribution with its distinct values for the characteristic parameters, the mean and standard deviation. The path can be viewed therefore as a serial connection of multiple elastic bodies as shown at Fig.5.

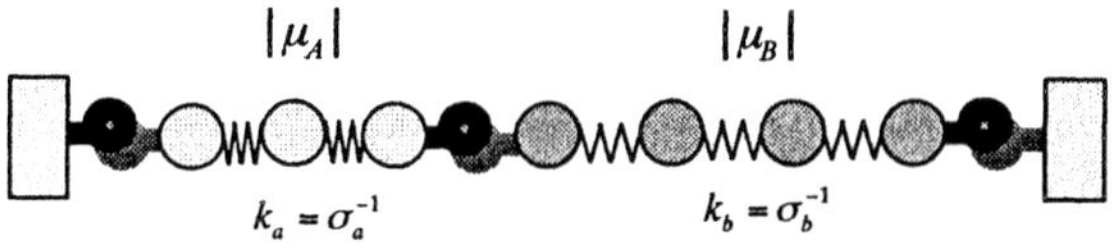

Figure 5. Coupling of Elastic Bodies of Time Resource

The multiple coupling of elastic bodies can be formulated as shown at Fig.6. The internal forces f_i's follow the Hooke's linear law, while their coupling behavior is generally non-linear. Their stable equilibrium solutions reveal the distribution of negative or positive strains in the scheduling under the external force F. Exertion of the external force here means controlling the whole period of a critical path in the project under various boundary conditions on its constituent activity elements whose moduli or inverse of standard deviations are allowed to vary due to various possible allocation of man x hour.

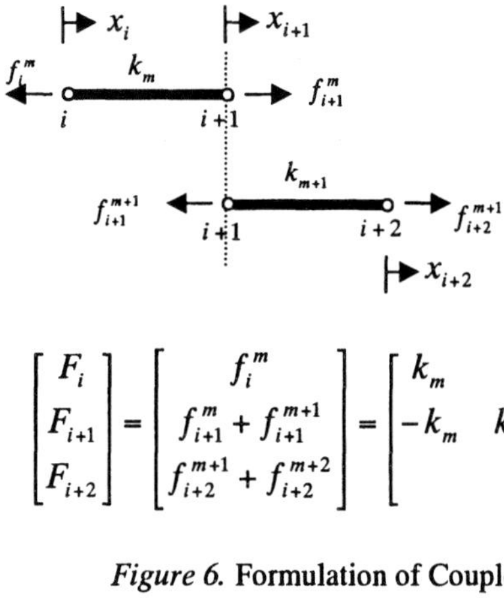

$$\begin{bmatrix} f_i^m \\ f_{i+1}^m \end{bmatrix} = \begin{bmatrix} k_m & -k_m \\ -k_m & k_m \end{bmatrix} \begin{bmatrix} x_i \\ x_{i+1} \end{bmatrix}$$

$$\begin{bmatrix} f_{i+1}^{m+1} \\ f_{i+2}^{m+1} \end{bmatrix} = \begin{bmatrix} k_{m+1} & -k_{m+1} \\ -k_{m+1} & k_{m+1} \end{bmatrix} \begin{bmatrix} x_{i+1} \\ x_{i+2} \end{bmatrix}$$

$$\begin{bmatrix} F_i \\ F_{i+1} \\ F_{i+2} \end{bmatrix} = \begin{bmatrix} f_i^m \\ f_{i+1}^m + f_{i+1}^{m+1} \\ f_{i+2}^{m+1} + f_{i+2}^{m+2} \end{bmatrix} = \begin{bmatrix} k_m & -k_m & \\ -k_m & k_m + k_{m+1} & -k_{m+1} \\ & -k_{m+1} & k_m + k_{m+1} \end{bmatrix} \begin{bmatrix} x_i \\ x_{i+1} \\ x_{i+2} \end{bmatrix}$$

Figure 6. Formulation of Coupled Elastic Time Resource

5 LOGICS OF BECOMING

The series of work activities are allocated on a critical path in scheduling. The elastic model for allocation of time resource over activities allows scheduling a dynamic analysis for planning and conducting schedule. Scheduling becomes analytic to examine stress distribution, where no stress indicates well-balanced schedule with neither waste nor over-burden of allocation of time resource and constant stress is the second best. Variety of appropriate boundary conditions can be imposed for examination of stress distribution and their effects are simulated. When over-stressed part is identified, reallocation of work load can be made before it happens.

Gantt chart is widely used for scheduling. It is however not an analytical tool, but only representational, for no internal structure is available. Non-locally distributed internal time made an analytical scheduling possible, where the modulus coefficient of elasticity is simply the inverse of temporal deviation. Temporal order is now re-examined closely to reveal it is the source of dynamics, where potential probability and force are conjoined.

5.1 Extended Dichotomy

The collection of internal times is a temporal potential space. The external time of events is an event space. The action operator acts on the temporal potential space independently at each moment to produce an event by integrating the latent potentials in its own way. The whole potentials get into each moment with its unique settlement. Or two streams of potential activities flow into the current activity at present, one from the past and the other from the future, convolute at present, bounce around their possible ways of settlement, come to a final settlement, and, in the end, transform the final settlement into an event at present as current optimal mixture of all the temporal potentials.

The series of moments is time. Every moment is distinct. Its distinctness however differs from an ordinary sense. It is a *distinct mixture of the same whole of the potentials*. Here is the point of departure from standard dichotomous logics. Moment A is now $A \times \overline{A}$, the direct product of A and $\overline{A}$, both of which are contained in A as potency. Moment $\overline{A}$ is likewise $\overline{A} \times A$.

$$\begin{aligned} (A \times \overline{A}) \cup (\overline{A} \times A) &= \Omega^{*} \\ (A \times \overline{A}) \cap (\overline{A} \times A) &= \phi \end{aligned} \qquad (4)$$

The law of excluded middle is preserved. However it is the logics of extended dichotomy for *becoming* [5]. It contrasts markedly with the standard dichotomous logic, logics of being.

5.2 Duality in Logics

These contrasted logics are in fact dual sub-logics of the even higher logics. The logical space Ω x Ω can be expanded as follows.

$$\Omega \times \Omega = [A \times A \,\dot{\cup}\, \overline{A} \times \overline{A}] \,\dot{\cup}\, [A \times \overline{A} \,\dot{\cup}\, \overline{A} \times A] = \Omega \,\dot{\cup}\, \Omega^{*} \qquad (5)$$

The self-referential logical space Ω x Ω contains both of the standard and extended dichotomous logics as shown at Fig.7 (a). These dual sub-logics are ceaselessly moving in and out each other at every moment via the action operator σ_{op} as shown at Fig.7 (b). The self-referential logical space maintains itself by this ceaseless movement.

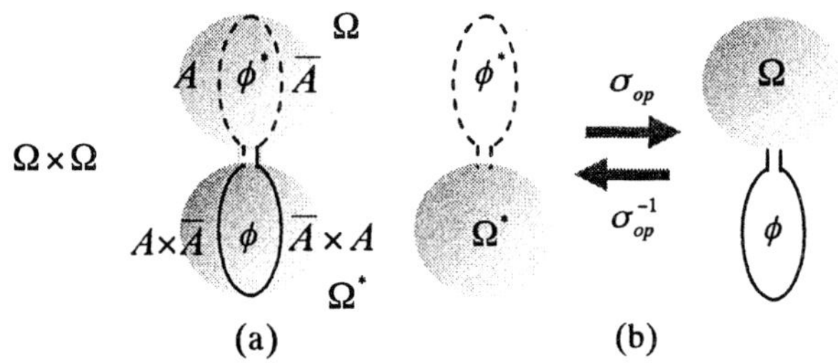

Figure 7. Logical Space and Internal Movement

6 CONCLUSION

It has been demonstrated that the reformulation of temporality more fitted to an actual decision process makes it possible to provide us with an analytical tool for scheduling. The elastic model was derived from the temporal probability, where the dynamics of elasticity and probabilistic fluctuation are conjoined in the theory. The union is rooted in temporality of becoming.

7 REFERENCES

1. Hilly. B.J., Peat, F.D: Quantum Implications, Routledge & Kegan Paul London and New York, 1991
2. Kano, Naruo: Construction Process Planning Methods, IAJ Award 1995, Institute of Architecture of Japan, 1995
3. Matsumoto, S. et al.: Scheduling of Construction Activities by Expert System, Operations Research of Japan, vol.32 No.11, 1987
4. Nagaoka et.al: Process Plan and Management for Repetitive Construction Projects, Institute of Architecture of Japan, No.492, 1997
5. Prigogine, Ilya: From Being to Becoming, W.H.Freeman And Company, 1980
6. Shoji.M, Arai, K: Research on Optimization of Process Plan Management (No.1), Annual Report of Kajima Research Institute, Vol.21, 1973
7. Yagi et al: Theory of Non-locally Distributed Time Resource, Institute of Architecture of Japan, July 2002

14

COMMUNICATION IN THE DIGITAL CITY AND ARTIFACT LIVES

Victor V. Kryssanov[1], Masayuki Okabe[1], Koh Kakusho[2], and Michihiko Minoh[2]
[1]*Japan Science and Technology Corporation*
[2]*Academic Center for Computing and Media Studies, Kyoto University, Japan*
e-mail: kryssanov@mm.media.kyoto-u.ac.jp

Abstract: This paper proposes a theoretical basis for the design and analysis of distributed information systems. A quantitative criterion is defined to estimate the efficiency of computer-mediated communication, and to monitor artifact lives as well. The theoretical concepts are discussed in a context of an example related to car use and servicing.

Key words: Distributed information systems, User interface design, Semiotics

1. INTRODUCTION

Recent advances in networked technologies and in performance of computing devices created the present situation, where the most critical problems in information system design concern with user interface. A user interface of an information system may be understood as a collection of symbols and interaction procedures that are to enable communication of the system users. The now generally recognized requirement of interface adaptability to both a particular user and dynamically changing system contents quickly made the task of interface design too complex to rely on hand-crafted decision procedures. Presently, however, no single theory exists that could provide comprehensive guidelines for the design and analysis of distributed information systems, such as digital cities, corporate memories, or the entire World-Wide Web [5].

In this study, we attempt to establish a theoretical basis – a new communication theory – by introducing fundamental axioms that would allow for the development of a science of information system design with its own explanatory and experimental apparatus. (Some applications of the proposed theory have been presented elsewhere [3-4].)

2. COMMUNICATION IN A DIGITAL CITY

A digital city usually comes as a distributed information system comprising a range of information resources associated with a certain geographical place or a specific human activity [5]. The principal function of a digital city is to support navigation of its users in the physical or abstract space by providing relevant information in a timely manner. To do this, the digital city enables various social interactions between users (e.g. through posting and retrieving documents), i.e. it enables communication of its users.

To illustrate major difficulties in designing and using digital city interfaces, let us consider a typical situation where a user of a digital city interacts with the system to find and, perhaps, go to "a good car-maintenance service station." First, the user must formulate her or his goal in terms, which can effectively and correctly be understood by the system. There may or may not be an appropriate pre-defined hyperlink (e.g. "car repair"), while submitting a query (e.g. "car maintenance") to a search engine will, as a rule, produce hyperlinks of arbitrary relevance to the user's goal (e.g. FAQ on the topic or any other document indexed with the query keywords). The user should yet unequivocally express her or his subjective notion of the "good" (e.g. some repair shop would be "good" in terms of service quality, but be too far from the user's location). To be effective, the interface has to develop and utilize a model of the user's goal and behavior.

Next, the user may access a component of the digital city (e.g. a site representing a "physical" car service-station). The semantics of the original query may then change, owing to the component design (e.g. some instructions on car maintenance or a "how-to-reach" map) and/or experience and practice currently prevailing in the society of the system's users (e.g. a particular shop may be considered "good" just because it is conveniently located and/or it is, at the moment, most strongly associated with the query keywords, owing to feedback from previous users).

After all, while browsing, the user may refine or even change her or his goal (e.g. to "to find a gas station"), based on the information learned from the different recourses (e.g. that the desired services are offered by gas stations). Therefore, to successfully communicate, the user has dynamically to adjust her or his behaviour and reconcile her or his subjective semantics with semantics implemented in the different parts of the system. It is understood that the efficiency of this adaptation process depends on both, the individual and the social dynamics of communication.

As a digital city is composed of a number of independently developed heterogeneous recourses – system components, its user interface usually consists of multiple parts that may employ different languages of different "digital genres" (e.g. text vs. graphics) and implement different and even conflicting semantics corresponding to different communities (e.g. "Western" *vs.* "Eastern") of the system users. To make the digital city interface capable of social adaptation, it is necessary to develop an integrated perspective on the communication process. Any of the current holistic approaches to modeling communication is, however, limited in the sense that it is based on an *ex ante* analysis, which assumes the (pre)existence of a society of the system users. In the case of digital cities, such a society does not pre-exist but emerges after a version of the digital city has been set up and made accessible. This emerged society is not stable, as not stable is the structure of the system: the users as well as the resources incorporated in the digital city have internal dynamics and are autonomous but mutually dependent when communicate. Being allied (functionally and/or communicatively), they create a complex system with a dynamics generally indefinable in terms of the classical communication theories. Thus, new approaches need to be found to provide guidelines for the design and development of distributed information systems, such as digital cities.

3. FROM QUANTITATIVE SEMIOTICS TO A NEW COMMUNICATION THEORY

3.1 Conceptual Framework

We will assume that all the systems (e.g. psychic) involved into communication are (higher order) autopoietic systems acting in the consensual domain (for details, see [5] and [4]). Each of these systems "belongs" to at least one self-organizing social system seen as a realization of the consensual domain. We will also assume that a psychic system engaged in communication is composed of interpretants (i.e. meanings) and is observationally equivalent to the totality of experientially effective behavior called objects. The social system is composed of signs and is equivalent to the totality of behavior maintaining the social system as a whole. (It should be noted that our treatment of the semiotic triad *object-sign-interpretant*, although does not generally contradict to the concept of infinite semiosis, moves forward from the canonical Peircean definition.)

We will consider communication as a partial sequence of interdependent semiosis processes $C=\{S_1,S_2,\ldots,S_K\}$, where $S_t=\{Object_t,Sign_t,Interpretant_t\}$ is a single semiosis process specified through its manifestation (that is an interpreted sign), and t is a discrete time-mark. The dynamics of the communication process is described as follows:

$$\begin{cases} \mathbf{Objects}_{t+1} = \text{Externalizing}(\mathbf{Objects}_t, \mathit{PsychicState}_t) \\ \mathit{PsychicState}_{t+1} = \text{Interpreting}(\mathit{PsychicState}_t, \mathbf{Signs}_{t+1}) \end{cases} \tag{1a}$$

$$\begin{cases} \mathbf{Signs}_{t+1} = \text{Authorizing}(\mathbf{Signs}_t, \mathit{SocialState}_t) \\ \mathit{SocialState}_{t+1} = \text{Evolving}(\mathit{SocialState}_t, \mathbf{Objects}_{t+1}) \end{cases} \tag{1b}$$

where "**Objects**" is a state vector representing behavior, i.e. psychic states as (self)observed, which are communicatively effective, and "**Signs**" is a state vector representing behavior socially valid. "Externalizing" and "Interpreting" are operators that represent the uttering and the understanding processes, respectively; likewise, "Authorizing" and "Evolving" represent the corresponding implied processes of social dynamics (see Fig. 1).

3.2 Quantitative Semiotics: Essential Definitions

To refine and make the conceptual model (1*a*-*b*) formal, the apparatus of algebraic semiotics can be used [1]. We will consider a sign system Ξ a logical theory that consists of ordered sets of symbols (for details, see [1]). We will call a semiotic morphism $f:\Xi\to\Xi'$ a translation from a sign system Ξ to a sign system Ξ'. We will also call a composition of semiotic morphisms $\mu_{t+1} : f_t[\Xi_t] \xrightarrow[P_{t+1}]{} \Xi_{t+1}$ as a basic semiotic component, where f_t is a single semiotic morphism on Ξ_t, and μ_{t+1} is a *probabilistic semiotic morphism* that specifies a set of L_{t+1} possible translations from Ξ_t to Ξ_{t+1} with probabilities $P_{t+1} = \{p_1, p_2, \ldots, p_{L_{t+1}}\}$, one for each translation.

Axiom I. A psychic system can be represented by a sign system Ξ. The state of the psychic system is completely described by a set of signs in Ξ. ♦

Definition 1. Two states of the psychic system, α and β, are called orthogonal, written $\alpha \perp \beta$, if α implies the negation of β, or vice versa. ♦

Definition 2. For a subset of states $A \subset \Xi$, its orthogonal complement is $A^{\perp} = \{\alpha \in A \mid \forall \alpha' \in A^{\perp} : \alpha \perp \alpha'\}$. ♦

Definition 3. $A \subset \Xi$ is orthogonally closed if $A = A^{\perp\perp}$. ♦

Definition 4. We will call *Object* an orthogonally closed set of psychic states with a single *Interpretant* understood as a distinction. ♦

An interpretant is a psychic state but also the result of interpretation. E.g., the experiencing of a "good repair shop" implies certain psychic states, and

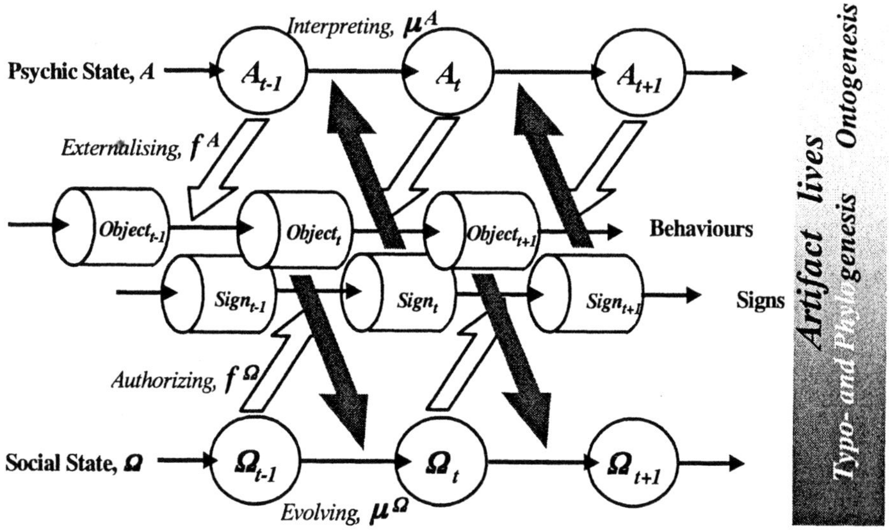

Figure 1. Semiosis of communication and artifact lives (also see [2])

interpretation of the signs "good repair shop" will result in a psychic state determined by the (past) experiences of the interpreter.

Axiom II. Right after an interpretation of an *Object* standing for some psychic states γ, which resulted in α, the psychic system is represented by α, i.e. the original states γ are translated to the *Interpretant* α by the interpretation. ♦

Similarly with a quantum system, the psychic system is in multiple states at once, and it cannot uniquely be interpreted: at every single moment, there can be made more than just one interpretation of the psychic system state. E.g. a repair shop can be "good" but "remote" for a particular customer.

Proposition I (Context Principle). For every two distinct psychic states $\alpha \neq \beta \subset \Xi$, there exists a context state $\alpha \cup \beta = \gamma \subset \Xi$ such that $\forall\delta \subset \Xi$, if $\delta \perp \alpha$ and $\delta \perp \beta$, then $\delta \perp \gamma$. ♦

Axiom III. Each interpreted psychic state can be represented in a unique way by a probabilistic semiotic morphism μ with $P = \{p_1, p_2, \ldots, p_L\}$ on Ξ. The probabilities of the morphism correspond to the possible interpretation results. ♦ (Probabilities *P* are determined with Axiom V, which is omitted in this paper, using the notion of semantic distance [4].)

An interpretant exists always only to the extent as the corresponding psychic states (i.e. the domain of μ) are accessible for interpretation. E.g., the words "a good repair shop" can hardly be associated with a unique psychic state unless an additional language context is provided.

Axiom IV. For a psychic system engaged into communication, the dynamics of the communication process is given by a pair of sequences of basic semiotic components defined recurrently as follows (see also Fig. 1):

$$\mathrm{A} = \{\mathrm{M}^{\mathrm{A}}, \mathrm{P}^{\mathrm{A}}, \mathrm{F}^{\mathrm{A}}, \Xi_{objects}\}, \tag{2a}$$

$$\Omega = \{\mathrm{M}^{\Omega}, \mathrm{P}^{\Omega}, \mathrm{F}^{\Omega}, \Xi_{signs}\}, \tag{2b}$$

where A is the model of the psychic system that includes M^{A} a set of semiotic morphisms μ_{t+1}, P^{A} a set of probabilities for each μ_{t+1} in M^{A}, F^{A} a set of semiotic morphisms f_t, $t=1,\ldots,K$, and $\Xi_{objects} = \bigcup_{m=1}^{M} Object_m$, M is the number of the interpretants by the psychic system prior to the communication. Ω is the model of the social system with analogously defined M^{Ω}, P^{Ω}, and F^{Ω}, and $\Xi_{signs} = \bigcup_{n=1}^{N} \Xi_{objects_n}$, where N is the number of psychic systems constituting the social system. ♦

3.3 Semiosis of Communication and Artifact Lives

(Closure) Theorem I. A communication is orthogonally closed *pragmatically* through the laws of nature in the sense that given an interpretant $Interpretant_t$, it is only the physical laws that determine its object $Object_t$ so that $Object_t = Object_t^{\perp\perp}$; *semantically* through the psychic system in the sense that $\Xi_{objects} = \Xi_{objects}^{\perp\perp}$; and *syntactically* through the social system in the sense that $\Xi_{signs} = \Xi_{signs}^{\perp\perp}$. ♦

It follows from the theorem (for the proof, see [4]) that psychic states corresponding to every physically possible *Object* should uniquely be determined as indicative of the given communication situation, but also that *Object* corresponding to a psychic state does not have to be unique. Besides, the theorem dictates that every single communication is orthogonally closed only to a degree. Indeed, given a communication situation, its pragmatic closure can be established if one considers all the possible *Objects*, which are to express the physical frames of the situation and to establish the interpretant (e.g. a perception or emotion). The latter is not a practical case (unless one considers learning by trial and error), and *Objects* are results of some relations (that are not necessarily conventions) developed from individual experience rather than exhaustive representations of the psychic state. It is obvious (due to Axiom IV) that semantic and syntactic closures are hardly reachable, too. Hence, *every single communication is uncertain*.

Lemma I. Given a communication situation with a pragmatic uncertainty *Const*, a natural limitation on the minimal requisite interaction for a psychic system engaged into the communication is determined by the degree of the communication closure $E_{O,M}$. The latter is inversely proportional to the communication uncertainty and can be estimated using the following formula ($E_{O,M} \in [0,1]$, $E_{O,M}=1$ is for the absolute certainty):

$$E_{O,M} = k_s \sum_{i=1}^{M} \left(\sum_{j=1}^{M} \frac{N_{Int}(Object_i \cap Object_j)}{N_{Int}(Object_i \cup Object_j)} - 1 \right) + k_c \sum_{i=1}^{O} \left(\sum_{j=1}^{O} \frac{N_{Int}(Sign_i \cap Sign_j)}{N_{Int}(Sign_i \cup Sign_j)} - 1 \right) - Const =$$
$$= k_s \sum_{i=1}^{M} \left(\sum_{j=1}^{M} \frac{N_{Int}(Object_i \cap Object_j)}{N_{Int}(Object_i) + N_{Int}(Object_j) - N_{Int}(Object_i \cap Object_j)} - 1 \right) + \qquad (3)$$
$$+ k_c \sum_{i=1}^{O} \left(\sum_{j=1}^{O} \frac{N_{Int}(Sign_i \cap Sign_j)}{N_{Int}(Sign_i) + N_{Int}(Sign_j) - N_{Int}(Sign_i \cap Sign_j)} - 1 \right) - Const,$$

where M is the number of *Objects* produced by the psychic system in the given communication, O is the number of *Signs* received by the psychic system, $N_{Int}(Object_i)$ is the number of interpretants by the psychic system for the same object $Object_i$, $N_{Int}(Object_i \cap Object_j)$ is the number of interpretants for both $Object_i$ and $Object_j$, $N_{Int}(Sign_i)$ is the number of interpretants of the same sign $Sign_i$ in the social system, $N_{Int}(Sign_i \cap Sign_j)$ is the number of common interpretants of $Sign_i$ and $Sign_j$, k_c and k_s are normalizing coefficients. Note that generally, $M \neq O$. ♦ (The proof can be found in [4].)

Const depends on pragmatics of the communication situation that may change throughout the lives of the artifacts (if any) in focus. In the case of a car repair shop, *Const* depends on the characteristics of the shop as a physical object, e.g. its organization. Having assumed that the uncertainty caused by the social dynamics converges to a constant for each particular topic in the given social system, one can (though indirectly) evaluate and assess the lives of artifacts-subjects of communication by monitoring the dynamics of $E_{O,M}$ (see Fig. 2). On the other hand, whenever *Const* is independent of time, $E_{O,M}$ can be used as a measure of the interface/information system quality or communication efficiency [3].

4. CONCLUDING REMARKS

In this paper, we have presented the new communication theory that offers a systematic basis for the description of communication phenomena. The proposed framework provides guidelines for the design and development of information systems capable of socio-cognitive adaptation, while the formalization permits us to qualitatively assess the efficiency of communication in each particular case. Furthermore, the theory explicitly defines and allows for exploring, both theoretically and empirically, the relationship between artifact lives and the efficiency of communication. All these are novel contributions of this study.

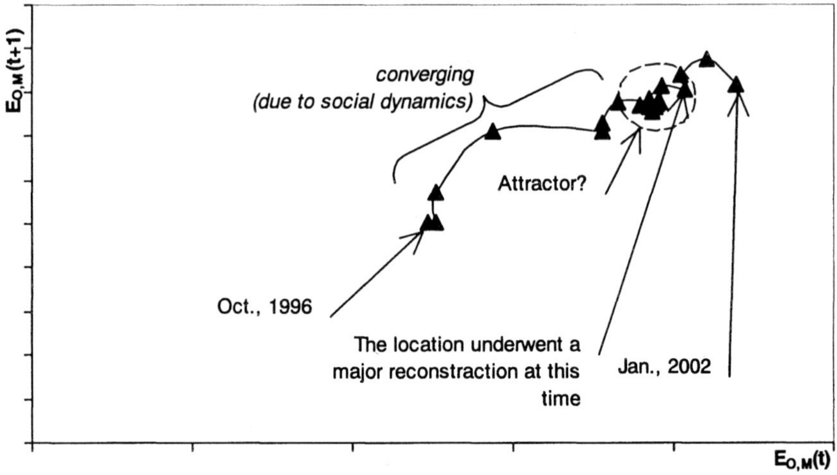

Figure 2. Monitoring artifact lives: Return map for $E_{O,M}$ (t) calculated for a (part of a) "Digital City" associated with a certain geographical location for the period from Oct., 1996, to Feb., 2002, with 1 month subinterval (65 months); the data obtained from the Web Archive http://www.archive.org

ACKNOWLEDGEMENTS

This work is part of the Universal Design of Digital City project funded by the Japan Science and Technology Corporation (JST).

REFERENCES

1. Goguen, J.A. (1999). Social and semiotic analyses for theorem prover user interface design. *Formal Aspects of Computing*, **11**, 272-301
2. Kryssanov, V.V., Goossenaerts, J.B.M. (2001). Modelling Semiosis of Design, In: J.P.T. Mo and L. Nemes (Eds), *Global Engineering, Manufacturing and Enterprise Networks*, pp. 111-118. Kluwer Academic Publishers
3. Kryssanov, V.V., Okabe, M., Kakusho, K., Minoh, M. (2002). A Theory of Communication for User Interface Design of Distributed Information Systems, In: C. Rolland, S. Brinkkemper, M. Saeki (Eds), *Engineering Information Systems in the Internet Context*, pp. 333-356. Kluwer Academic Publishers.
4. Kryssanov, V.V., Okabe, M., Kakusho, K., Minoh, M. (2002). A Theory of Communication for User Interface Design, In: R. Jorna, H. Gazendam, R. Cijsouw (Eds), *Proc. Fifth Workshop on Organizational Semiotics*, pp.145-161. Delft, The Netherlands
5. Kryssanov, V.V., Okabe, M., Kakusho, K., Minoh, M. (2002). Communication of Social Agents and the Digital City – A Semiotic Perspective, In: M. Tanabe, P.V. den Besselaar, T. Ishida (Eds): *Digital Cities II, Computational and Sociological Approaches*. LNCS **2362**, Springer, 56-70

15

VALIDATING MEDIQUAL CONSTRUCTS

Reliability, Empathy, Assurance, Tangibles, and Responsiveness

Sang-Gun Lee and Jae H. Min
School of Business, Sogang University, Korea
e-mail: sglee@unlserve.unl.edu or jaemin@sogang.ac.kr

Abstract: *In this paper, we validate MEDIQUAL constructs through the different media users in help desk service. In previous research, only two end-users' constructs were used: assurance and responsiveness. In this paper, we extend MEDIQUAL constructs to include reliability, empathy, assurance, tangibles, and responsiveness, which are based on the SERVQUAL theory. The results suggest that: 1) five MEDIQUAL constructs are validated through the factor analysis. That is, importance of the constructs have relatively high correlations between measures of the same construct using different methods and low correlations between measures of the constructs that are expected to differ; and 2) five MEDIQUAL constructs are statistically significant on media users' satisfaction in help desk service by regression analysis.*

Key words: SERVQUAL, MEDIQUAL, Media Choice, Help Desk

1. INTRODUCTION

Recently, IT support for end-users emerged as one of the leading concerns in an organization [1]. Continuous adopting and updating of communication technologies have driven most organizations to develop more effective help desk or customer service centers. Organizations are always looking for new ways to provide better help desk service in order to satisfy the growing demands and expectations of customers [2].

In recent help desk research, following the growing demands of new technology support, a number of commercial products have started using advanced techniques, such as self-service customer-center problem resolution, the instant help function, and desk automation. Remote on-line troubleshooting and Internet-based products are examples of more advanced technologies that are currently being used to support customer centers [3].

The main purpose of this study is twofold. First, we extend MEDIQUAL constructs in previous research [4] to reliability, empathy, assurance, tangibles, and responsiveness, based on the SERVQUAL theory. Second, we validate MEDIQUAL constructs as new belief criteria on media users' satisfaction.

2. LITERATURE REVIEW

Most early studies on media choice have paid attention to social presence and media richness theory. The researchers define social presence as "the degree to which a medium permits communicators to experience others as being psychologically present" or "the degree to which a medium is perceived to convey the actual presence of the communicating participants" [5]. According to the social presence theory, communication media are perceived as rating in social presence. Social presence is then determined by the degree to which one medium transmits information about facial expression, direction of looking, posture, dress and nonverbal, vocal cues. For example, conventional media such as face-to-face and group meetings are perceived as ranked high in social presence. By contrast, electronic media such as early e-mail and computer-based written documents are poorly perceived in terms of social presence. Therefore, social presence theorists argued that conventional media are more appropriate for tasks requiring high social presence, whereas electric media and written documents are more appropriate for tasks with low social presence requirement.

Similarly to the social presence theory, the media richness theory focuses on the nature of media characteristics, but on their match with task characteristics [6, 7, 8, 9]. This theory is based on task variety and task analyzability: Task variety is "the frequency of unexpected and novel events that occur in the conversation process" and task analyzability refers to "the degree to which tasks involve the application of objective, well-understood procedures that do not require novel solution" [8, p. 554]. Media richness theorists suggested that rich media, such as face-to-face and telephone, facilitate the immediate exchange of a wide range of communication cues, while leaner media such as e-mail, written roles and regulations, letters, and written notes allow exchange of a restricted range of such cues over a longer period. Then, the media richness theorists posed that richer media are more appropriate for unanalyzable tasks such as resolving disagreement, making important decision, generating ideas and exchange of confidential/sensitive information, whereas leaner media are more appropriate for analyzable tasks such as exchanging routine information, clarifying confusing viewpoints, and exchanging urgent/timely information. They suggest that when equivocality is high, organizations allow for rapid information cycles among managers, typically face-to-face and telephone, and prescribe fewer rules for interpretation [7].

Although these two theories tried to explain end users' media choice, authors of many empirical studies have suggested that media choice cannot be logically explained or predicted by considering only the inherent richness or social presence of medium and the characteristics of the task [10, 11, 12, 13, 14, 15]. For instance, Ngwenyama and Lee [16] found that electronic media are increasing their richness through messenger service such as call and page functions.

Since media richness theory is only partially supported by empirical tests, it is likely that other factors or dimensions might affect end-users' media choice. In order to overcome the criticisms media richness theory, we adopted the service quality (SERVQUAL), created by Parasuraman et al. [17], as new characteristics that are associated with medium. They also presented a 22-item scale consisting of five service quality dimensions and theorized that regardless of type of service, customers use basically similar criteria in evaluating service quality and that these criteria span virtually all aspects of service [18]. Those dimensions are:

- Tangible: physical facilities, equipment, and appearance of personnel.
- Reliability: ability to perform the promised service dependably and accurately.
- Responsiveness: willingness to help customers and provide prompt service.
- Assurance: knowledge of employees and their ability to inspire trust and confidence.
- Empathy: the provision of caring individualized attention customer.

3. HYPOTHESES DEVELOPMENT

To investigate the relationship between media choice and end-user belief on help desk service, a research model was developed as shown in Figure 1. In this model, we use five constructs from SERVQUAL [19, 20] that are thought of as closely related to end-user satisfaction with customer service: reliability, empathy, assurance, tangibles and responsiveness. This study contains, as already mentioned, adapted SERVQAUL constructs as MEDIQUAL constructs instead of commonly used equavocality and uncertainty, or analyzability and diversity.

To extend end-user belief criteria such as reliability, empathy, assurance, tangibles, and responsiveness, we first need to validate that these criteria measures. On the basis of SERQUAL theory, the defined MEDIQUAL constructs are follows. **Reliability** is defined as the degree of feeling that customer service shows sincere interest in customer service, and that they will solve the problem correctly the first time and in a timely fashion. **Empathy** consists of the degree to which customer service shows personal interest and politeness, and is perceived by customers as displaying personal attention. **Assurance** refers to the degree of confidence customer service gives, staffs are knowledgeable about transactions, and customers can feel safe about the privacy of transacted information. **Tangibles** include modern-looking equipment, visually appealing, and increasingly aesthetic attention to materials. **Reliability** is customer service's perceived ability to provide service in a timely fashion, show sincere interest, and correctly diagnose and solve the problem on the first try. **Responsiveness** is customer service's perceived ability to provide prompt service, readiness to respond to requests, and to never be too busy to respond to requests.

The concept of end user satisfaction has been widely used by researchers [21, 22, 23, 24, 25] as a surrogate of system success. For example, DeLone and McLean [21] introduced a comprehensive taxonomy to organize diverse research that posed six major dimensions or categories of information system success: system quality, information quality, use, user satisfaction, individual impact, and organizational impact. Ives et al. [26] developed a 13 items instrument, which was later confirmed by Baroudi and Orlikowski [27]. They defined end user satisfaction as felt need, system acceptance, perceived usefulness, MIS appreciation, and feelings about a system. Bailey and Pearson [23] evaluated overall satisfaction, which they suggested is affected by 38 items, measured by the 7-point Likert scale. Recently, Lee and Ulgado [20] investigated the relationship between perceived SERVQUAL value and customers' overall satisfaction in fast-food industry.

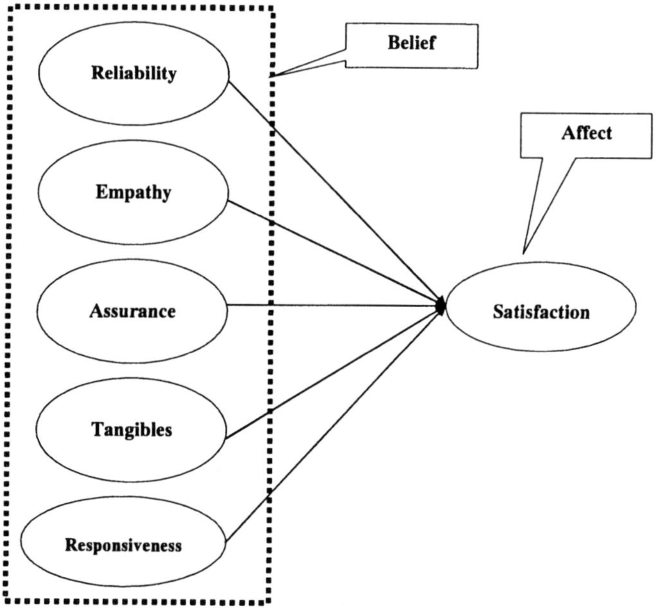

Figure 1. Research Model

On the basis of previous research, we suggest a new relationship between MEDIQUAL constructs and end-user satisfactions, and hypothesize that the assessed MEDIQUAL constructs strongly affect end-users' overall satisfaction. The degree of overall customers' satisfaction with the media customers will be measured within the five MEDIQUAL constructs (empathy, assurance, tangibles, reliability and responsiveness). From this general statement, we can propose a series of hypotheses to explore in this study.

TAM and TAM2 [28, 29] assumed perceived usefulness (as belief) as a direct predictor of attitude, and Bhattacherjee [30] created PAM, which adapted TAM and showed that users' perceived usefulness is positively associated with their satisfaction, which is defined as affect (evaluated attitude). Hence our hypotheses will be:

H1: Service satisfaction comes from reliability, empathy, assurance, tangibles, and responsiveness.
H1a: Reliability positively affects end-user's satisfaction.
H1b: Empathy positively affects end-user's satisfaction.
H1c: Assurance positively affects end-user's satisfaction.
H1d: Tangibles positively affects end-user's satisfaction.
H1e: Responsiveness positively affects end-user's satisfaction.

Table 1. User Perceptions Questionnaire items adapted by SERVQUAL

		Modification	*Resulting Participative perception*
Tangibles (Parasuraman, et al., 1991)	Material associated with the service (such as pamphlets or statement) will be visually appealing in excellent telephone company	Replaced to	The medium you use the most will increase visual attention using materials such as pamphlets or public relations.
	The physical facilities at excellent telephone companies will be visually appealing	Replaced to	The medium you use the most makes you feel customer service is visually appealing.
	Excellent telephone companies will have modern-looking equipment	Replaced to	The medium you use the most makes you feel customer service has modern-looking equipment
	Employees of excellent telephone companies will be neat-appearing	Dropped	
Responsive. (Parasuraman, et al., 1991)	Employees of excellent telephone companies will never be too busy to respond to customer requests	Replaced to	The medium you use the most makes you feel customer service is never too busy to respond to your requests.
	Employees of excellent telephone companies will always be wiling to help customers	Replaced to	The medium you use the most makes you feel customer service is always ready to respond your request.
	Employees of excellent telephone companies will give prompt service to the customers	Replaced to	The medium you use the most makes you feel customer service provides a prompt service
	Employees of excellent telephone companies will tell customers exactly when service will be performed	Dropped	
Assurance (Parasuraman, et al., 1991)	Customers of employees of excellent telephone companies will feel safe in their transactions	Replaced to	The medium you use the most makes you feel customer service make you feel safe with your transactions
	The behavior of employees of excellent telephone companies will instill confidence in customers	Replaced to	The medium you use the most makes you feel customer service gives you a confidence about the transaction
	Employees of excellent telephone companies will have the knowledge to answer customer questions	Replaced to	The medium you use the most makes you feel customer service has a knowledge about your transactions.
	Employees of excellent telephone companies will be consistently courteous with customers	Dropped	
Reliability (Parasuraman, et al., 1991)	When excellent telephone companies promise to do something by a certain time, they will do so	Integrated into	The medium you use the most makes you feel that customer service provides the service just in time.
	Excellent telephone companies will provide their services at the time they promise to do so		
	Excellent telephone companies will perform the service at the first time	Replaced to	The medium you use the most makes you feel that customer service will solve the problem right the first time.
	When customer have a problem, excellent telephone companies will show a sincere interest in solving it	Replaced to	The medium you use the most makes you feel that customer service shows sincere interest in customer service.
	Excellent telephone companies will insist on error-free records	Dropped	
Empathy (Parasuraman, et al., 1991)	Excellent telephone companies will give customers individual attention	Replaced to	The medium you use the most makes you feel that customer service shows an personal interest
	Excellent telephone companies will have employees who give customers personal attention	Replaced to	The medium you use the most makes you feel customer service tries to pay personal attention.
	Excellent telephone companies will have operating hours convenient to all their customers.	Integrated into	The medium you use the most makes you feel that customer service is polite to you.
	Excellent telephone companies will have The customers' best interests at heart.		
	Excellent telephone companies will understand the specific needs of their customers.		
Total	22		15
Satisfaction (Lee and Ulgado, 1997)	The food and service offered by McDonald's are very good value for the money	Integrated into	Please specify the degree of overall satisfaction with the media you use most often
	The food and service offered by McDonald's are very good bargain, considering the prices		
(Bhattacherjee, 2001)	How do you feel about overall experience of OBN use:		
Total	2		1

4. RESEARCH METHOD and ANALYSIS RESULTS

To gather the data for this study, we incorporated a 7-point Likert-type scale to measure end-users' overall belief of service in rating the research variables. That is, the following set of statements (i.e., variables) relate to customers' feelings about the medium they use most. Each statement shows the extent to which the user feels the medium's feature described by the statement: "1" means users strongly disagree the medium has that feature, and "7" means users strongly agree.

To improve the design of survey, we conducted a pilot test employing a web-based survey. The questionnaires were distributed to 214 end-users at two organizations during the summer of 1999 and then they were revised. The revised questionnaires were redistributed to 1,000 MBA students at five universities and six companies throughout the Republic of Korea during the spring of 2001. Study finally tests the aforementioned hypotheses with 222 subjects (a return rate of 22.2 percent).

To investigate hypotheses, we performed a factor analysis to validate end-users' service perception of the customer service center, while validity and reliability of ender users' perception is widely recognized in the IT field [31, 32]. Factor analysis is commonly used to reduce a set of variables to underlying factors. In general, linear combinations of original variables create a clear structure, and suggest clear discriminant validity for these constructs. Table 2 shows the factor analysis results. As can be seen the five factors account for more than 78.36 percent of the observed variance.

The loading of each of the 15 measures on their respective factors is well over 0.40, and the eigenvalues of all the constructs are above 1. The results of factor analysis show the importance of the constructs, in that they have relatively high correlations between measures of the same construct using different methods and low correlations between measures of construct that are expected to differ [33]. Based on this result, we validated the MEDIQUAL constructs.

To further investigate relationships between MEDIQUAL constructs (belief) and satisfaction (affect), we utilized multiple regression analysis. Since the primary objective of this step is to test the comparative influence of several sets of predictors on customer satisfaction, we use them as independent variables in the regression model. The equation representing the model developed in this study can be expressed as:

$$SP = \beta_0 + \beta_1 \, TA + \beta_2 \, RP + \beta_3 \, AS + \beta_4 \, RL + \beta_5 \, EM,$$

Where: SP = Overall Satisfaction
TA = Tangibles RP = Responsiveness
AS= Assurance RL = Reliability EM = Empathy

Table 3 summarizes the results of multiple regression analysis to end-users' overall satisfaction. The result shows that H1 is partially supported from the fact that assurance and reliability turn out to be statistically significant level of 0.01.

Table 2. Factors Analysis of independent Variables

	Tang.	*Resp.*	*Assu.*	*Reli.*	*Emp.*
The medium you use the most will increase visual attention using materials such as pamphlets or public relations.	.854				
The medium you use the most makes you feel customer service is visually appealing.	.838				
The medium you use the most makes you feel customer service has modern-looking equipment	.780				
The medium you use the most makes you feel customer service is never too busy to respond to your requests.		.906			
The medium you use the most makes you feel customer service is always ready to respond your request.		.813			
The medium you use the most makes you feel customer service provides a prompt service		.774			
The medium you use the most makes you feel customer service make you feel safe with your transactions			.839		
The medium you use the most makes you feel customer service gives you a confidence about the transaction			.784		
The medium you use the most makes you feel customer service has a knowledge about your transactions.			.598	.420	
The medium you use the most makes you feel that customer service provides the service just in time.				.834	
The medium you use the most makes you feel that customer service will solve the problem at the first time.				.774	
The medium you use the most makes you feel that customer service shows sincere interest in customer service.			.447	.594	
The medium you use the most makes you feel that customer service is polite to you.			.454	.476	.420
The medium you use the most makes you feel that customer service shows an personal interest					.847
The medium you use the most makes you feel customer service tries to pay personal attention.					.771
Eigenvalue	2.506	2.476	2.462	2.404	1.906
Percentage of Variance Explained	16.708	16.504	16.416	16.028	12.705

Table 3. Results of Regression

Hypotheses	***Condition*** (β and p)
Hypothesis1a (Tangibles)	β_1=0.186 and p=0.099
Hypothesis1b (Responsiveness)	β_2= 0.216 and p=0.056
Hypothesis1c (Assurance)	β_3= 0.420 and p=0.000
Hypothesis1d (Reliability)	β_4= 0.337 and p=0.003
Hypothesis1e (Empathy)	β_5= 0.108 and p=0.337
Model Summary	R=0.366, R^2= 0.112, p=0.000

5. FINDINGS

The hypotheses test whether the MEDIQUAL constructs affect end-users' satisfaction or not. Table 3 summarizes the results from the multiple regression. The adjusted R^2 scores of model indicate that overall the five variables constitute good predictors of end-users' satisfaction (R=0.366, R^2= 0.134, p= 0.000). Specifically, the assurance and reliability constructs appear to be the significant predictor at the level of 0.005 (β_3= 0.420, p= 0.000, β_4= 0.337, p= 0.003), and responsiveness and tangibles constructs are acceptable predictors at the level of .10 (β_2= 0.216, p = 0.056, β_1= 0.186, p = 0.099). Although the empathy construct is not statistically significant, the predictor value turns out to be positive (β_5= 0.108, p= 0.337). Even though empathy is not significant, our empirical test supported the original theory of SERVQUAL; and thus, we can assume that SERVQUAL constructs are strongly related to end-users' satisfaction.

6. CONCLUSION

As we hypothesized, end users' satisfaction is influenced by MEDIQUAL constructs, that is, different media users feel satisfaction from reliability, empathy, assurance, tangibles, and responsiveness. The results of this study have some important theoretical implications. First, we applied the concept of media choice to the domain of customer service center. Second, our research incorporated SERVQUAL theory into media selection mechanism.

The practical implication of this study is that by understanding the media choice mechanism, we may be able to provide a better help desk (customer service center), which is one of the critical success factors in the competitive business world.

ACKNOWLEDGEMENT

This work was supported by the Brain Korea 21 Project in 2003.

REFERENCES

1. Whiting, R. (1997). "Panic at the Help Desk," *Software Magazine*, (17:11), pp. 92-95.
2. Dabolkar, P. A., Shepard C. D., Thorpe, D. I. (2000). "A Comprehensive Framework for Service Quality: An investigation of Critical Conceptual and Measurement Issues Through a Longitudinal Study," *Journal of Retailing*, (76:2), pp. 139-173.
3. Oxford, C. (1998). "Strategies for Resolving the Help Desk Crisis," *White Paper*, Help Desk 2000™
4. Lee, Z., Kim, Y., Lee, S. (2001). "The Influences of Media Choice on Help Desk Performance Perception," *Proceedings of the Thirty-Fourth Annual Hawaii International Conference on Systems Sciences*, Maui, Hawaii, January 2001.
5. Short, J., Williams, E., Christie, B. (1976), *The Social Psychology of Telecommunications*, John Wiley, New York, NY.
6. Hiltz, S. R., Turoff, M. (1981). "The Evaluation of User Behavior in a Computerized Conferencing System," *Communications of the ACM*, (24:11), pp. 739-751.

7. Daft, R. L., Lengel, R. H. (1984). "Information Richness: A New Approach to Managerial Behavior and Organization Design," *Research in Organizational Behavior,* (6:2), pp. 191-233.
8. Daft, R. L., Lengel, R. H. (1986). "Organization Information Requirements, Media Richness and Structural Design," *Management Science*, (32:5), pp. 554-571.
9. Rice, R. E. (1992). "Task Analyzability, Use of New Media, and Effectiveness: A Multi-Site Exploration of Media Richness," *Organization Science*, (3:4), pp. 475-500.
10. Markus, L. (1988). "Information Richness Theory, Managers, and Electronic Mail," *Proceedings of the Academy of Management National Meeting*, Anaheim, CA.
11. Rice, R. E., Shook D. E. (1990). "Relationships of Job Categories and Organizational Levels to Use Communication Channels, including Electronic Mail: A Meta-Analysis and Extension," *Journal of Management Studies*, (27:2), pp. 195-229.
12. Zmud, R. W., Lind, M. R., Young F. W. (1990). "An Attribute Space for Organizational Communication Channels," *Information Systems Research*, (1:4), pp.440-457.
13. Yates, J., Orlikowski, W. J. (1992). "Genres of Organizational Communication: A Structurational Approach to Studying Communication and Media," *Academy of Management Review*, (17:2), pp. 299-326.
14. Trevino, L. K., Webster, J. (1992). "Flow in Computer-Mediated Communication," *Communication Research*, (19:5), pp. 539-573.
15. King, R. C., Xia, W. (1997). "Media Appropriateness: Effects of Experience on Communication Media Choice," *Decision Science*, (28:4), pp. 877-910.
16. Ngwenyama O. K., Lee, A. S. (1997). "Communication Richness in Electronic Mail: Critical Social Theory and Contextuality of Meeting," *MIS Quarterly*, (21:2), pp. 145-167.
17. Parasuraman, A., Zenithal, V., Berry, L. (1985). "A Conceptual Model of Service Quality and Its Implications for Future Research," *Journal of Marketing*, (49), pp. 41-50.
18. Parasuraman, A., Zenithal, V., Berry, L. (1988). "SERVQUAL: A Multiple-Item Scale for Measuring Consumer Perceptions of Service Quality," *Journal of Retailing*, (64:1), pp. 12-37.
19. Parasuraman, A., Zenithal, V., Berry, L. (1991). "Refinement and Reassess of the SERVQUAL Scale," *Journal of Retailing*, (67:1), pp. 420-450.
20. Lee, M. and Ulgado, F. M. (1997). "Consumer Evaluations of Fast Food Service: A Cross-National Comparison," *The Journal of Service Marketing*, (11:1), 1997, pp. 39-52.
21. DeLone, W., McLean, E. (1992). "Information System Success: The Quest for the Dependent Variable," *Information Systems Research*, (3:1), pp. 60-95.
22. Ginzberg, M. J. (1979). "A Study of the Implementation Process," *TIMS Studies in the Management Sciences*, (13), pp. 85-102.
23. Bailey, J., Pearson, S. W. (1983). "Development of a Tool for Measuring and Analyzing Computer User Satisfaction," *Management Science*, (29:5), pp. 530-545.
24. Melone, N. P. (1990). "A Theoretical Assessment of User Satisfaction Construct in Information Systems Research," *Management Science,* (36:1), pp. 76-91.

25. Klenke, K. (1992). "Construct Measurement in Management Information System: A Review and Critique of User Satisfaction and User Involvement Instruments," *INFOR*, (30:4), pp. 325-348.
26. Ives, B., Olson, M. H., Baroudi, J. J. (1983). "The Measurement of User Information Satisfaction," *Communications of the ACM*, (12:10), pp.785-793.
27. Baroudi, J. J., Orlikowski, W. (1988). "A Short-Form Measure of User Information Satisfaction: A Psychometric Evaluation and Notes on Use," *Journal of MIS*, (4:4), pp. 44-59.
28. Davis, F., Bogozzi, R. P., Warshaw, P. R. (1989). "User Acceptance of Computer Technology: A Comparison of Two Theoretical Models," *Management Science*, (53:8), 1989, pp. 982-1003.
29. Venkatesh, V., Davis, F. (2000). "A Theoretical Extension of the Technology Acceptance Model: Four Longitudinal Field Studies," *Management Science*, (46:2), pp. 186-204.
30. Bhattacherjee, A. (2001). "Understanding Information Systems Continuance: An Expectation-Confirmation Model," *MIS Quarterly*, (25:3), pp. 351-370.
31. Kettinger, W. J., Lee, C. C. (1994). "Perceived Service Quality and User Satisfaction with Information Service Function," *Decision Science*, (25:5), pp. 737-766.
32. Jiang, J. J., Klein, G., Crampton, S. M. (2000). "A Note on SERVQUAL Reliability and Validity in Information System Service Quality Measurement," *Decision Science*, (31:3), pp. 725-744.
33. Cronbach, L. J. (1971). "Test Validation," in *Education Measurement*, 2nd Edition, Thorndike, R. L.(ed.), American Council on Education, Washington, D.C., pp. 443-507.

PART II

External Collaboration

16

DISTRIBUTED ENGINEERING ENVIRONMENT FOR INTER-ENTERPRISE COLLABORATION

Koji Kawashima[1], Koichi Kasahara[1], and Yasuyuki Nishioka[2]
[1]*Mitsui Engineering & Shipbuilding Co., Ltd., Japan*
[2]*Dept. of Industrial and Systems Eng., Hosei Univ., Japan*
e-mail: kkawashi@mes.co.jp

Abstract: This paper focuses on collaboration of engineering information among globally distributed companies and describes DEE (Distributed Engineering Environment) of design and production on inter-enterprise network. For DEE, the integration framework and methodology to execute engineering are described in the paper. As for the forms of information description, XML, PSLX (Planning and Scheduling Language for XML) and Web browsing technologies are used. In DEE, XML technology prescribes the communication data form that contains hierarchical functional data structure on a common specific Web browsing space. And PSLX technology prescribes the dynamic processes and their controls of the connection among the distributed business activities of the creation of product data and documents. A prototype of DEE for engineering collaboration is also introduced to evaluate the feasibility. This prototype has been developed as a part of IMS/GLOBEMEN of international project.

Key words: Virtual Enterprise, Collaborative Engineering, PLC, XML, PSLX, WWW

1. INTRODUCTION

For the total product lifecycle and by the development of the global information communication environment, each enterprise has growing recognition of the importance of the international business models, ideal and practical models, and of more close world-wide relationships with the related companies and with companies that possess characteristic products and technologies.

In IMS/GLOBEMEN (GMN), we have been studying "Global Engineering and Manufacturing in Enterprise Network" by VME (Virtual Manufacturing Enterprises) that consists of the companies with the core competence. The target of R&D in GMN, is the development of methodologies and tools necessary to the lifecycle business activities that change dynamically. This paper focuses on the DEE (Distributed Engineering Environment) in the management of the dynamic engineering business and the collaboration of the information sharing among VME.

At first, the framework and the second, DEE model to realize the environment of the information sharing among VME are mentioned. And then, a prototype of DEE for the engineering collaboration is introduced to evaluate the feasibility.

2. DEE FRAMEWORK

The framework of the requirement and the function description that shows in Fig.1 is proposed in GMN to deal with information integration in the VME [1].

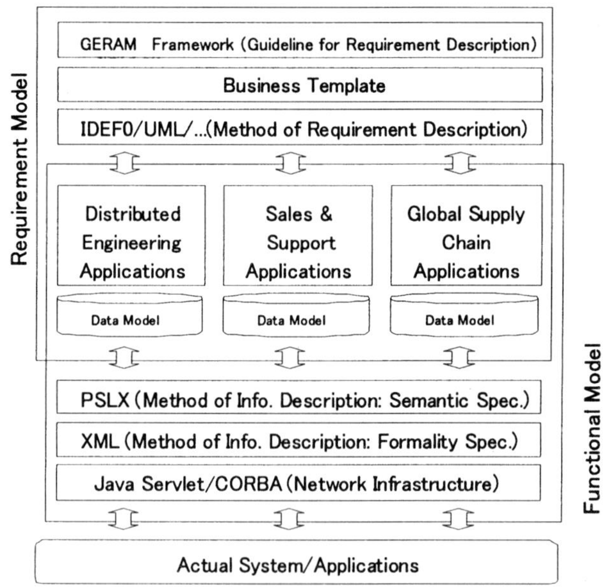

Figure 1. DEE Framework of Requirement and Functional Description

In the figure, the requirement model describes the present condition (as-is model) and the goal (to-be model) for each business and the function model shows the implementation form to realize the requirement in the VME. By the requirement model, in each business, which part is equivalent, and related, or roles of each part in the whole business model of the VME can be defined.

The function model shows implementation forms of the requirement from the requirement model in the VME on the IT infrastructure. The specifications described in the requirement model are developed into the executable model in the function model.

DEE is a system that operates and integrates the business information dispersed around the world based on the framework shown in the figure 1. It becomes the following concretely.

- The description of the requirements in each business is classified and reified by the modeling tool such as IDEF0 or UML. The abstract function of the design requirement is disassembled by the function elements into the detailed executable level for the actual production by the classification and reification. The hierarchical function structure from the abstraction to the business activities is constructed.

- The engineering is carried out by the process flow of the function elements with executable resources. The process is constructed among the process elements due to the flows of data exchange. These processes have the dynamic connections that contain the feedbacks among the process elements. The product data is created by applications in each element.

- The function of a business process element is equivalent to the requirement function element described hierarchically. And the product data is created in each process element. Therefore, function, process and product data are dependent on the mutuality and compose all the systems.

DEE is the platform to manage the mutual dynamic connections of function, process and product data on Information and Communication technology (ICT) base.

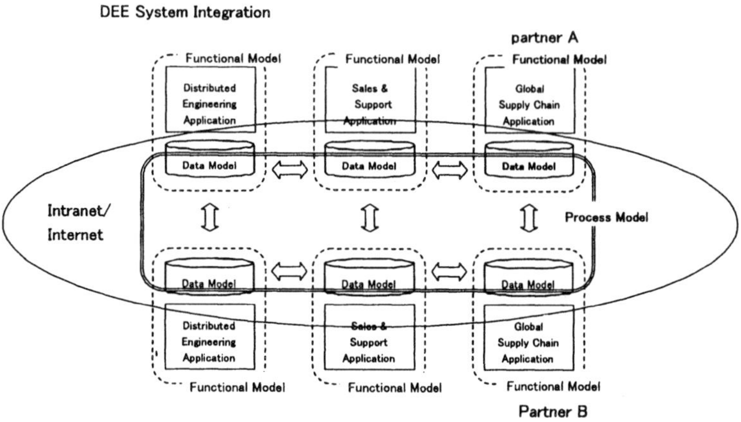

Figure 2. DEE System Integration

3. DEE MODEL

DEE has been studied as the activities of Work Package3 in the GLOBEMEN international joint research. The system concept of WP3/DEE is shown in the following.

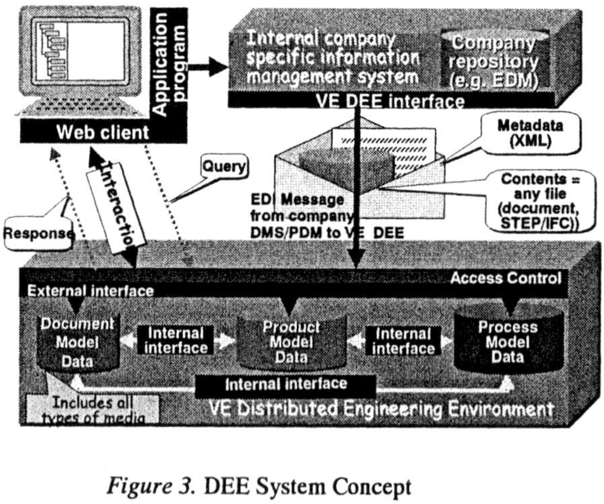

Figure 3. DEE System Concept

DEE manages document, process and product data elements independently and integrates their elements according to each engineering phase of product life cycle. The connection model of the engineering information in DEE is shown in the following.

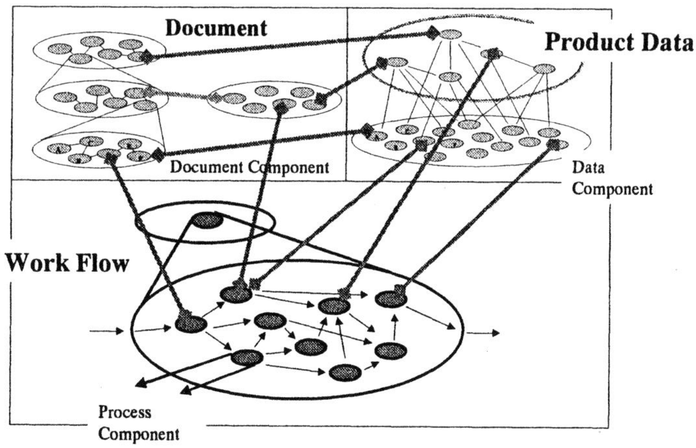

Figure 4. DEE Intercommunication Model

The workflows that show the processes of business functions are connected among groups having functional roles in VME. The functional roles have static hierarchical function layers. The business activities are connected as time-dependent flow in the execution. This concept is shown in Figure5 and Figure6.

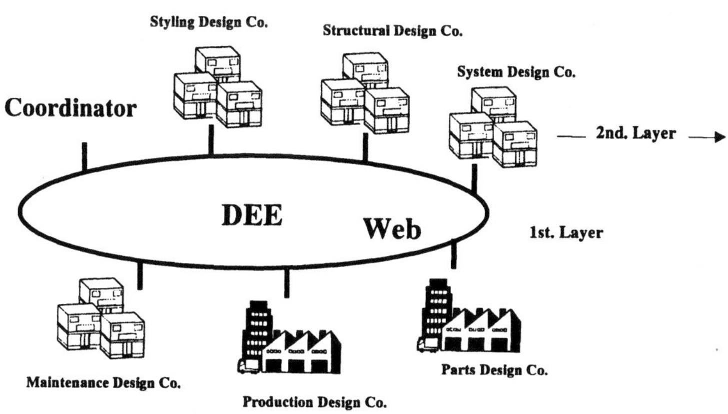

Figure 5. VME Layered Model

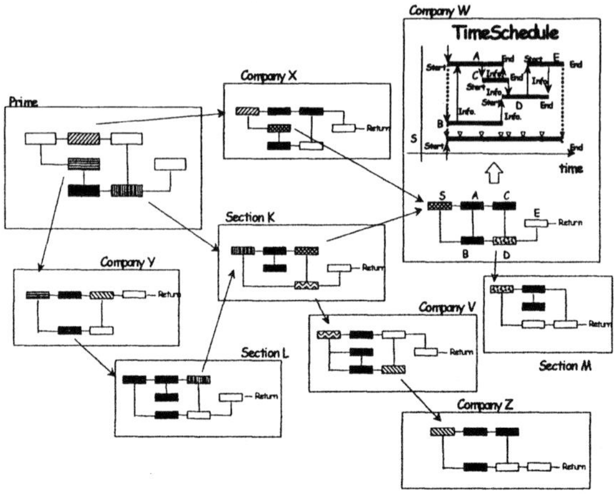

Figure 6. DEE Process Integration Model

The multimedia product data is created in each process element in the distributed engineering environment. In the process element, the product data is created by the application software (resources) using the output data from documents outputted by other process elements. The output data is included in the document description in the process element. The document elements that created in each process element are integrated according to the functional roles. That means the document can be browsed as the document of independent part, hierarchical group and all. And also, the multimedia data is seen independently. This concept is shown below.

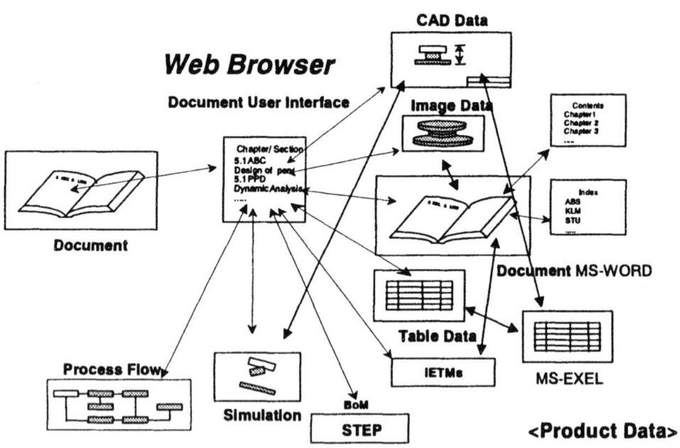

Figure 7. DEE Document Integration Model

As for the characteristics of DEE, mutual related links are constructed among process, document and product data elements by using XML. By the link structure, engineering collaboration and business management are possible on time as for the control of the whole business, the design change, and so on. And, each element made a part becomes possible to reuse to all the product lifecycle by constructing them in the hierarchical class level and by the purpose. And, PSLX [2] is used for the dynamic business control.

4. DEE PROTOTYPE

A prototype DEE system platform has been developed to evaluate DEE model. The result is shown in Fig. This environment has a common 4D virtual space in VME on the Web browser that shows business functions, individual/integrated documents and multimedia product data in the document. The PSLX based scheduler controls the dynamic processes. The document includes HTML and word processing document, and product data such as SVG drawing and 3D XVL (eXtensible Virtual world description Language) simulation [3] and so on are called from Web data warehouses. XVL is the compression technology of 3D model transmission on the Web and has animation execution capability.

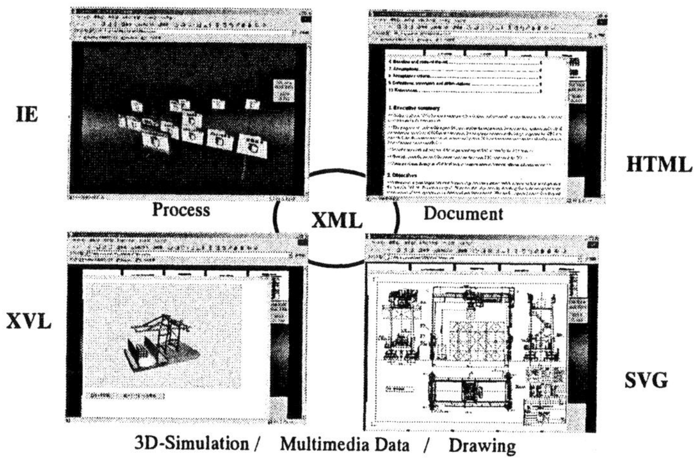

Figure 8. DEE 4D Virtual Space

5. CONCLUSION

This paper describes the collaborative engineering environment among VME for the lifecycle information integration. It is shown that such a distributed engineering environment is implemented by element parts of process, product data and document for the business and their dynamic integration. And the functional integrated information is re-usable to each product lifecycle phase such as design, manufacturing, maintenance and customers services. Moreover, we developed the prototype platform of Web P2P browsing space based on ICT and verified the suitability and effectiveness of the environment.

Our next step will be the knowledge-based engineering of the Semantic Web RDF (Resource Description Framework) approach of OWL (Web Ontology Language) of W3C for DEE.

REFERENCES

1. Nishioka Y., Kamio Y, Kawashima K. and Fukuda Y (2001). A Booking Type Production System as a Collaboration Method for Virtual Enterprises, Global Engineering, Manufacturing and Enterprise Networks, Kluwer Academic Publishers, 164-172
2. Nishioka, Y (2001). XML-based Standardization of Planning and Scheduling Problems for Virtual Enterprise Collaboration, Proceedings of ICPR-16, Czech Prague
3. K.Morita et. al. "Web-based maitenance manual with three-dimensional simulation model", DIISM2000, Nov.15-17, 2000, Melbourne, Australia

17

AGENT BASED MANUFACTURING CAPABILITY ASSESSMENT IN THE EXTENDED ENTERPRISE USING STEP AP224 AND XML

Svetan Ratchev and Omar Medani
School of Mechanical, Materials, Manufacturing Engineering and Management, University of Nottingham, UK
e-mail: Svetan.Ratchev@nottingham.ac.uk

Abstract: Data exchange in the extended enterprise is one of the most critical tasks in supporting collaborative decision-making. Companies often rely on geographically distributed suppliers for efficient product design and manufacture. Early design assessment can substantially reduce the cost of product development and production. This research proposes a new STEP AP224 EXPRESS based data model to facilitate the exchange of part and process data between distributed key agents in the early design process. The approach is illustrated using a prototype XML/CORBA environment to support the information exchange between collaborating design and manufacturing agents.

Keywords: Early Design Assessment, Collaborative Design, STEP AP224 and XML

1 INTRODUCTION

In today's competitive manufacturing world, companies constantly seek to optimise their manufacturing processes and improve their overall performance. One of the key factors for success of new products is how efficient is the utilisation of the product development time and how well the early design decisions are coordinated between all major stakeholders. Early design assessment is an effective method for cutting costs throughout the product development phase. Performing the early design assessment in a collaborative environment is a technique used to concurrently review the design from different points of view, thus reducing the gap between the design and the production stages in product development.

One of the critical developments for improved communication of product and processing data in machining has been the development of the STEP AP224. Feature extraction routines utilising STEP AP224 were proposed by Bhandarkar and Nagi[1] The research focuses on the extraction of the feature information by converting low level geometry into higher level manufacturing information. Ming et al.[2] developed an information model for CAPP by using object-oriented model based on STEP AP224 and product data exchange. The model consisted of part information model, process plan information model and production resource information model. Zha and Du [3] proposed a prototype system for integrated design and assembly planning utilising an EXPRESS model representing STEP information for the part representation. Pena-Mora and Hussien[4] reported a PDM system with integrated early design system using AP 224 for feature representation. Moreover, the system supported parameters for manufacturability assessment and process planning generation. Due to the complexity of STEP AP224 representation, there is still a gap between the capabilities of the AP224 and its use and the great potential for data integration offered by AP224 has been mostly unutilised.

Another important drive for developing collaborative decision making applications in manufacturing has been the development of web enabling tools and data structures that can support distributed environments. `Other reported distributed design support systems include the distributed agent-based design negotiation and co-ordination environment CAIRO [5]; the web-based electronic design tool WELD[6]; and the internet-based computer-aided design evaluation tool WebCADET [7].

With the advent of internet-based communication XML (eXtensible Mark-up Language) has been used increasingly to transmit structured information in a universal way across different architectures [8][9] reported a new Product Data Markup Language (PDML) creating a unified use of XML for the purpose of product data exchange. The PDML used STEP as a Product Data Exchange (PDE) technology.

Despite the advances in developing distributed design systems there is still a lack of integration between the design and facility planning activities at early development stages [10][11]There is also insufficient understanding of companies own production capabilities, their distribution between different company sites and the production capabilities which major subcontractors and suppliers can offer which results in inefficient use of available resources and development of more expensive designs. There is a further need to develop generic methods and tools to capture the capabilities of manufacturing processes and map the parameters used for their representation to the machines and processing systems participating in the extended manufacturing environment.

The paper reports on a STEP AP224 compliant product data model and pilot environment for rapid product manufacturability assessment in extended enterprises using collaborative, autonomous design and manufacturing agents. The approach is based on applying the product data model using XML for exchange of requests and information between design, processing, and facility planning agents. The decision making process is supported by multilevel manufacturability domain and inference models.

2 DATA MODELS – AN OVERVIEW

The decision making at the early stages of design manufacturability assessment is based on matching product requirements to processing capabilities and available resources in the extended enterprise.

2.1 Component Data Model

The request for design manufacturability assessment is based on a component data model representing the target components constituting the new product. Components are described using form features and their relationships[12].

STEP AP224 is an ISO application protocol offering a framework for using machining features in process planning. It specifies the requirements for defining the machining features and the parameters needed to correctly and sufficiently represent them. It would therefore be beneficial if AP224 could be used for representing the product model in a way that would allow structured feature information to be directly used for downstream planning decisions. A major barrier for this is the fact that the standard STEP file even for a very simple part feature is long and difficult to process or understand by the planner. Traditionally a STEP files contains a variety of data about the part and its features that is not closely related to process planning. Here a supplementary part feature model is proposed to reduce the complexity of representation and provide a meaningful part feature definition closely related to AP224 which can be used in process planning activities.

2.2 Process and Facility Data Models

The approach adopted in this work is to relate process capability to three fundamental levels. The first is a "form generating schema" to be used for describing process knowledge at a level that is independent of the machine tool and machining facility used for process execution. Form generating schema (FGS) is defined as a technologically meaningful combination of tool of specific geometry, set of relative motions between a part and the tool and the typical levels of technological output that can be associated with using that combination of tool and relative motions FGS are machine independent but can be used to provide a generalised description of machine tool capabilities.

The second level of abstraction is the "facility" level that is used to describe the distribution of manufacturing capabilities in each facility belonging to the extended enterprise. The overall capabilities of a manufacturing facility are described by a set of resource elements (RE) [12]. Each RE represents a collection of form generating schemas which uniquely define the exclusive machine tool capability boundary and the shared boundaries between machine tools. At extended facility level, the distribution of manufacturing capabilities between different geographically distributed facilities is described by technological elements (TE). TEs are collections of form generating schemes which uniquely define the exclusive and shared machining capability boundaries between the machine tools of different business entities/facilities[13].

3 STEP AP224 COMPLIANT DATA MODEL

The product model is described using EXPRESS and EXPRESS-G representations. The STEP AP224 file was analysed and a technique for data extraction was introduced. The extracted information was represented using XML as the main data communication format for the proposed prototype manufacturing system. The model starts with the PRODUCT entity where basic information about the product such as name, ID, description and components are held (see Figure 1a). The four attributes are used to differentiate the part as all the part information is stored in a database and these can help in proper retrieval of information. The components attribute of the PRODUCT is defined in terms of a set of instants of PART entity. This way sets the coherency between the product and its components.

The PART entity holds the specific machining data. (See Figure 1b). The product could be made of one or several parts and this is not in the focus of this research. This research concentrates on how each part is defined in terms of machining features based on STEP AP224. The PART entity again contains attributes to differentiate it from other parts, and these are ID and Name. The description attribute holds text containing any extra data the user wants to state about the part. The remaining three attributes are the most related to the process planning. The first one is the material; this attribute is an instance of the MATERIAL entity. This entity holds information about the type and the properties of the proposed material.

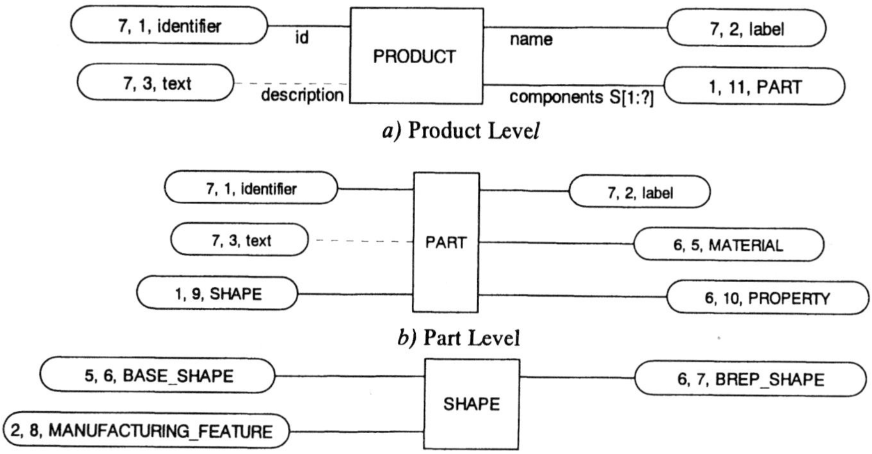

c) Shape Level

Figure 1. Product and Part level description

The shape entity holds all the information about the geometry of the part and as well as the part's features (see Figure 1c). The shape is represented by the *Base_shape* entity representing the shape of the block raw material; the *Brep_shape* entity representing the boundary representation definition of the part in terms of constituting faces and the *Manufacturing feature*, where the high-level shape information is held. A machining feature is a subtype of manufacturing feature and it is a supertype of one of a number of possible types (see Figure 2a). The figure shows just four of these subtype features. Each of these features has its own set of attributes needed to be conforming to the standard. The model is illustrated with two entities: a hole and a slot (see Figure 2b&c).

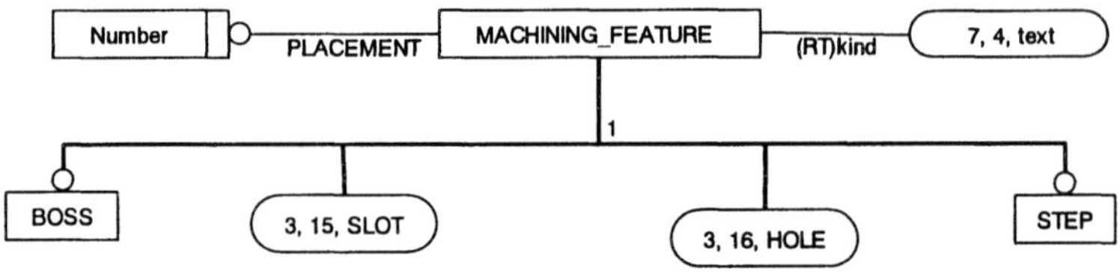

a) Machining Feature DataModel

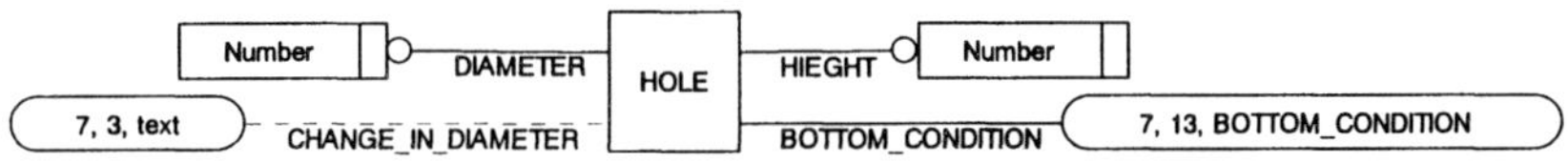

b) Machining Feature - Hole

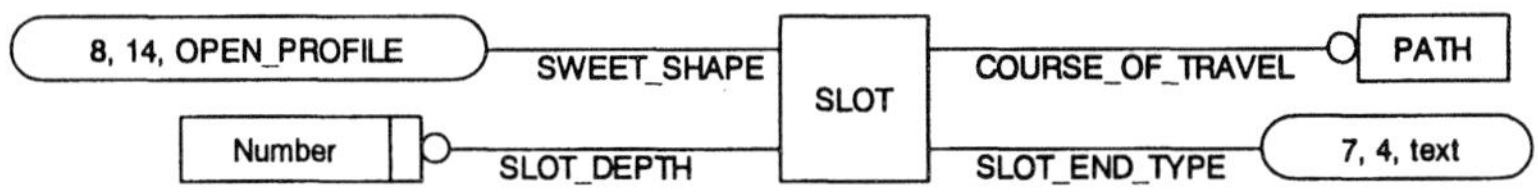

c) Machining Feature – Slot

Figure 2. Machining Feature Data Representation

4 AGENT BASED SYSTEM ARCHITECTURE

In an extended manufacturing environment the decision making is based on collaboration between different geographically distributed facilities with a high level of autonomy in the extended enterprise organisation. Such autonomy can be effectively described by using a multi-agent representation of the main decision making modules within the distributed design environment. In the context of the reported research an agent is considered to be autonomous collection of object oriented decision making algorithms, software tools, and data structures to support human-centred decision making process in a well defined domain and provide communication to other agents. The experimental product and facility prototyping system is developed as a distributed multi-agent CORBA environment (see Figure 3).

The decision-making agents are implemented using CORBA objects with IDL (Interface Definition Language) interfaces allowing application independent specification of available member functions to a client. To call member functions a client needs to know only the object's IDL definition without any details such as programming language used, object location, or operating system. The communication between different agents is supported by a CORBA interface database that stores information about the IDL interfaces implemented within the environment. Each agent is based on communication layer (ORB), parser using KQML/XML message representation and a user interface.

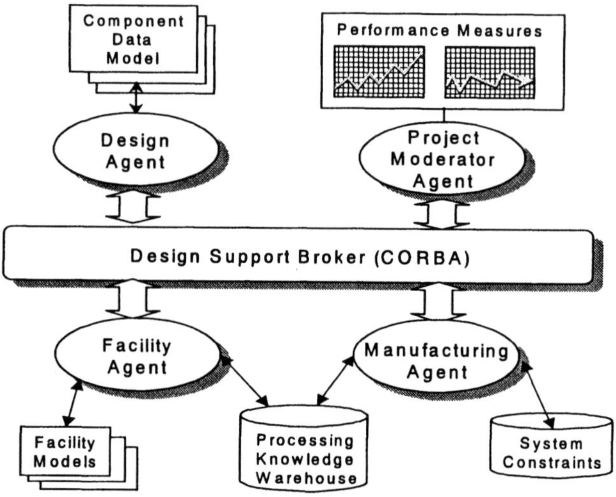

Figure 3. Integrated product and facility prototyping environment – an overview

The sequence of decision-making activities is summarised in Figure 4. The design agent passes the design model to the manufacturing agent with a request for process and facility prototyping. The manufacturing agent communicates with the corresponding facility agents for developing planning alternatives and establishing initial facility prototype. The project moderator is an agent that supports the decision making process by providing periodic evaluation and guidance based on a set of performance indicators. The prototype agent-based modules have been implemented within a distributed computer environment consisting of three workstations linked together within a local area network (LAN) and using TCP/IP protocol.

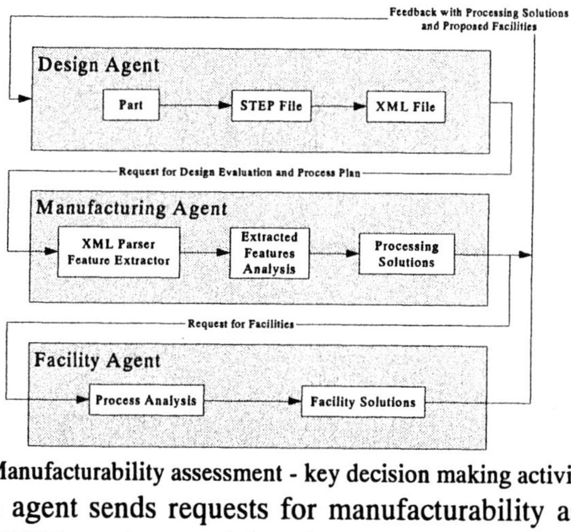

Figure 4. Manufacturability assessment - key decision making activities and actors

The design agent sends requests for manufacturability assessment based on a STEP AP224 product model and a query message. The STEP file is represented using XML based on AP224 application protocol (REF). Once the request is accepted by the process planning agent the critical part features are extracted from the XML file. The geometry of the part can be viewed using the STEP file and if any additional data is needed, it could be obtained from the STEP file. After analysing the part, process planning agent facilitates the selection of processing alternatives. As a result, a number of possible solutions are generated that are communicated to the facility agent.

The facility agent attaches technological elements to each processing alternatives and provides support for optimisation of the overall set of technological elements and final machining equipment selection. Each proposed processing and equipment solution is evaluated by the project moderator agent (not shown) and performance measure values attached. The preferred solution is then sent to the design agent and issued to the designer who initiated the request. An example of the results from the manufacturability assessment process is shown in Figure 5 illustrating the initial design model, the selected processing solutions and facility resources.

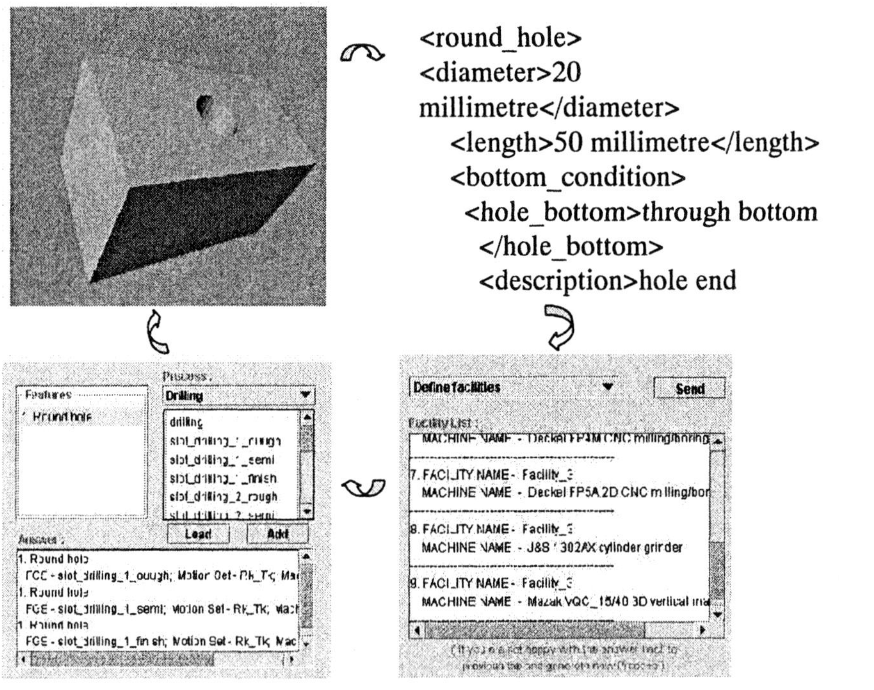

Figure 5. Manufacturability assessment process - an example

5 CONCLUSIONS

The reported research aims at providing the designers with a rapid manufacturing feasibility assessment tool to be used at different design and planning stages in extended manufacturing enterprises. The design evaluation framework provides an integrated platform to support the decision making in distributed design teams. It is based on matching the design requirements to the facility capabilities in an extended enterprise environment using generic models for representation of the processes and resource capabilities in an extended ('virtual') manufacturing environment.

REFERENCES

1. Bhandarkar M. P., Nagi R.2000 STEP-Based Feature Extraction from STEP Geometry for Agile Manufacturing. *Computers in Industry*; pp 3-24.
2. Ming XG, Mak KL, Yan J. Q.1998 A PDES/STEP-Based Information Model for Computer-Aided Process Planning. *Robotics and Computer Integrated Design*; 14: pp 347-61.
3. Zha and Du 2002 A PDES/STEP-based Model and System for Concurrent Integrated Design and Assembly Planning. *Computer Aided Design*; 34: pp 1087-1110.
4. Sharma R. and Gao J. X. 2002 A Progressive Design and Manufacutring Evaluation System Incorporating STEP AP224. *Computers in Industry*; 47:155-167.
5. Pena-Mora F., Hussein K., 1998, Proactive meeting management for distributed collaborative design, *Advances in Engineering Software*, Dec, Vol.29, No.10, pp. 839-849
6. Chan F., Spiller M., Newton A., 1998, WELD - An Environment for Web-based Electronic Design, *Proceedings of Design Automation Conference*, pp. 146-151
7. Caldwell N., Rodgers P., 1998, WebCADET: Facilitating Distributed Design Support, IEE Colloquium (Digest), No.307, pp. 9/1-9/4
8. OMG, Object Management Group Inc., 1999 XML & CORBA.
9. Burkett W. C. 2001 Product Data Markup Language: A New Paradigm for Product Data Exchange and Integration. *Computer Aided Design*; 33:489-500.
10. Hale M. A., Marvis D. N.2000 A Lean-Server Foundation For Collaborative Design. *Advances in Engineering Software*; pp 679-685.
11. Shyamsundar N, Gadh R. 2001 Internet-based Collaborative Product Design with Assembly Features and Virtual Design Spaces. *Computer Aided Design*, 33: pp 637-651.
12. Gindy, N. N. Z. Ratchev, S. M., 1997, Component grouping for Cellular Decomposition of Manufacturing Facilities using Resource Elements, *Integrated Manufacturing Systems*, 8, pp 215-222.
13. Ratchev, S. M., 1998, Resource Capability Modelling for Responsive Process Planning in Extended Manufacturing Facilities, *Globalisation of Manufacturing in the Digital Communications Era of the 21st Century*, Kluwer Academic Publishers, pp. 389 – 400.

18

INTER-ENTERPRISE PLANNING OF MANUFACTURING SYSTEMS APPLYING SIMULATION WITH IPR PROTECTION

Kai Mertins and Markus Rabe
Fraunhofer IPK Berlin, Germany
e-mail: markus.rabe@ipk.fraunhofer.de

Abstract: Discrete Event Simulation is a well-proved method to analyse the dynamic behaviour of manufacturing systems. However, simulation application is still poor for external supply chains or virtual enterprises, encompassing several legal entities. Most conventional simulation systems provide no means to protect intellectual property rights (IPR), nor methods to support cross-enterprise teamwork. This paper describes a solution to keep enterprise models private, but still provide their functionality for cross-enterprise evaluation purposes. Applying the new modelling system, the inter-enterprise business process is specified by the user, including a specification of the objects exchanged between the local models. The required environment for a distributed simulation is generated automatically. The mechanisms have been tested with a large supply chain model.

Key words: Distributed Simulation, HLA, Supply Chain, Manufacturing Engineering, IPR Protection

1. INTRODUCTION

Simulation is a well-proved method to analyse the dynamic behaviour of manufacturing systems. Today, simulation systems for manufacturing are basically closed systems, which always run the complete scenario within one simulation execution machine. This brings several disadvantages, especially for the efficiency of the team work, the reusability of simulation models and the protection of the intellectual property rights (IPR). IPR are a specific issue if the system under consideration is an external supply chain or a virtual enterprise, and includes several legal entities.

A consortium of companies and research institutions from the European Union, Japan and USA have faced this problem, and joined for the common IMS project " Modelling and Simulation Environments for Design, Planning and Operation of Globally Distributed Enterprises " (MISSION) [1]. Methods have been developed to efficiently structure the simulation model into several separate federates, which may run on different computers within different simulation execution machines, even from different software vendors.

The European Module of the MISSION project has put specific effort to introduce methods to keep the interfaces flexible, and at the same time to support common enterprise models shared by several enterprises. Specification methods and tools provide the necessary environment to model the federation on the interface level and to generate the required interfaces.

2. SIMULATION AND IPR

Simulation models contain detailed strategies of the company. Engineers might not wish to make these known to others, especially not outside of the enterprise. Therefore, mechanisms for knowledge protection are required.

Knowledge is distributed, as local engineers know the local rules and environment best. Therefore, best modelling results are expected with decentralised models, which are built and maintained locally. For these decentralised (partial) models, there is no IPR problem, as none of these models has to leave the originating enterprise. However, in order to achieve results about the relationships within the supply chain, these partial models need to be combined.

Within classical simulation systems, all parts of a large simulation model have to be physically joined, before the simulation can be run [2]. Therefore, the detailed description of the enterprises' business processes has to be handed out to an external company, in the shape of the simulation model. There might be a chance to define a neutral third party, e.g. a simulation service provider, for this purpose. Specific IPR protection regulations between this service provider and the co-operating enterprises might help to secure the business process information.

However, enterprises might wish to make sure, and not provide their internal processes to any other firm. Separate distributed models are adequate to fulfil this requirement. They are built and maintained locally, and joined for evaluation purposes, only. The interface description, which is necessary to evaluate the complete supply chain, can be used for purposes of process engineering, evaluation by simulation and specification of the supply chain control system.

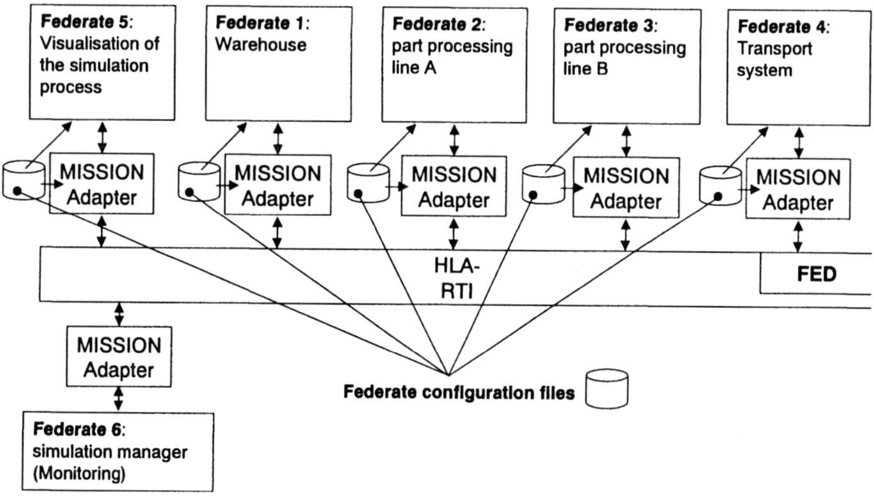

Figure 1. Execution of a Simulation Scenario

3. MODELLING TECHNOLOGY

There is no super-model describing the supply chain as one large simulation model. In contrary, separate models of the chain elements are communicating in a way, which is rather similar to their communication in reality. As a consequence, the interface descriptions generated by the MISSION process description can be directly used as specification of the supply chain interfaces.

3.1 Connection Platform

The High Level Architecture (HLA) [3] was selected as the base architecture. The HLA satisfied many requirements for distributed simulations [4], e.g. time synchronisation, communication between independent simulation models etc. [5]. Woerner [6] states, that HLA requires exchange of detailed model information. However, experiments have demonstrated that the exchange of interface information is sufficient, and all model details may be hidden when applying HLA.

In fig. 1 the runtime environment of the model is shown. The separate simulation models are called “federates” and may be enhanced by other federates for purposes like monitoring or visualisation. The run-time infrastructure (RTI), which provides the basic mechanisms, requires a specific set of Federation Execution Data (FED). The interface between RTI and federates is provided by adapters. These adapters encompass a generic part (bridge to RTI) and a second part bridging to the simulation systems, which has to be implemented once for each additional simulation system.

3.2 Federation Specification

Before an evaluation can be started, the manufacturing system under consideration has to be planned and modelled. This is done within the Manufacturing System Engineering (MSE) process, and the result is the process model of the Manufacturing System (MS) [7].

A central supporting structure is the template library [8]. It enables a flexible definition of classes, attributes and objects. The library content is not fixed. The user can add classes and attributes at any time.

The specification task is done in two steps. First, the process itself is modelled. This process describes the steps for production and logistics, but it does not define the systems implementing these steps. Then, the template library is searched for suitable production and logistics systems, and those are added to the model, connecting from the bottom to the respective process steps. Parameters are set for the single templates. For example, if a specific warehouse is selected as part of the chain, it should be specified how much space for pallets should be reserved there.

Exchange Objects, which are described for each template, specify the classes used for the communication with the distributed simulation environment.

3.3 Adapters

A new software package includes all the necessary interfacing components as well as special sets of building blocks for specific commercial simulation systems (Taylor and ARENA available, will be extended). The major task of these new elements is to replace the sources and sinks of the original models by connectors to the outside world. Using these building blocks, engineers can model the interface of the simulation model just in the same way they are used to model the manufacturing system itself.

A major problem of the HLA detected when developing the adapters is the ownership mechanism. HLA requires that exactly one federate "owns" an attribute at any point of time, which avoids conflicts and ensures that there is always exactly one valid instantiation. However, when applying to manufacturing system simulation, a federate releases ownership of an exchange object, when it has fulfilled all necessary tasks with this object. Sometimes there is a time gap before an other simulation system is ready to take this ownership. In these cases the exchange object can be completely lost [9]. For the adapters described, workarounds have been implemented, which collect dangling ownership and provide it later to a federate requiring it. A more sophisticated approach is under development.

3.4 Configuration Files

The HLA-RTI expects a Federate Execution Data (FED) file. It contains the class names of common objects as well as the names of their attributes. This file can be easily generated from the federation specification. The FED information is sufficient to run the RTI, but not adequate to control the interfaces to the RTI. HLA proposes to implement specific interfaces based on the specification. This means, that most changes in the classes will induce changes in the adapter implementations.

The approach described in the previous section, however, requires that the interface between the commercial simulation system and the HLA-RTI is somehow "self-configuring". This is done by separating all model-specific information from the interface code, and storing it in a separate XML file, the *Federate Configuration File* (Table 1). All information which is contained in this file has already been specified by the engineer within the template library or the process model before, as described above. Therefore, it was possible to automate the generation of these files.

The Federate Configuration Files (FCF) are generated using the process model (including the parameters of the single templates), the Template Library and information about the input and output segments, specified with building blocks (fig. 2). Further details about the FCF, the interface and the underlying mechanisms are reported by Rabe and Jäkel; the technical implementation is given by Rabe and Gurtubai [7, 10].

Table 1. Federation Execution Definition File (FED) vs. Federate Configuration Files

Content of the Federation execution definition (FED) file used by the HLA-RTI	Content of the federate configuration files (FCF) used for simulation interface configuration
• necessary class names of exchanged object classes • attribute names	• class names of exchanged objects • attribute descriptions of the classes including: + names, + type information + default values • initial values of attributes or parameters • parameter settings (e.g. max vehicle speed) • Information about sequences (process flow) • Information about output and input relations

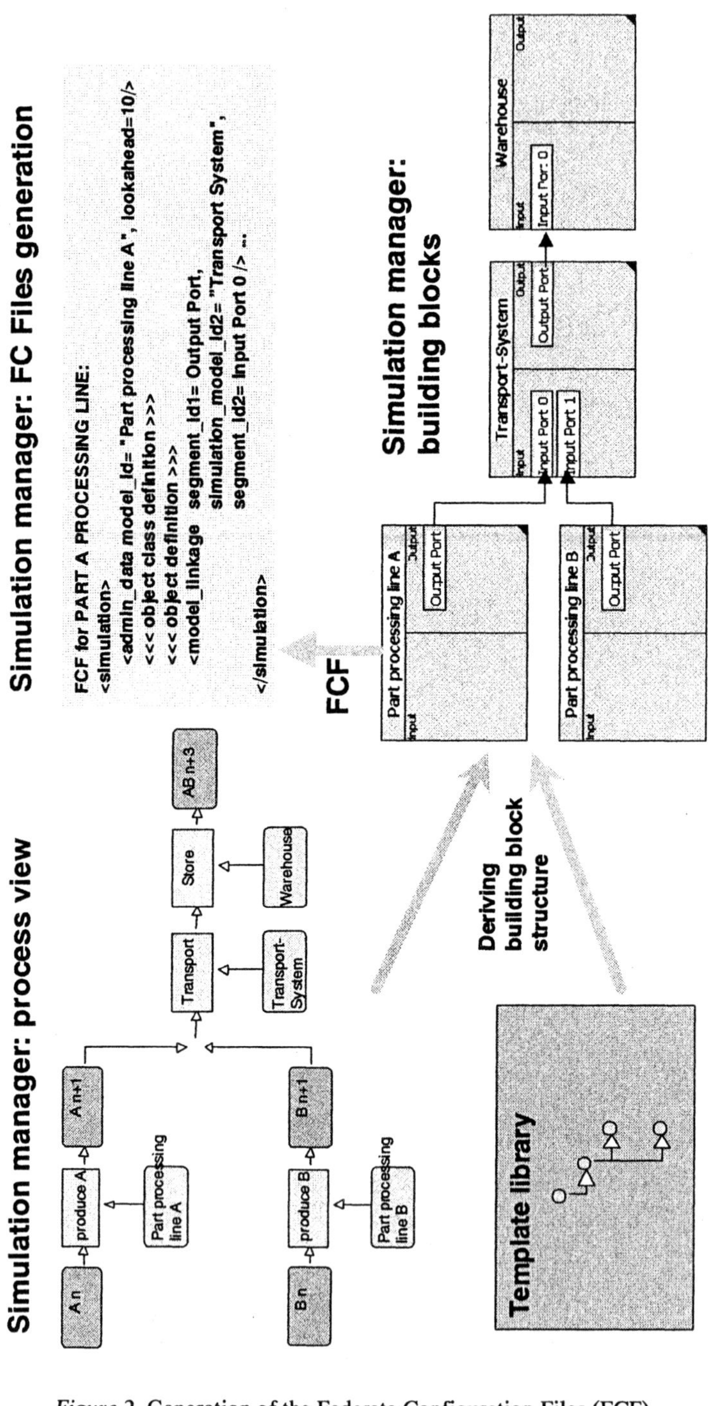

Figure 2. Generation of the Federate Configuration Files (FCF)

4. SUPPLY CHAIN APPLICATION

The MISSION methods and software have been applied within a large supply chain scenario. Specific focus has been set on the realistic representation of the order flow. Research conducted at the U.S. National Institute of Standards and Technology (NIST) in the framework of the IMS MISSION project [11] has been the used for the Exchange Objects representing this order flow.

5. BENEFITS ACHIEVED AND CONTINUED WORK

The approach described has proved to work with large applications, including real-world simulation models. Modelling is efficient, and can be performed without significant additional skills compared to an engineer applying simulation today. The major advantages achieved are:

- IPR protection, as models can be run on any server with specific access rights. Only the template has to be published.
- Reusability, even when changing simulation tools, as the single federates can be implemented by any system with a suitable interface.
- No need to agree on common simulation tools within the supply chain members, for the same reason.
- Clear and structured teamwork on large simulation models, as the single parts can be developed, tested and maintained separately.
- For large models, performance improvement is achieved by distributing computer power together with the models.

The mechanisms have been tested with a large supply chain model, which now serves as an adaptable system to configure and test supply chains very fast. However, the specification has been done for any kind of manufacturing system, and tests have been performed with a common Japanese-European example during the development work. Therefore, the mechanisms can be used for any large manufacturing simulation model, in order to improve the reliability and efficiency of the work.

Any component of this model may at any time be replaced by an application-specific detailed model. Furthermore, the interface information can be amended without the need to change any interface software. This has been proved with one application from the automotive supplier company Bosch (Germany).

The components described are ready to use. An additional approach is ongoing to apply the MISSION technology for training purposes, in cooperation with the University of Limerick (Ireland). An advanced adapter has been designed and implemented work in co-operation with the University of Bari (Italy).

6. CONCLUSION

The European module of the MISSION project has provided a modelling technology, which sets up the structure of a distributed simulation model and automatically generates all the necessary interfaces. Specific template libraries support the distributed generation and maintenance of the simulation models, and support flexibility as well as some standardisation. . As only the interfaces are published, IPR protection is given with respect to the internals of the model.

REFERENCES

1. Rabe, M.; Jäkel, F.-W. (2000). Simulation for Globally Distributed Enterprises, *12th European Simulation Symposium* (ESS), Hamburg, September 2000, pp 322-327.
2. Rabe, M.; Jäkel, F.-W. (2002). Decomposition of Large Simulation Systems for Supply Chains and Manufacturing Systems, *9th European Concurrent Engineering Conference* (ECEC), Modena, pp. 96-102.
3. IEEE (2000). IEEE P1516.1/D5. Draft Standard for Modeling and Simulation (M&S) High Level Architecture (HLA) - Federate Interface Specification. IEEE.
4. Kuhl, F.;Weatherly, R.; Dahmann, J. (1999). *Creating computer simulation systems:* an introduction to the high level architecture. Upper Saddle River (USA, NJ): Prentice-Hall.
5. Schulze, Th.; Straßburger, St.; Klein, U. (1999). Migration of HLA into Civil Domains: Solutions and Prototypes for Transport Applications, *Simulation* 73 (1999) 5, pp. 296-303.
6. Woerner, J.; Laengle, T.; Woern, H. (2002). Corporate planning and Simulation of plant production facilities in the virtual world, *18th International Conference on CAD/CAM, Robotics and Factories of the Future*, Porto (Portugal) 2002. Porto: INESC Porto, 2002, pp. 109-116.
7. Rabe, M.; Jäkel, F.-W. (2001a). Non Military use of HLA within Distributed Manufacturing Scenarios, *Simulation und Visualisierung 2001*, Magdeburg, pp. 141-150.
8. Mertins, K.; Rabe, M.; Jäkel, F.-W. (2000). Neutral template libraries for efficient distributed simulation within a manufacturing system engineering platform, *Winter Simulation Conference*, Orlando (USA), pp. 1549-1557.
9. MISSION Consortium (2001). Public Deliverable 21 Input to standardisation working groups. Internet publication www.ims-mission.de.
10. Rabe, M.; Garcia de Gurtubai, G.; Jäkel, F.-W. (2001b). Modelling and Simulation for Globally Distributed Enterprises, *Proceedings of the EUROSIM 2001 conference*, Delft (CD ROM Publication).
11. Lee, Y.T.; Umeda, S. (2001). Information Model of the Supply Chain Simulation, *Proceedings of the EUROSIM 2001 conference*, Delft (CD ROM Publication).

19

A STUDY ON SUPPORT SYSTEM FOR DISTRIBUTED SIMULATION SYSTEM OF MANUFACTURING SYSTEMS USING HLA

Hironori Hibino[1] and Yoshiro Fukuda[2]
[1]*Technical Research Institute of JSPMI, Japan*
[2]*Engineering Faculty of Hosei University, Japan*
e-mail:hibino@tri.jspmi.or.jp

Abstract: In this paper, we firstly describe outlines of the developed distributed simulation system using HLA and the developed synchronization method in the past. We clarify requirement contrivances of the user support system. We explain requirement functions of the user support system that is premised on using the developed distributed simulation system and the developed synchronization method. We propose and develop the user support system. The case study was then carried out to evaluate the performance of the cooperative work.

Key words: HLA, distributed simulation, manufacturing system, user support system, XML

1. INTRODUCTION

A simulator plays an important role in designing new manufacturing systems [1][2][3][4]. As manufacturing systems are being created on larger and more complicated scales than ever before, it is increasingly necessary to design a manufacturing system with several persons concurrently. Then each manufacturing system designer uses manufacturing system simulator. At present, it is very difficult to combine different models which are made by different simulators because there are no standard descriptions for simulation models and a standard language for simulation programs has not been developed. In order to evaluate the whole system, a distributed simulation technology begins to attract industries. The distributed simulation technology is defined as executing a simulation while connecting and synchronizing several different simulators [5].

Concerning the distributed simulation technology, there are many research papers in academe [5][6]. But there are a few cases where industries have used the ideas presented in those papers. In order to use the distributed simulation technology in actual designs of manufacturing systems, it was necessary to develop the following items.

1. A distributed simulation system which is easy to use in various industries using commercial based manufacturing system simulators.
2. A method to synchronize commercial based manufacturing system simulators using characteristics of manufacturing system designs.
3. A user support system to define several relationships such as network relationships, simulation model relationships and so on using visual user interfaces while consistently keeping defined parameters along a definition procedure and to execute a distributed simulation remotely.

We have already proposed and developed the distributed simulation system using High Level Architecture (HLA) and the developed synchronization method [1][2][3][4]. Therefore we focus on the user support system.

In this paper, we firstly describe outlines of the developed distributed simulation system and the developed synchronization method in the past. We clarify requirement contrivances of the user support system. We explain requirement functions of the user support system that is premised on using the developed distributed simulation system and the developed synchronization method. We propose and develop the user support system. The case study was then carried out to evaluate the performance of the cooperative work.

2. DISTRIBUTED SIMULATION IN ACTUAL DESIGNS OF MANUFACTURING SYSTEMS

2.1 Manufacturing adapter of distributed simulation systems using HLA

HLA is only one standard for the distributed simulation system architecture as IEEE 1516 [7]. Currently there are several commercial based HLA Runtime Infrastructures (HLA/RTIs) which provide many kinds of services which are defined in IEEE1516.1 interface specifications [7].

However in order to use the commercial based HLA/RTIs for distributed simulations of manufacturing systems, the manufacturing designer needed to implement interface modules combing program methods based on the specifications by himself. Therefore it was difficult for the designer to use HLA.

In order to solve the problem, an adapter intermediately between manufacturing simulators and HLA/RTI has been developed [1][2]. This adapter is called a manufacturing adapter. Figure 1 shows the proposed manufacturing adapter.

Concerning communication between the manufacturing adapter and the simulators, XML format messages have been developed [1][2]. Using the manufacturing adapter, the simulator can access the services of HLA/RTI by only sending the XML format messages without implementing any program methods.

We have also proposed a Relational Data Definition (RDD) file which defines exchanged message contents between the simulators [1][2]. We have developed XML format definitions in the RDD file. The manufacturing adapter has been developed to get the message exchange services by reading parameters of the RDD file when the manufacturing adapter starts. The contents of the RDD file describe exchanged message contents. Therefore the RDD file is important to connect with simulation models.

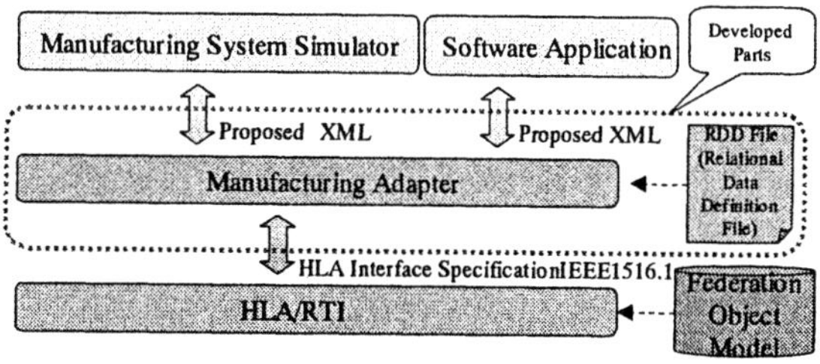

Figure 1. Outline of the manufacturing adapter.

2.2 Synchronization method for distributed manufacturing simulation systems

In actual manufacturing systems, relationships between the subsystem and the other subsystems are defined as input and output of material flow. Storage function units such as warehouses and buffers are usually located intermediately between the input part of a subsystem and the output part of other subsystems. The model of the each subsystem on the suitable simulator has an input part and an output part of the material flow. Therefore a system model can be made by connecting the input part of the subsystem model and the output part of the other subsystem models. To connect different simulators, the focus of our study has been on the storage function units as association points between the subsystems. In simulators used in this study, the storage function units are defined as storage models.

Figure 2 shows a fundamental structure to connect two different simulators by using the storage model. Each subsystem is modeled by different simulators (A and B). The warehouse is modeled as a storage model in each simulator. The simulators are synchronized by the proposed synchronization method using conditions of amounts of stock in the storage models [3][4]. This method can be adapted for synchronizing distributed simulation clocks without the rollback function which is to return the simulation clock to passed time to synchronize events among the simulations [3][4]. This method is necessary for actual designs of manufacturing systems because commercial based manufacturing system simulators do not have the rollback function.

Figure 2. Outline of the proposed synchronization method.

2.3 User support system for distributed manufacturing simulation systems

In a definition procedure to execute the distributed simulation system, the manufacturing system designer needs to define many parameters. If only a definition contradiction occurs, the designer does not have inconsistent simulation results, but also loses a lot of time to detect the definition contradiction. Therefore it is necessary to prevent careless mistakes through the definition procedure such as definition contradicts of messages among the simulators, definition omissions, double definitions and so on. In order to prevent careless mistakes through the definition procedure, it is necessary to support the definition procedure consistently while storing and reusing the definition parameters.

Particularly in a case that a relationship among a message sender and message receivers is in the ration of one to many and also there are many kinds of messages such as very large manufacturing systems, there are probabilities to make definition mistakes for the relationships because the relationships become complicated. In such a case, it is effective to use a visual user interface (VUI) which provides modeling functions to define and express the relationships.

On the other hand, in a case that each simulator is located away from the other simulators, it is necessary for only a manufacturing system designer to initiate and execute the distributed simulation system remotely.

In this paper, we propose a user support system to solve the above issues. The user support system is premised on using the developed distributed simulation system (op. cit. chapter 2 section 1) and the developed synchronization method (op. cit. chapter 2 section 2). Figure 3 shows an outline of the user support system and relationships among the user support system, the distributed simulation system, and the synchronization method.

The user support system includes the following contrivances.

1. The user support system has a contrivance to support the definition procedure consistently while storing and reusing the definition parameters.
2. The user support system has a contrivance to provide visual user interfaces (VUI). For example, a visual interface provides modeling functions to make and express message relationships among a message sender and message receivers.
3. The user support system has a contrivance for only a manufacturing system designer to execute the distributed simulation remotely.

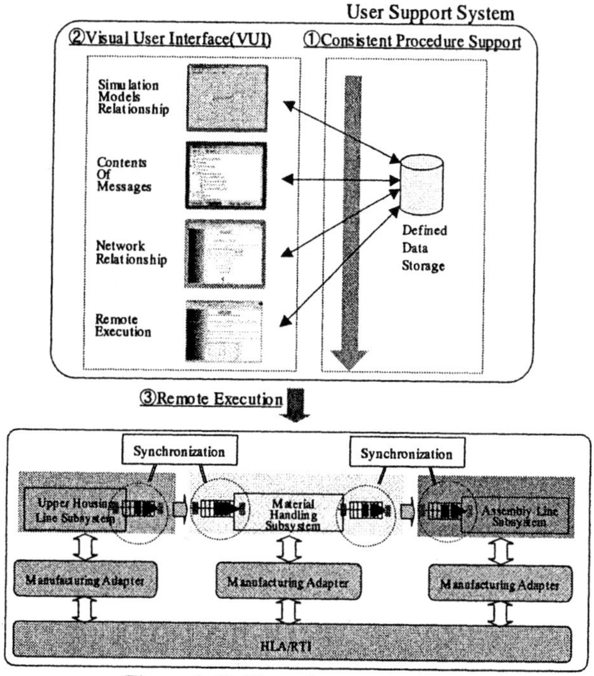

Figure 3. Outline of user support system.

3. REQUIREMENT FUNCTIONS OF USER SUPPORT SYSTEM

We analyzed procedures of the definitions and execution for the distributed simulation system using the synchronization method. We clarified requirements for the user support system. We classify the requirements into seven functions for user support system as a project management function, a modeling function, a network definition function, a RDD file generation function, a FOM file generation function, a file transfer function, and a remote execution function. Figure 4 shows an outline of the functions of the user support system.

Firstly, main roles of the project management function are to declare a project as the first step in the definition procedure and to manage storing and reusing the defined parameters by the designer.

Secondly, the modeling function includes three functions. At first, in order to define the storage models as the association points among the simulation models, the modeling function provides a function to define a relationship between a name of a federate which means a simulator in HLA and a name of the storage model which is modeled on the federate. The modeling function also provides a function to make message relationships between a storage model as a message sender and the other storage models as message receivers. Using these two functions, it is possible to express the message relationships among the federates visually by arrows from the federate as the message sender to the federates as the message receiver. The modeling function provides such a visual expression function.

Thirdly, main roles of the network definition function are to define a network environment of each federate and to define parameters of the socket communication between the manufacturing adapter of each federate and HLA/RTI. In this function, the designer can reuse data that have been defined in the modeling function by the project management function.

Fourthly, the RDD file generation function generates data of each RDD file using the defined parameters in the modeling function and the network definition function. Contents of the RDD file are formatted by XML. The RDD file includes the defined messages contents, the defined network environment, the defined parameters of the socket communication and so on using data of the modeling function and the network definition function.

Fifthly, the FOM file generation function generates data of the FOM file using all of the definition messages in the modeling function.

Sixthly, the file transfer function includes two functions. At first, the file transfer function provides a function to define exact locations where the RDD files and the FOM file are located in personal computers (PCs). The file transfer function provides a function to send the RDD files and the FOM file to the defined locations.

Finally, the remote execution function provides a communication function of a client-server type between a client side PC which sends indications to initiate and execute the federate and a server side PC which receive and serve the indications to initiate and execute the federate.

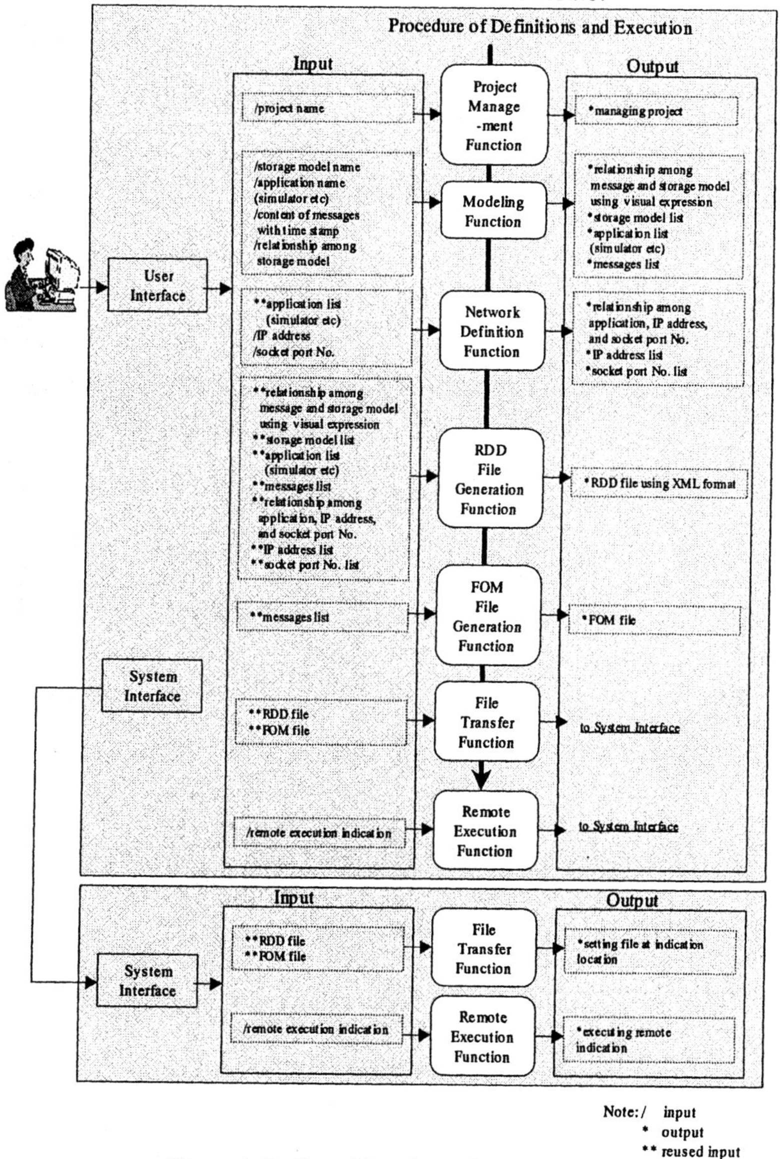

Figure 4. Outline of functions of user support system.

4. DEVELOPMENT OF USER SUPPORT SYSTEM

In order to implement classified seven functions, we propose and develop Distributed Simulation Modeling and Execution (DISMO) as the user support system. In DISMO, the Web server technology, the servlet technology, and the Java program technology are used for the designer to get services of seven functions using an Internet browser. Apache version 3.20 as the Web server and Tomcat version 3.1 as the servlet are used in DISMO.

DISMO consists of a DISMO Design Server (DISMO-DS) and DISMO Simulation Servers (DISMO-SS). DISMO-DS is a core server to support the designer through the Internet or Intranet technology. DISMO-DS provides seven requirement functions. DISMO-SS is a serve where an application such as the manufacturing system simulator is implemented. DISMO-SS provides the file transfer function and the remote execution function.

5. CASE STUDY

The developed user support system is confirmed using a case to evaluate a hypothetical manufacturing system which produces motors. Figure 5 shows a layout of the manufacturing system and relationships among simulation model using the storage models. With this manufacturing system, it is assumed that three companies design the manufacturing system separately and concurrently. Company B designs the upper housing line subsystem. Company C designs the assembly line subsystem. Company A designs the other subsystems which include the storage and the material handing subsystem. In this case study, three manufacturing system simulators as QUEST [8], eMPlant [9], and GAROPS [10] are used.

Three simulators are located on three places as Tokyo Metropolitan, Saitama Prefecture, and Aichi Prefecture. It is technically possible to connect three simulators using Internet. However this is not reality because of security problems. Therefore we chose Intranet to connect the simulators. Three simulators are connected by Integrated Service Digital Network (ISDN) using Remote Access Server (RAS).

At first, this case study was registered as a new project in DISMO. The designer defined message contents and the relationships between the storage models using the VUIs which are provided in the modeling function. Figure 6 shows the definition of the message contents using the VUI. Figure 7 shows the message relationships using the VUI by arrows from the federate as the message sender to the federates as the message receiver. After the modeling function, the network parameters for each manufacture adapter were defined. After these definitions were finished, the RDD files and the

FOM file were generated by DISMO. These files were translated to the defined locations. Then the designer indicated to initiate and execute the distributed simulation system. After the procedure to define and execute the distributed simulation, we confirmed that the distributed simulation system using the synchronization method could be executed exactly. Therefore the case study of the developed user support system was carried out to evaluate the performance of the cooperative work.

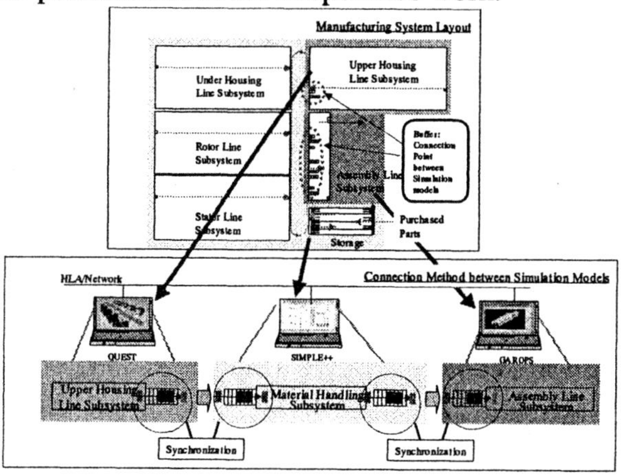

Figure 5. Layout and relationships among simulation models.

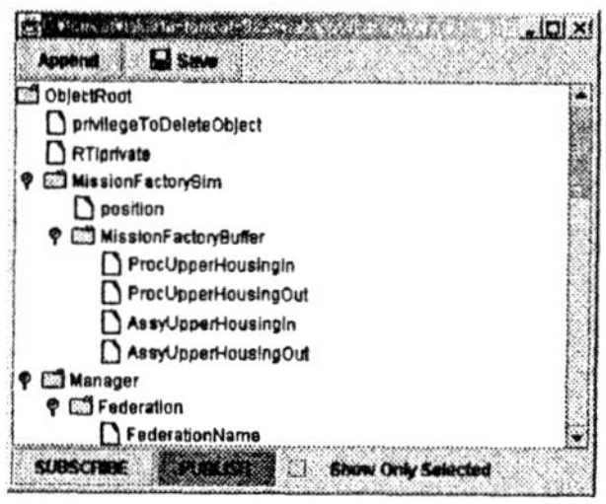

Figure 6. Definition of message contents by DISMO.

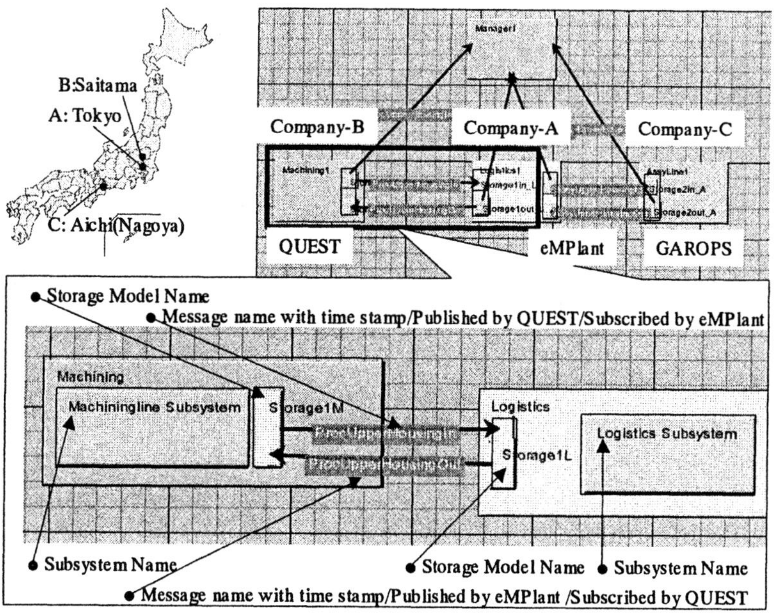

Figure 7. Definition of message relationships visually by arrows using DISMO.

6. CONCLUSION

In this paper, we firstly described outlines of the developed distributed simulation system using HLA and the developed synchronization method in the past. We clarified requirement contrivances of the user support system. We explained requirement functions of the user support system that is premised on using the developed distributed simulation system and the developed synchronization method. We proposed and developed the user support system. The case study was then carried out to evaluate the performance of the cooperative work.

ACKNOWLEDGEMENTS

This research is a part of the "Study on Manufacturing System Using Open-Oriented Information Technologies" research project of JSPMI. This project was supported by funding from the Japan Keirin Association.

REFERENCES

1. Hibino H., Fukuda Y., Yura Y., Mitsuyuki K., Kaneda K. (2002). Manufacturing Adapter of Distributed Simulation Systems Using HLA, *Proceedings of the 2002 Winter Simulation Conference*, 1099-109
2. Hibino H., Fukuda Y., Nakano M., Yura Y., Fujii S. (2002). A Study on Distributed Simulation Systems to Evaluate Manufacturing Systems Using HLA, *Proceedings of 2002 Japan-U.S.A. Symposium on Flexible Automation*, 923-30
3. Hibino H., Fukuda Y., Nakano M. (2001). A Synchronization Mechanism for Distributed Manufacturing Simulation Systems, *Proceedings of the International Conference of Production Research*, CD-ROM
4. Hibino H., Fukuda Y. (2002). A Synchronization Mechanism without Rollback Function for Distributed Manufacturing Simulation Systems, *Transactions of the Japan Society of Mechanical Engineers*, **68**(672), 2472-8
5. Fujimoto R. (1990). Parallel Discrete Event Simulation, *Communication of ACM*, **33**(10), 30-53
6. Nicol D., Johnson M., Yoshimura A. (1997). The IDES Framework: A Case Study in Development of a Parallel Discrete-event Simulation System, *Proceedings of the 1997 Winter Simulation Conference*, 93-9
7. IEEE 1516 (2001). http://www.demso.mil/
8. QUEST (2003). http://www.delmia.com/
9. eMPlant (2003). http://www.tecnomatix.com/
10. GAROPS (2003). http://www.tpyotamacs.co.jp/products/catalog13/

20

METHOD AND TOOL FOR DESIGN PROCESS NAVIGATION AND AUTOMATIC GENERATION OF SIMULATION MODELS FOR MANUFACTURING SYSTEMS

Masaru Nakano[1], Fumiko Kubota[1], Yutaka Inamori[1] and Keiji Mitsuyuki[2]
[1]*Toyota Central R&D Labs., Inc., Japan*
[2]*DENSO Corporation, Japan*
e-mail: nakano@robotics.tytlabs.co.jp

Abstract: Manufacturing system designers should concentrate on designing and planning manufacturing systems instead of spending their efforts on creating the simulation models to verify the design. This paper proposes a method and its tool to navigate the designers through the engineering process and generate the simulation model automatically from the design results. The design agent also supports collaborative design projects among different companies or divisions with distributed engineering and distributed simulation techniques. The idea was implemented and applied to a factory planning process.

Key words: Manufacturing System, Concurrent Engineering, Simulation, Design Agent

1. INTRODUCTION

Global competition forces industrial companies to reduce the cost and time needed for the development of new manufacturing systems. To achieve this goal, manufacturing system simulators have been employed. However, manufacturing system designers should be concentrating on designing and planning the systems instead of spending their efforts on creating the simulation models in order to verify the results. The more time designers need to build simulation models, the less time they have to develop the manufacturing system. On the other hand, if consultants, vendors or software engineers build and verify simulation models for the manufacturing system engineers, many problems such as incorrect input data, misunderstanding of the model and misleading of the decision from the results may occur.

To overcome the problem, this paper proposes a method and its tool, referred to as a design agent system [1][2], to navigate the designers through the engineering process and generate the simulation model automatically from the design results to evaluate the system performance. Data exchange from CAD systems to simulators was considered [3]. This paper deals with not only the exchange but also the building of manufacturing system models. The Product Process Resource (PPR) [4] approach is also a promising approach to build manufacturing system models. Though it is a supporting tool to manage a large amount of data in a standard engineering process and collect the appropriate data for the simulation, the approach in this paper focuses on modelling the engineering process linked to the system data and directly generates the simulation model from the design specifications.

The design agent supports not only a designer in a company or division but also collaborative design projects among different companies or divisions. The manufacturing system model is built concurrently and translated into a simulation model or separate simulation models if different simulators are employed. In the latter case, the models are simulated in a distributed environment using the High Level Architecture (HLA) [5]. The proposal has been discussed under Intelligent Manufacturing Systems (IMS) Modelling and Simulation Environments for Design, Planning and Operation of Globally Distributed Enterprises (MISSION) international project including institutes and companies from Japan, Europe and USA [6]. The design agent system was applied to a factory planning process involving four companies to develop a factory including assembly lines, processing lines and a transportation system.

2. DESIGN PROCESS NAVIGATION

The structure of the design agent system for navigating the design process to build manufacturing systems is shown in Figure 1. It has two categories of agents: client agents which interact with engineers and server agents which manage data base for the agent system (referred to as Design Agent Database), deal with communications among client agents and provide mobile agents such as collecting data in a distributed network. The client agents are downloaded as JAVA applets. The Design Agent Database includes Project Case Library to accumulate and reuse project cases and Manufacturing System Component Library (referred to as MS Component Library) to define the structure of components in manufacturing systems such as enterprise, factory, production line, cell and equipment in an object-oriented manner. The MS Component Library provides unique names, standard values and design rules of design parameters in each component. All agents communicate through the Manufacturing System Engineering (MSE) Network such as CORBA and DCOM, distinguishing the distributed simulation network described later.

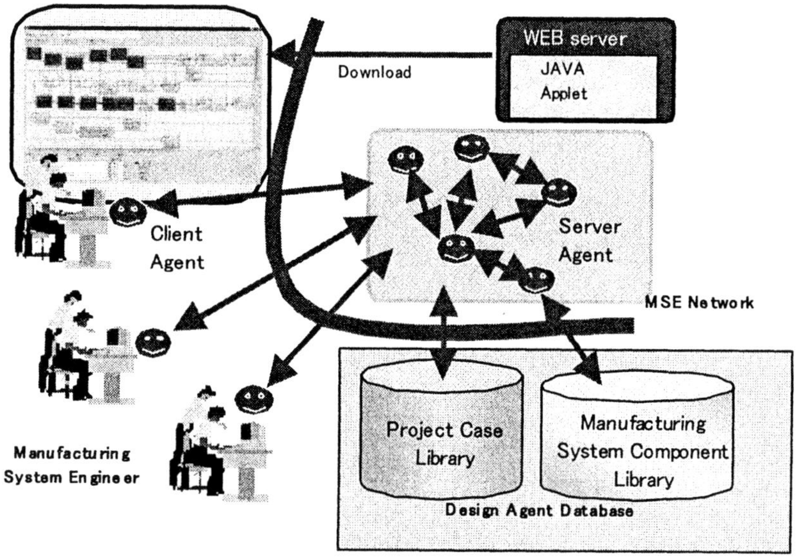

Figure 1. Architecture of the design agent system.

Design Process Agent is a client agent that has a central role to navigate engineers in the design of manufacturing systems as shown in Figure 2. At first, engineers give the requirements of a manufacturing system to the Design Process Agent. The agent helps the engineers search for an appropriate project case in the Project Case Library, and routinely support them to build the system model by showing a list of appropriate components in the MS Component Library. As the design process improves from an initial rough-cut model to a detailed one, the Design Process Agent creates the design process and the manufacturing system model that consists of the manufacturing process, the system components such as facilities and workers and the layout.

A requirement, design process, layout and manufacturing system model are described in eXtensible Markup Language (XML) so that software vendors can easily develop the interface to the design agent system.

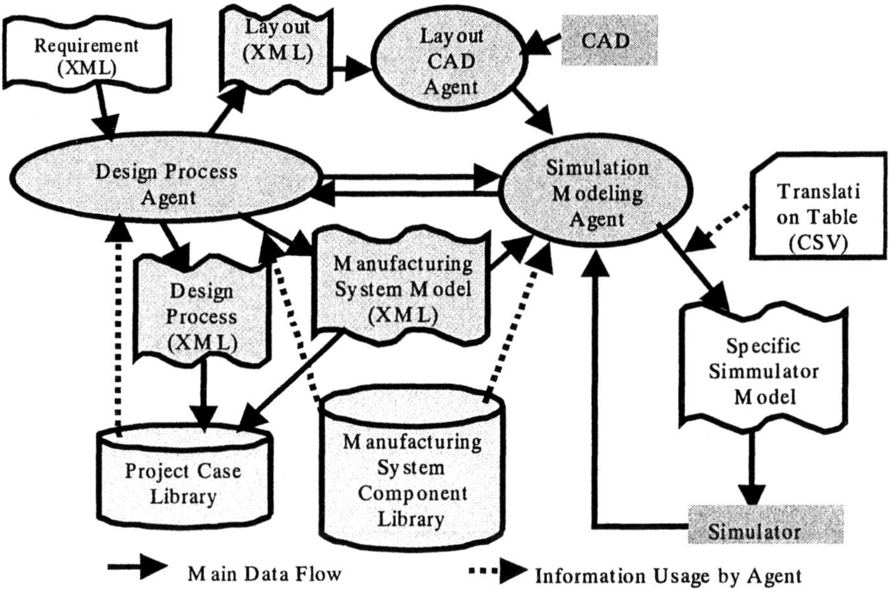

Figure 2. Client agents and their roles

3. AUTOMATIC GENERATION OF SIMULATION MODELS

Tool Agents are client agents that support engineers to use engineering tools. Simulation Modeling and Layout CAD Agents in Figure 2 are the Tool Agents. The Simulation Modeling Agent has the following functions:

- Know what simulation packages are available and how to select them [2].
- Know how to translate a manufacturing system model created by the Design Process Agent into specific simulation models.
- Know how to get simulation results from simulators and give the results to the Design Process Agent.

The agent translates the system model to a model for a simulator to evaluate or animate the model. The Translation Table in Figure 2 is defined to translate the attributes of a class in the MS Component Library to the corresponding attributes in the factory facilities used in specific simulators. For example, the class "storage" in the MS Component Library may be called "Buffer" in simulator A, and "Accumulator" in simulator B. There are specific parameters such as a facility color and simulation length in simulation study. If those parameters are not found in the MS Component Library, the Simulation Modeling Agent asks users to provide appropriate parameters. The Translation Table is described in Comma Separated Values (CSV) format so that engineers can easily change the table. The Simulation Modeling Agent provides the Design Process Agent with the simulation results such as the system performance and the utility rates of machines.

A simulator can be employed before or after a CAD system as shown in Figure 3 because the simulation model usually serves for both evaluation and animation. The design agent system provides an initial layout file for a CAD system and the Simulation Modeling Agent can use either a layout from the manufacturing system model to evaluate the system performance or a layout from a CAD system to animate with higher fidelity. Because CAD systems do not usually output XML files, the Layout CAD Agent is employed to translate the layout data for CAD systems into the XML files.

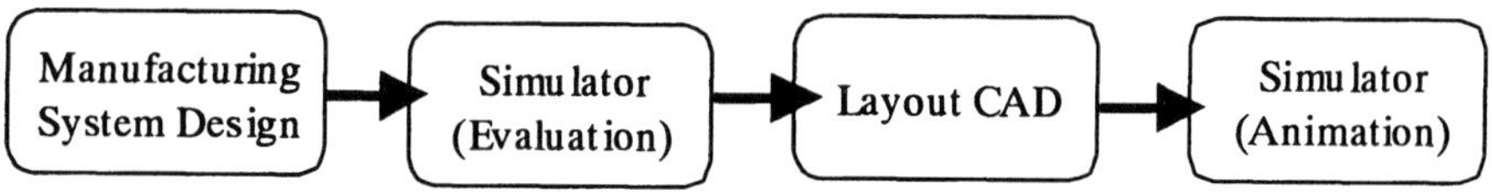

Figure 3. Roles of CAD and simulator

4. IMPLEMENTATION OF DESIGN AGENT SYSTEM

The design agent system was implemented by using JAVA language. We designed a virtual factory that produced motors for automobiles and consisted of processing lines, assembly lines and a logistic system where automated guided vehicles transported parts between the production lines and a storage system. Figure 4 shows the main procedure used with the agent system.

The requirements include a design order sheet including a rough specification such as production volume and work schedule, product structure usually given in a CAD system, and an initial layout for information. The manufacturing model of an appropriate case is shown up as a template to the engineer in three views: the design process, the manufacturing process and the layout. The Design Process Agent navigates engineers to change the design process that in turn produces a new manufacturing process and a new layout. The design parameters of the facilities are provided in the design process. A simulator [7] was employed to evaluate the system performance because it provides a language for easy development of the interface to the Translation Table.

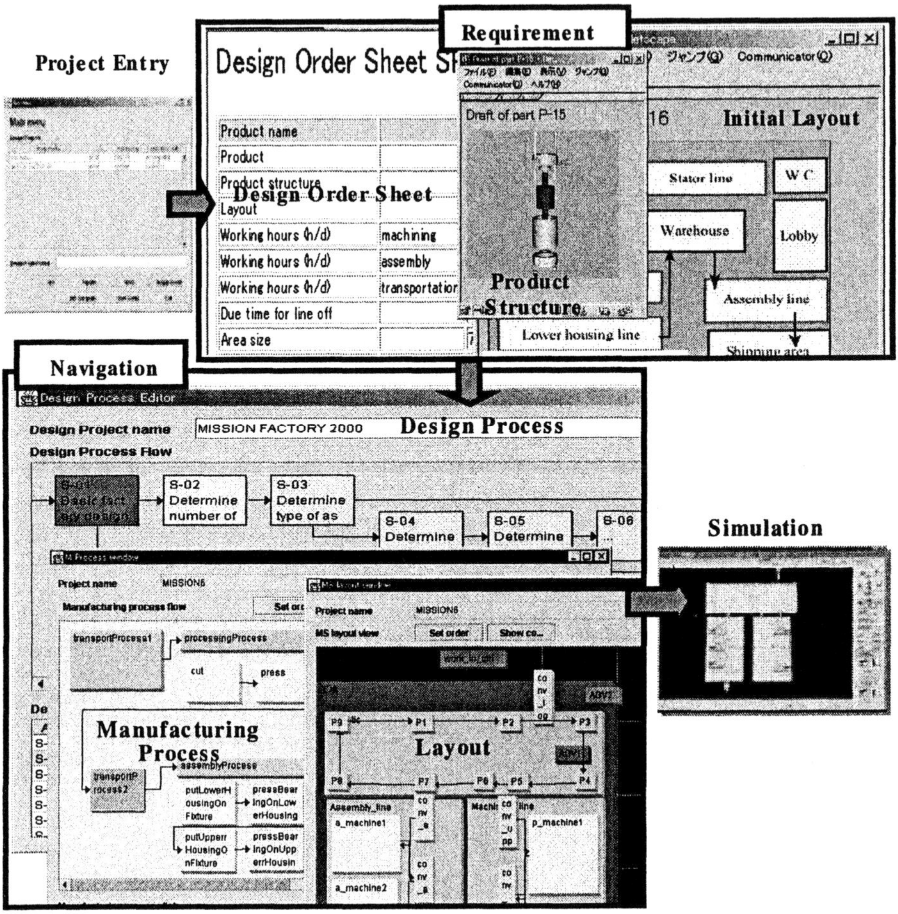

Figure 4. Main procedure of the design agent system.

5. DISTRIBUTED ENGINEERING AND DISTRIBUTED SIMULATION

The design agent system can be employed in collaborative projects to design not only a manufacturing system but also a manufacturing enterprise including sales, logistics and planning for manufacturing in a globally distributed enterprise. The design agent system can also be employed to design a virtual enterprise in which several companies do business together. As a result of distributed engineering, a distributed simulation is desired to evaluate or animate the entire model [8]. Although techniques of distributed simulation have been studied [5], the combination of the distributed engineering and distributed simulation has not yet been studied in depth. In this solution, the MS Component Library has an important role to integrate the manufacturing system models into the simulation model correctly.

Four companies in Japan collaborated in a study to build the virtual factory described in Section 3. In this virtual project, the owner company gave three companies the requirements to build assembly lines, processing lines and a transportation system respectively as shown in Figure 5. The architecture of the distributed simulation should have the capability to develop a control policy to proceed simulation time. The proposed system is based on the HLA. The data model communicated among simulators is based on the MS Component Library but data specific to the distributed simulation such as a time stamp, a source ID and a destination ID are added. The Distributed Simulation Modeling Agent is proposed to generate the data model in the Federation Execution Data (FED) file. The HLA adapter has an interface to the individual simulators and manages the simulation data under the HLA. The simulators have the function to pause a simulation at a certain time in order to simultaneously proceed with other simulators. The HLA manager monitors the HLA adapters and gives the feedback of the results to the Design Process Agent. The virtual collaboration project was successfully studied by the cooperation of the four companies [2].

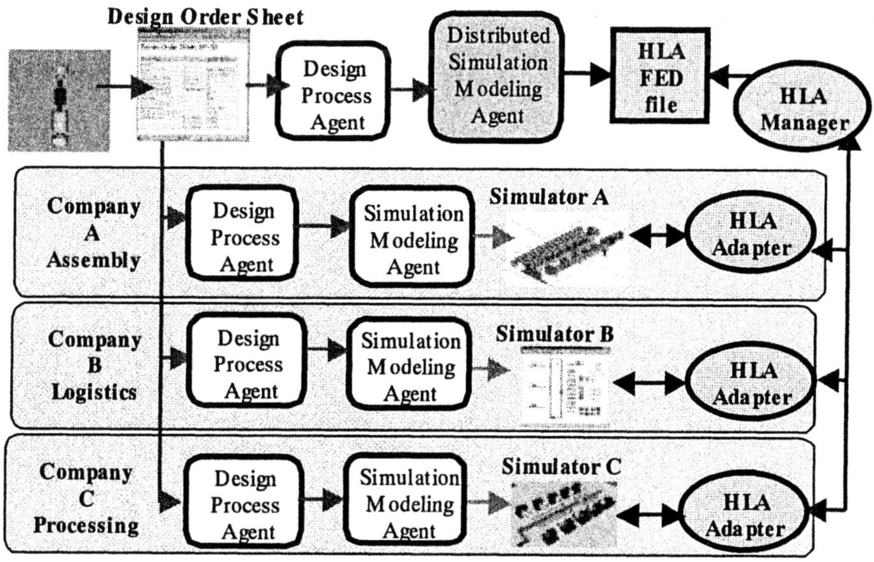

Figure 5. Distribution engineering and simulation

6. CONCLUSION

The paper proposes the method and its tool to navigate engineers to develop manufacturing systems. The manufacturing system models from the design system are translated automatically to the simulation models to evaluate the performance or animate. The combination of distributed engineering and distributed simulation is also proposed for global enterprises, demonstrating that the design agent system can be applied to the enterprise planning beyond manufacturing systems.

ACKNOWLEDGEMENTS

This research is sponsored by the IMS program in Japan. Useful discussions with members in the IMS/MISSION international project are much appreciated. The authors also would like to thank Dr. Roser in TCRDL for his help in the preparation of the paper.

REFERENCES

1. Nakano, M., Kubota, F., Yura, Y., and Arai, E. (2002). Design Agent System for Collaborative Design Process in Manufacturing Systems, *Proceedings of Eurosim 2001 Congress,* pp.1-6
2. Nakano, M., Sato, S., and Arai, E. (2002). Collaborative Design Supporting System for Manufacturing Systems, *Japan Society of Mechanical Engineering (JSME) International Journal, Series C.*, Vol.45, No.2, pp.575-580
3. Moorthy, S. (1999), Integrating the CAD Model with Dynamic Simulation: Simulation Data Exchange, *Proceedings of the 1999 Winter Simulation Conference*, pp.276-285
4. Brown, R. G. (2000), Driving Digital Manufacturing to Reality, *Proceedings of the 2000 Winter Simulation Conference*, pp.224-228
5. Hibino, H., Fukuda, Y., Nakano, M., Yura, Y., and Fujii, S. (2002). A Study on Distributed Simulation System to Evaluate Manufacturing Systems using HLA, *Proceedings of 2002 Japan-USA Symposium on Flexible Automation*, pp.923-930
6. Yagi, J., Fujii, S., Nakano, M., Hibino, H., Arai, E., Kojima, F., Fukuda, Y., Sugimura, N., and Isago, S. (2000). IMS MISSION PROJECT -MISSION Modelling Platform, *Proceedings of Sixth International Conference on Manufacturing (PCM' 2000),* pp.961-969
7. Nakano, M., Sugiura, N., Tanaka, M. and Kuno, T. (1994). ROPSII: Agent-Oriented Manufacturing Simulator on the Basis of Robot Simulator, *Proceedings of JAPAN-U.S.A. Symposium on Flexible Automation,* pp. 201-208
8. Kubota, F., Sato, S., and Nakano, M. (1999) Enterprise Modeling and Simulation Integrating Manufacturing System Design and Supply Chain, IEEE *Proceedings of International Conference on System, Man, and Cybernetics*, Ⅳ, pp.511-515

21

KNOWLEDGE MANAGEMENT IN BID PREPARATION FOR GLOBAL ENGINEERING AND MANUFACTURING PROJECTS

Mingwei Zhou [1], John Mo [1], Laszlo Nemes [1], and William Hall [2]
[1]*CSIRO Manufacturing & Infrastructure Technology, Australia*
[2]*Strategy and Development, Tenix Defence, Australia*
e-mail: Mingwei.Zhou@csiro.au

Abstract: The core competence of a global engineering and manufacturing enterprise increasingly depends on the quality of its intellectual resources and how these resources are used in critical missions such as bid preparation. This paper discusses the knowledge management issues in the development of VIEWBID, a web-based system for supporting online bidding document preparation for global engineering and manufacturing projects. The VIEWBID system aims to supports inter-enterprise collaboration for compiling accurate bids. The enterprise engineering architectures and methodologies, such as VERA and iRoadmap, have been used to analyze the bidding process to capture different levels of procedural knowledge. A set of component-based technologies has been developed using XML and Java to capture, configure and compose the bidding documents.

Key words: Knowledge Management, Enterprise Engineering, Global Manufacturing

1. INTRODUCTION

With the globalization of business activities, the core competence of a manufacturing enterprise increasingly depends on the quality of its intellectual resources and how these resources are used in critical missions such as bid preparation.

Research in the technologies to support the bidding process has led to the development of the VIEWBID (Virtual Enterprise Workbench for

Worldwide Business Integration and Development) [1], a web-based system for supporting online bidding document preparation for global engineering and manufacturing projects.

The VIEWBID system provides three layers of functionality. The inner layer is a corporate knowledge management platform that serves as the corporate memory. The middle layer is a team collaboration environment for supporting the design and operation of a virtual enterprise, and the management of cross-corporate boundary bidding project team. The outer layer is the bidding workbench that enables the bidding team to prepare the cost estimation and technical proposal for the bidding document.

This paper discusses the knowledge management issues in the development of VIEWBID system that aims to support inter-enterprise collaboration for compiling accurate bids with managed risks. The bidding team needs extensive and up-to-date knowledge about the company's engineering capability and capacity in order to make an inform decision on the technology baselines and costing strategies. The knowledge created in the bidding process will also be captured and utilized as the corporate memory for supporting preparation of following bids. If the bid is successful, the knowledge captured in the bidding phase can also be further used in the following contracting, engineering & procurement, and manufacturing phases.

2. BIDDING PROCESS

To compete in the global business environment, companies need to collaborate across their company boundary to form a virtual enterprise and bid for new projects together. This distributed bidding team involves members from the leading bidding company as well as its partners and possibly the client.

During the bidding process, information entered into the process will need to be captured in a specific structure that can be utilized as part of the corporate memory for supporting the preparation of bids quickly. In addition, the knowledge captured in the bid can also be used for contracting and total product lifecycle support.

A bidding document usually consists of the technical proposal to client's requirements, and the associated terms and conditions. Preparation of the technical proposals for large engineering projects is usually very time consuming. The technical proposals are quite different for different projects or products, but many components of the content are often the same within each proposal. The same components, such as the manufacturing capabilities of the company, may be used in many different bidding documents. Also, a bidding document developed for one project can also serve as the basis for later projects having similar requirements. Re-use of information based upon similarities between projects is a major goal for knowledge management in bid preparation. On the other hand, different kinds of contextual and procedural knowledge need to be captured during the bid preparation.

The bidding process usually begins with the reception of a client's Request for Proposals (RFP) outlining the capabilities sought from the project. The essential contexts are knowledge relating to the client's statement of capabilities, engineering decisions, notes, standards, correspondence, and a variety of documentation from lower tier suppliers. The bidding team needs to distil this kind contextual information into preliminary technical proposal.

Based on information provided by preliminary technical proposal, the client normally issues a Request for Tender (RFT) to a short list of suppliers. The bidding team of the suppliers then rework and extend the documentation developed in their proposals and include a detailed commercial response to the RFT.

Technical proposals and RFT responses are required to be completed within very limited time periods, such that if documentation is not delivered by the specified due date the proposal or bid will not be considered. Development of proposals and bids are highly competitive and place major pressure on to minimize documentation production cycle times and to maximize the quality of information/knowledge they contain.

3. KNOWLEDGE MODEL

Knowledge is the internal state of an agent following the acquisition and processing of information, here the agent could be a human being or a computer system. To categorize human knowledge, many knowledge models have been proposed. Polanyi [2,3] identified that human knowledge has two major components: the tacit and explicit knowledge. Following Polanyi's concept, Nonaka [4,5] further proposed his theory that tacit knowledge consists of personal relationships, practical experience, shared values and explicit knowledge consists of formal policies and procedures.

Nickols further clarified the intrinsic meanings of various knowledge terms by proposing a testable knowledge model [6] that includes: explicit, tacit, implicit, declaration, and procedural knowledge. As shown in Figure 1, Nickols' model also depicted a testable procedure to distinguish the relationship among different classes of knowledge.

In our approach, we classified the knowledge involved in the bid preparation process into three categories based on Nickols' framework.

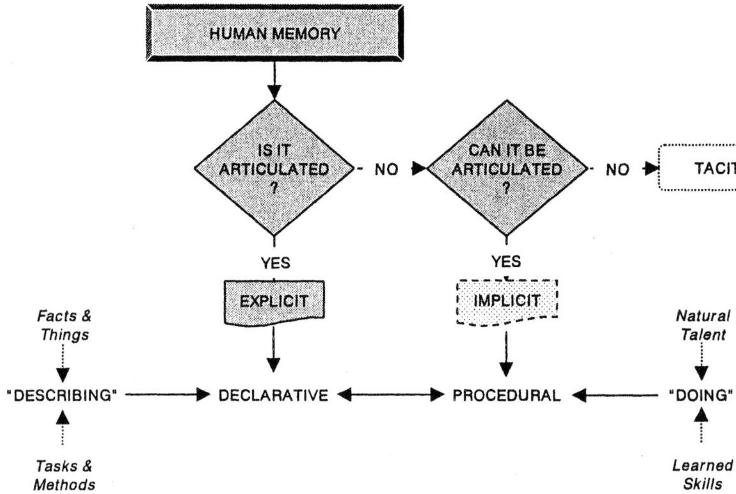

Figure 1. Nickols' Knowledge Model [6]

The first category is the "direct knowledge" or facts, which are the explicit or declarative knowledge. This category of knowledge is visible, written, transferable, sharable and reusable. It is usually documented and stored and transmitted externally to a human brain. In the bidding process, engineers assimilate their information and turn it into bidding documents conveying distilled knowledge. A knowledge management system should help the engineers capture, validate, and preserve knowledge; and assist discovery, reuse, retrieval and transmission of that knowledge.

The second category is the "procedural knowledge" or the best practices, which are usually implicit, and context sensitive. This category of knowledge is related to processes, methods, practices in groups and professions. This type of knowledge needs to be identified, captured and made explicit in the way that can be shared. However, it is not always well documented.

The third category is the "tacit knowledge", which is the most difficult to understand and represent. It is indirect, embedded in experience, owned by individuals and cannot be articulated in words.

Our research presented in this paper focuses on the first two categories of knowledge, aiming to capture and manage the direct and procedural knowledge. Our previous work has attempted to interpret and convert the tacit knowledge into the direct and procedural knowledge [7]. We found that the direct and procedural knowledge is more critical to the bid preparation process.

4. MANAGING DIRECT KNOWLEDGE

The direct knowledge used in the bid preparation is mainly the factual information about the company and its engineering and manufacturing capabilities to fulfill the client's requirements. This class of information is often available in various forms, such as text documents, diagrams, spreadsheets, and engineering drawings. Traditionally, engineers have been using the "cut and paste" technique to re-use this kind of information.

Basically, managing direct knowledge is to develop an advanced replacement of the "cut and paste" technology. The key issue here is how to "cut" the direct knowledge into "components" with a right size, so that they can be both easily found and "pasted" into the new bidding documents.

Based on our analysis of the bidding documents, and especially the technical proposals, we found it is feasible to segment the re-usable parts of an engineering document according to their functions in the document. These functional segments can further be classified into four categories:

1. Basic Document Component (BDC): this kind of component contains self-explanatory facts. For example, the description of a specific material or equipment.
2. Associated Document Component (ADC): this kind of component is usually associated with a BDC, and used to further depict the content of the BDC. For example, an illustration diagram or a table of operational parameters for a machine that is described by a BDC. The existence of an ADC is depending on the existence of a BDC. If a BDC is modified, its ADCs may also need to be modified accordingly. However, a BDC and its ADCs can be modified independently, as long as they are consistent.
3. Derived Document Component (DDC): this kind of component is usually derived from a BDC or an ADC either automatically or manually. For example, a bar chart (DDC) of a table (ADC) or an image (DDC) of a 3D model (BDC). The existence of a DDC depends on the BDC or ADC from which it derived. Whenever a BDC or an ADC is modified, its derived DDC must be re-created.
4. Composite Document Component (CDC): this kind of component is usually a combination of a BDC with its ADCs and DDCs. For

example, a manufacturing facility can be described using a CDC that consists of a description of the machine (BDC), its major parameters (ADCs), and some charts and images (DDCs).

This classification scheme facilitates us to capture, re-use and update the direct knowledge used in the bid preparation.

5. MANAGING PROCEDURAL KNOWLEDGE

The procedural knowledge for bid preparation is the best practices that lead to a winning bid. We classified those best practices into three levels according to their coverage.

The top-level procedural knowledge is the information roadmap, or iRoadmap [8]. The iRoadmap is developed based on the Virtual Enterprise Reference Architecture (VERA). Along the lifecycle of a virtual enterprise (VE), extensive amount of critical information is created, shared, used, modified and disposed. The iRoadmap attempts to trace the content and flow of such information in the whole life of a VE. The iRoadmap for the bid preparation in a global engineering and manufacturing network environment provides a guideline for the bidding team, and enables the development of knowledge management system to support the bidding process.

The middle-level procedural knowledge is the bidding decision-making and bidding document configuration processes. These best practices will help the bid manager in analyzing the business opportunities, and adopt a suitable bidding strategy according to market, technology, or alliance considerations. With the selected bidding strategy, the engineering baseline, cost estimation, and risk exposures can be established. Furthermore, the structure of the bidding document, including the technical proposal and terms and conditions can be defined. Most re-usable components for the bidding document can be identified and included shortly after the configuration of the bidding document is determined.

The detailed-level procedural knowledge is the know-how of how technical, financial, and legal information is used in the bid preparation. These best practices may include the selection of manufacturing equipment, the specification of design or operation parameters, trade-off in production processes, and selection of corresponding testing and validation methods. It may also include advise to the bidding team about alternative clauses of terms and conditions as well as their impacts and consequences.

The top-level procedural knowledge could be developed and tailored to suit any bidding team in a virtual enterprise setting. However, the middle-level procedural knowledge could only be developed for a specific industry sector or a specific group or network of companies. The detailed-level procedural knowledge could only be developed for a specific company. This is because the best practice in one company could be disastrous for another.

6. IMPLEMENTATION AND CASE STUDIES

A proof-of-concept VIEWBID bidding workbench has been implemented. The Lotus QuickPlace [9] has been used as a web-based collaboration platform to support the whole bid preparation process, from receiving the RFP or RFT, through to the online co-authoring of technical proposal, to the final delivery of the bidding document.

The various document components are defined using XML DTDs, which are then used to create forms and templates in the QuickPlace to enable the capturing and re-use of direct knowledge. A Java based modeling and execution tool has also been implemented to capture the procedural knowledge in the bidding process, however, only selected a few of them have been implemented in QuickPlace as agents.

Two case studies have been conducted. The first case study looks at the bidding process for the design and build of an airframe component. The second case study investigates the feasibility of implementing SMART 2000 [10], the Australian Defence Industry guideline for bidding and contracting in bid preparation for defence projects.

Our case studies indicate that development of a knowledge management system for bidding requires careful analysis of the bidding process, and the effectiveness and efficiency of such a system largely depends on the easiness of capturing and reusing existing knowledge. Managing knowledge as intangible assets in the dynamic virtual enterprise environment is still a challenging research topic [11].

7. CONCLUSION

This paper presented an investigation on the knowledge management issues in the development of VIEWBID, a web-based bid workbench for global engineering and manufacturing projects. We classified knowledge involved in the bid preparation process into three categories, and focused our research on the capturing and management of the direct and procedural knowledge, while trying to interpret and convert the tacit knowledge into the direct and procedural knowledge.

The enterprise engineering architectures and methodologies, such as VERA and iRoadmap, have been used to analyze the bidding process to capture different levels of procedural knowledge. A set of component-based technologies has also been developed using XML and Java to capture, configure, and compose the bidding documents.

REFERENCES

1. Zhou M., Zheng J., Williams A., Alexander R. (2000). "A Web-based Bidding Workbench for Global Manufacturing", Global Engineering, Manufacturing and Enterprise Networks (Proceedings of 4th International Conference on Design of Information Infrastructure Systems for Manufacturing), 15-17 November, Melbourne, Australia, pp.206-211
2. Polanyi, M. (1969). Knowing and being, The University of Chicago Press
3. Polanyi, M. (1997). "Tacit Knowledge," Chapter 7 in Knowledge in Organizations, Laurence Prusak, Editor. Butterworth-Heinemann: Boston
4. Nonaka, I. (1991). "The Knowledge Creating Company." Harvard Business Review, November-December: 96-104
5. Nonaka, I. (1995). The Knowledge Creating Company. Oxford University Press.
6. Nickols, F. (2000). "Knowledge in Knowledge Management." Knowledge Management Yearbook, Butterworth-Heinemann: Boston.
7. Zheng J., Zhou M., Mo J.P.T., Tharumarajah A. (2000). "Background and Foreground Knowledge in Knowledge Management", Global Engineering, Manufacturing and Enterprise Networks (Proceedings of 4th International Conference on Design of Information Infrastructure Systems for Manufacturing), 15-17 November, Melbourne, Australia, pp.332-339
8. Hartel I, Sonderegger S, Kamio Y, Zhou M (2002). The Information Roadmap of Virtual Enterprises, e2002 - eBusiness and eWork Conference, Prague (CS), 16-18 October.
9. Morrison D, et al, (2002) Deploying QuickPlace, http://ibm.com/redbooks
10. SMART 2000 Handbook, http://www.defence.gov.au/dmo/lsd/smart2000
11. Mo J.P.T., Zhou M. (2003) Tools and Methods for Managing Intangible Assets of Virtual Enterprise, Computers in Industry, Vol.51, Issues 2, pp197-210

22

SUPPLY CHAIN ENGINEERING AND THE USE OF A SUPPORTING KNOWLEDGE MANAGEMENT APPLICATION

Frank Laakmann
Chair of Factory Organization (LFO), University of Dortmund, Leonhard-Euler-Strasse 5, D-44227 Dortmund, Germany, laakmann@lfo.uni-dortmund.de, www.lfo.uni-dortmund.de

Abstract: The future competition in markets will happen between logistics networks and no longer between enterprises. A new approach for supporting the engineering of logistics networks is developed by this research as a part of the Collaborative Research Centre (SFB) 559: »Modeling of Large Networks in Logistics« at the University of Dortmund together with the Fraunhofer-Institute of Material Flow and Logistics founded by Deutsche Forschungsgemeinschaft (DFG). Based on a reference model for logistics processes, the process chain model, a guideline for logistics engineers is developed to manage the different types of design tasks of logistics networks. The technical background of this solution is a collaborative knowledge management application. This paper will introduce how new Internet-based technologies support supply chain design projects.

Key words: Logistics Networks, Knowledge Management, Supply Chain Engineering, Supply Chain Management, Business Process Reengineering

1. INTRODUCTION

Currently the different disciplines of logistics and several companies are looking for new instruments and applications to manage their distributed knowledge about the design of business processes, the planning of logistics facilities and the determination of other organizational aspects, like network strategies and policies, organizational structures, information flows, performance measurement metrics etc.. Examining the engineering process of physical products and software applications a new engineering methodology for logistics networks is developed to reduce the complexity of the design process of logistics networks. This is done by using an existing process reference model, by developing a project reference model and by collecting construction catalogues for several organizational aspects as described above.

2. CHALLENGES OF THE DESIGN OF LOGISTICS NETWORKS

»Large Logistics Networks« can always be found in domains where different goods or products are transformed and transported over several stages by different cooperating partners. They do not only include material and information flow but also the organizational framework, the required resources, policies for planning and control as well as the people acting in this environment [1]. Here, supply chains (SC) and value added chains are used as synonyms for logistics networks. The organizational framework for example covers strategic alliances, extended enterprises and virtual or multinational companies. Within this framework it is possible to identify supply networks, distribution networks, transportation networks, communication networks, reusable package pools, knowledge networks or other communities. All partners in these networks provide multiple but also complementary services or competencies. The components of these large logistics networks (organizations, resources, goods, information, knowledge etc.) are connected by numerous different relations [2]. The number of connections is growing permanently, because there is an increasing demand for more competitive products or services that can only be satisfied by cooperating with powerful partners within these logistics networks.

3. KNOWLEDGE MANAGEMENT AS A INVEVITABLE COMPONENT OF SUPPLY CHAIN ENGINEERING

The management of logistics processes as parts of large logistics networks needs the integration of all partners starting at the beginning of the design phase. By concentrating on core competencies OEM depends more and more on suppliers and service providers that have the knowledge and experience in providing integrated solutions for their customers. This means that for example logistics service providers are not only operators of transportation networks anymore. It also includes the integration of all partners and the management of the information and communication flow along the supply chain. Therefore, the management of knowledge has become a strategic resource for producers and logistics service providers.

Currently the Western industrial societies undergo a change from the industrial age to the knowledge society. In several industries, especially in high tech and in the service industry, knowledge has become a major part of the total industrial value added and often that share amounts to more than 60 %. At the same time, the business partners face a growing lack of guidance through all available information. The today's challenge is to determine which information is relevant for all acting people and how to provide it in different situations of given projects.

Knowledge Management (KM) is the explicit and systematic management of vital knowledge – and its associated processes of creation, organization, diffusion, use and exploitation. Knowledge is a critical asset of current business activities. It is the development of information products, business processes and business scenarios for the application of concepts and ICT (information and communication technology) for data and information management. KM can be divided into two areas: internal and external. Internal KM targets reflect the own company and internal organisational units. External KM focuses on the operation of knowledge products for external or public groups or organizations.

A leading edge KM in logistics networks integrates the adjustment of KM strategies to the different logistics tasks, the experience of detecting relevant information, the design of knowledge acquisition workflows and the configuration of technical solutions for information management.

KM means the availability of information and knowledge for the design of logistics processes as well as for management and execution tasks.

- **KM for Design Tasks:** Participants are decision makers, strategists and logistics planners designing and implementing systems and processes for material flow and information management. Knowledge for logistics design processes covers methodological know-how, modern design concepts and knowledge in best practices. Benefits are more excellent design alternatives and improved ramp-ups. This area is only been very few supported by ICT systems yet. The solution that has been developed in this project targets this area.
- **KM for Management Tasks:** This includes the monitoring of the operating logistics systems (capacity, inventory, orders, positions etc.) the management of historic data for analysis and the management of workflows (forecasting, dispatching, bidding etc.) for the collaboration. Additional training and qualification offers are essential. Here, target groups can be found on the management and on the administrative level

being prepared for a flexible alert management and skilled to perform improvement and innovation processes.

- **KM for Execution Tasks:** For the execution of logistics processes the work force, which often belongs to different companies, has to be trained and educated for their tasks.

Available solutions for KM put their emphasis either on internal or on external participants. Traditionally, requirements of knowledge management can be fulfilled by existing enterprise resource management systems (ERP), SCM-systems and other systems for communication and data analysis. Specific solutions also based on internet technology are internet portals, extended services of trading exchanges, integration platforms of competence networks and e-learning solutions. These new solutions allow fulfilling future requirements of modeling specific knowledge.

To meet all requirements, a deep understanding of the possible use and existing IT-structures is crucial. Beside this, a cooperative and communicative staff is necessary for the successful implementation of KM in logistics networks.

4. BUILDING-UP PROCESS OF A KNOWLEDGE BASE

As a normal information and communication, technology system (ICT) the Workbench at first only provides a knowledge base containing data objects, which can be linked together by the implemented works. As the knowledge base is empty at first it has to be filled or built up in a process of input. The users, either logistics planners or experts, who carry out and supervise this process, will be called here "editors".

During the building up of the knowledge base, an analysis of real application fields of logistical networks by these editors is necessary. Either they analyse the logistical network itself or the planning process, which led to its realisation. The ideal state of affairs is the combination of both. As mentioned above, it surely will be possible in future to complement the knowledge contents by direct usage of the workbench in planning projects or by storing of intermediate and final results. This knowledge doesn't have to be entered by editors but only, if necessary, has to be processed and checked.

The greatest problem for the editors is the identification of relevant knowledge contents. The following general procedure for logging knowledge has proved successful in the past. It is structured into several steps and does not exclude iterations.

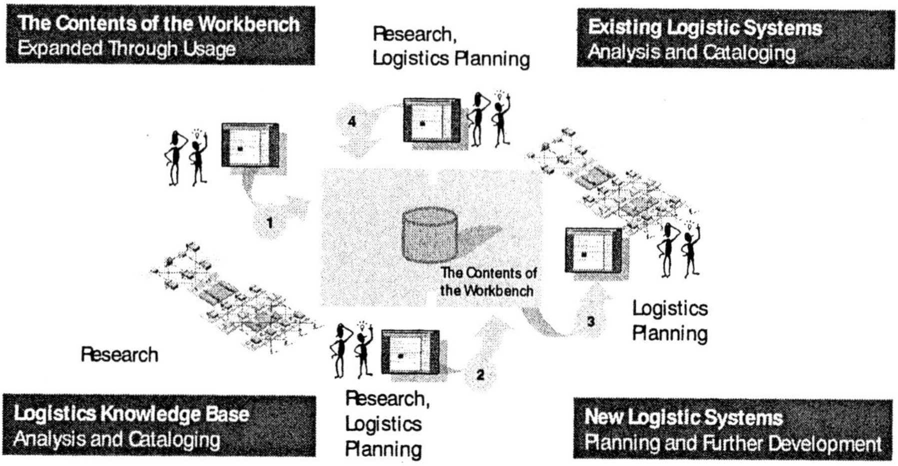

Figure 1. Use Cases for a Collaborative Knowledge Management Application for Design Projects of Logistics Networks

- **Selection of a subject:** the knowledge contents of the workbench can only be building up during a long-term process. Therefore it is functional that the editors concentrate on single subjects for which they should enter utmost complete knowledge into the workbench. One example for reactively complete knowledge would be the "Sourcing Process" as an example for processes. Another subject would be "Packaging Material" as an example for resources.
- **Delimitation and cut away:** the selected subject and, if necessary, the individual case of application are the range of analysis which has to be delimited sharply from things not to be taken into consideration. This is done by a mental border between those facts which are to be displayed in the following and those which are disregarded for the time being. A key word list which is structured corresponding to this pattern is helpful as well.
- **Pre-analysis of every single component:** the analysis of the analysis range has to identify roughly the knowledge elements which can be mapped in the workbench. For further details see chapter "data input" below. Concerning the subject Sourcing process we have to consider Sourcing processes, Sourcing concepts, transportation means, material flow systems, planning projects for the delivery chain and so on. This step is still done without using the workbench.
- **Collection of knowledge objects:** In the next step, after a rough pre-analysis, the knowledge objects in detail have to be collected. The matter in this case is the completeness of the collection of elements. A distinct

structure is not yet necessary here. For example all identifiable processes concerning procurement can be collected. In the first step the names of the processes and textual descriptions are put together. This information can either be directly entered into the workbench or can be registered in tables in standardised form.

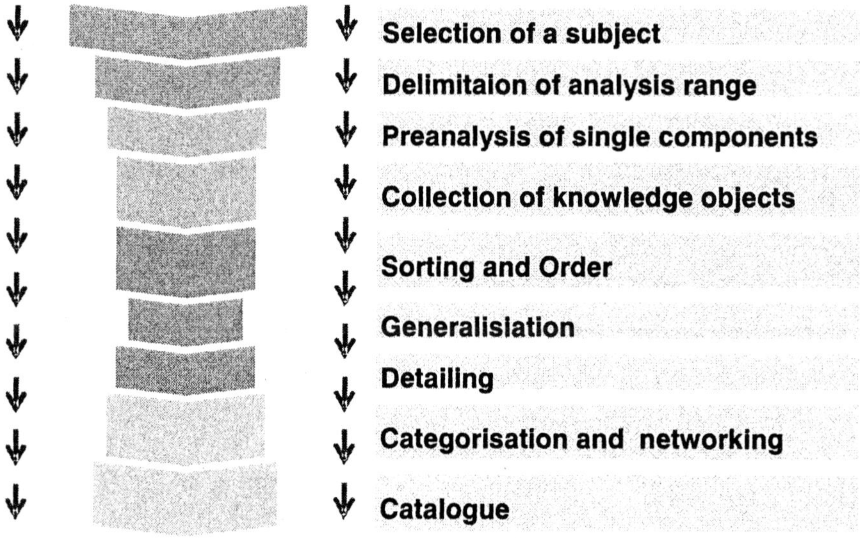

Figure 2. Procedure for building up the knowledge base

After this step a substantial intermediate result of the documentation of the knowledge concerning this subject is reached. Much relevant basis information has been registered as knowledge objects in structured form and is available for further working. This working can be done by the same editor or by other users.

- **Sorting and order:** If on the preliminary stage of collecting knowledge objects a certain number of similar ones have been put together it is practical to sort, structure and file them afterwards. This simplifies the finding of stored objects as well as the linking to other objects later on (for instance the linking of planning data to planning methods). The found procurement processes for example are subdivided in processes concerning information flow and processes concerning material flow. Main processes are set over partial processes.

This step introduces a number of other possible working stages. Yet superficially these are only further steps.

- **Generalisation:** During the collection of knowledge objects there may be equal or similar facts registered separately. These are put together to a comprehensive object during the sorting process. This leads to a consolidation and reduction of the collected knowledge on relevant information. Typical example for this are registered processes for which no general mapping rule can be developed. Single partial processes are put together to a comprehensive process and the contents are integrated.

- **Detailing of identified knowledge contents:** a model of knowledge can only develop by abstraction of real facts while the essential structures and features are supposed to be kept. In this case the task of the editor is to find a suitable middle course between closeness to reality and mapping expenditure. An example for this is the logging of planning projects and the description of the planning steps carried out. Surely one can find many recurring processing series. Here the editor has to recognise these steps as independent planning method, to abstract them as such and to insert them into the collection of methods.

The workbench provides several techniques for structuring and sorting the knowledge contents. The most important techniques are categorisation and cataloguing.

- **Categorisation**: the category system consists of several roots, the so called main categories, and the categories. Thus, the category system looks like a tree with several stems. But also possible are associations between categories so that single branches can be subordinated flexibly to other categories.
- **Cataloguing**: Thematically coherent knowledge objects can also be put together in catalogues. For this purpose first of all a catalogue for a particular subject is set up. Then already stored objects are attached to this catalogue. Within the catalogue, there are functions to sort and to typify the embedded objects.

An important component of this procedure is the review of the collected and stored information to ensure a high quality of the information and knowledge. It is obvious that at the first time the quality is a question of the design of the process of collecting and structuring the information. Secondly it is a question of additional features of the IT-System used for the knowledge management. Thirdly, it is a question of the motivation of the users and stakeholders of this system. These aspect will be the focus of the further steps in the project.

5. CONCLUSION AND OUTLOOK

In this research project the use of the Workbench will give further requirements for the design of additional features of the workbench.

However, the most interesting aspect of the use of this software pilot is the experience in using a knowledge workbench for sharing information between logistics planners in projects.

Successfully proven modeling elements like methods, construction elements and design process patterns [3], [4] will extent the knowledge base of the workbench. The Process Chain Model outlines the powerful combination between interdisciplinary approaches coming out of information technology, engineering, industrial engineering, economics and logistics.

This research is part of the Collaborative Research Centre (SFB) 559: »Modeling of Large Networks in Logistics«, Project M6: »Construction Rules« at the University of Dortmund (http://www.sfb559.uni-dortmund.de) together with the Fraunhofer Institute of Material Flow and Logistics founded by Deutsche Forschungsgemeinschaft (DFG).

REFERENCES

1. Beckmann, H. (1996). Theorie einer evolutionären Logistik-Planung. Praxiswissen.
2. Kuhn, A. (1995). Prozessketten in der Logistik. Entwicklungstrends und Umsetzungsstrategien. Praxiswissen.
3. Beckmann, H. (2000). Supply Chain Method Handbook. Praxiswissen.
4. Laakmann, F. (2002). Knowledge Management and Applications for Supply Chain Design. *Proceedings of the 6th International Conference on Engineering Design and Automation (EDA) – „Engineering Practice and Education". Maui, Hawaii, USA, August 7-8*

23

A PLANNING FRAMEWORK FOR THE DEPLOYMENT OF INNOVATIVE INFORMATION AND COMMUNICATION TECHNOLOGIES IN PROCUREMENT

Robert Alard[1], Martin Gustafsson[2], and Jörg Nienhaus[1]
[1] *ETH-Center for Enterprise Sciences (BWI), Swiss Federal Institute of Technology Zurich (ETH), Zurichbergstrasse 18, 8028 Zurich, Switzerland*
[2] *Unisys (Schweiz) AG, Commercial Services, Zurcherstrasse 59, 8800 Thalwil, Switzerland*

Abstract: The management of buyer-supplier relations is a major topic for many enterprises today. Modern Information and Communication Technologies (ICT) offer interesting perspectives on opportunities and implementation approaches. Today, logistics and procurement departments of numerous enterprises are evaluating the possibilities and opportunities of new ICT solutions and especially of internet-based electronic procurement solutions for the optimisation and re-engineering of their buyer-supplier relationships. Due to the highly innovative character of the new ICT solutions and the scarcely available operational examples in the industry, only little guidance exists to support responsible managers during the evaluation, planning and designing of internet-based electronic procurement solutions. This paper describes a framework for the strategic evaluation and planning of the deployment of internet-based procurement solutions for direct materials. The presented approach supports enterprises in the analysis of procurement objects and procurement structuring, in the definition and management of buyer-supplier-relationships, in the requirements analysis of ICT solutions as well as the assessment of the potential to support procurement with innovative ICT and internet-based electronic procurement solutions.

Key words: Planning Framework, Procurement, Supplier Relationship Management, Internet-based Electronic Procurement

1. INTRODUCTION

One of the most significant trends influencing the strategic procurement today is the increased usage of Information- and Communication Technologies (ICT). Since some time now ICT have had a major impact on the design of business processes. The software applications focused initially on internal business processes. With Electronic Data Interchange (EDI), Supply Chain Management (SCM)-applications and electronic business (e-business) solutions the focus shifted to trans-company approaches. For Small and Medium-sized Enterprises (SME) e-business applications provide good access opportunities to SCM concepts [1]. Internet-based electronic procurement solutions have claimed an especially important role in the e-business and SCM environment [2, 3].

Following Arnold´s definition of procurement [4], internet-based electronic procurement can be understood as electronic support, based on Internet standards, of company and/or market activities that aim to procure objects that the company requires but does not produce itself. Electronic procurement provides support for strategic and operational procurement processes and is able to adapt to the trans-corporate context and design aspects of the supplier-side of the value chain.

2. A PROCUREMENT PLANNING FRAMEWORK

There is a lack of research in the field of internet-based electronic procurement of direct materials and limited experience when it comes to usage of such applications [5, 6, 7]. Therefore, enterprises are challenged when taking strategic decisions on implementation and usage of internet-based procurement solutions. The planning framework presented below structures procurement objects and groups and thereby provides support in planning and evaluation of procurement solutions.

In the literature there are plenty of suggestions for the support of strategic decisions. In the procurement theory and practice the portfolio approach has proved useful [7, 8, 9].

In preparation for the implementation of internet-based electronic procurement solutions in the industry, important questions must be answered in terms of procurement objects and sources in the Value Chain. An instrument for analysis and structuring of procurement objects and groups is the analysis framework described below, figure 1 [10, 11]. It is used in the evaluation and planning of electronic procurement solutions aiming to create a differentiated view on buyer-supplier solutions.

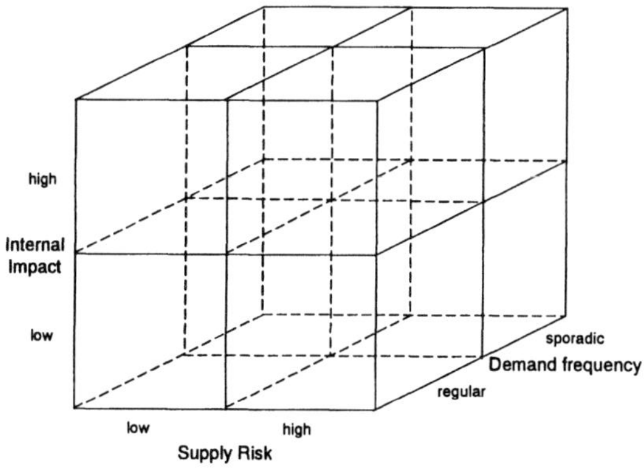

Figure 1. Analysis Framework for the procurement

The dimension Internal Impact is a crucial parameter in shaping the procurement relationships and processes and is quantified according to ABC-analysis with purchasing volume per procurement object/group and qualitatively assessed through expert-groups. Qualitative and quantitative criteria in the assessment of the internal impact are e.g. price of procurement object, costs of the purchase of the procurement object in a defined time period (volume), impact on quality, functionality and image of the end product or possible penalties caused by the procurement object.

The Supply Risk can be interpreted as the technical and logistical complexity and the resulting insecurity of the procurement market. Should the procurement object be easily reachable on the market, it is labeled with low supply risk. Criteria used to assess the supply risk are e.g. number of potential suppliers, level and sophistication of supplier-buyer-cooperation necessary, level of standardization, technical complexity, competition situation of suppliers and the existence of substitution products.

The demand frequency shows the demand structure of the procurement object from the internal standpoint. The objective of this dimension is to assess the ability to make exact prognoses of demand. The XYZ-analysis is useful in the assessment of this dimension. For objects with a continuous demand a prognosis level of 90-95% each week is expected [11, 14].

Even though objective and quantifiable criteria are used in single dimensions of the analysis framework it should be noted that the overall assessment is strongly dependent of enterprise specific and partly subjective criteria. This should not be considered something negative [8]. The presented planning framework (see figure 1) provides a grouping of the procurement objects/groups and assigns them to their respective procurement reference relationships (PRR) (characteristic buyer-supplier-relationships). These should be understood as procurement design guidelines which aim to support optimised concepts. The design possibilities are vast and include organisational as well as information technical aspects [13, 15].

The positioning of procurement objects and procurement groups in the analysis framework as well as the design of procurement reference relationships should be managed in the context of commodity management, i.e. in cross-organisational teams with representation from logistics, production, research and development, construction, quality and strategic as well as operational procurement. Figure 2 presents five characteristic procurement reference relationships which have been identified in the machine and plant industry.

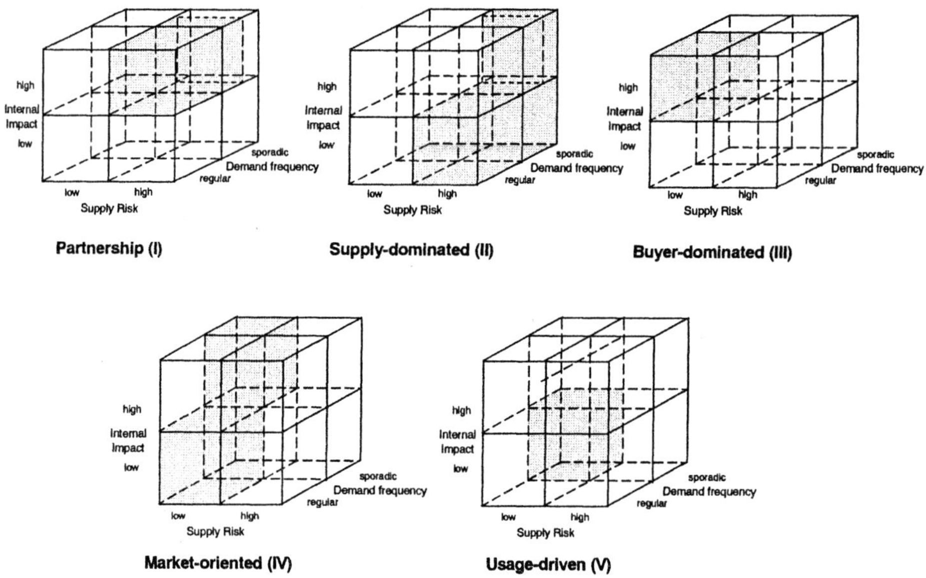

Figure 2. Analysis framework for structuring procurement items and strategic procurement units and characteristic procurement reference relationships

The identified procurement reference relationships are: Partnership (I), Supply-dominated (II), Buyer-dominated (III), Market-oriented (IV) and Usage-driven (V). In most enterprises only a sub-set of the procurement reference relationships are relevant [12, 16]. The design of procurement reference relationships is influenced by market strength of suppliers and buyers and should be assessed with respect to the cooperation willingness and ability of the suppliers within the enterprise-specific context of the enterprise.

3. SUPPORT AND DESIGN POTENTIAL THROUGH ELECTRONIC PROCUREMENT SOLUTIONS

Even considered the wide spectrum of support potential offered by internet-based electronic procurement solutions, not all strategic and operative procurement processes can be supported. In strategic procurement for example the potential is focused on the phases of procurement market research, supplier choice and contracting.

Table 1 presents procurement process phases and the corresponding support potential with examples. The internet-based electronic procurement solutions are structured according to their institutional ownership of the software solution [2]: Sell-side solutions, Buy-side solutions, neutral electronic marketplaces and buyer-supplier direct connections.

Table 1. Procurement process phases in the strategic procurement and support potential by internet-based electronic procurement solutions / Examples

	Type of solution		Examples	PRR
Procurement market research	Sell-Side Solutions		▪ Product catalogue ▪ Product specification and configuration	I-V
	Buy-side Solutions	Classic buy-side solution	▪ Supplier and product search in catalogue (restricted choice of products)	IV, V
		Buyer-controlled Electronic Marketplaces	▪ Offer submissions (passive supplier search) / Call for tenders	I-III (IV-V)
	Neutral Electronic Marketplaces		▪ Product catalogue ▪ Supplier catalogue ▪ Register / Data bases ▪ Offer submissions (passive supplier search) / Call for tenders ▪ Bulletin Board Services / News groups / Discussion services ▪ Search services / Information brokers	I-V
Supplier choice	Sell-Side Solutions		▪ Product catalogue ▪ Product specification and configuration	I-V
	Buy-Side Solutions	Classic buy-side solutions	▪ Supplier choice from catalogue (restricted number of suppliers)	IV, V
		Buyer-controlled Electronic Marketplaces	▪ Offer submissions / Call for tenders ▪ Auctions	I, III, IV, V, (II)
	Neutral Electronic Marketplaces		▪ Offer submissions / Call for tenders ▪ Auctions	I, III, IV, V, (II)
Contracting	Sell-Side Solutions		▪ Auctions ▪ Contract preparations	-
	Buy-Side Solutions	Buyer-controlled Electronic Marketplaces	▪ Auctions ▪ Contract preparations	I, III, IV, V, (II)
	Neutral Electronic Marketplaces		▪ Auctions ▪ Contract preparations	I, III, IV, V, (II)

( ): limited suitability

The preference of the internet-based electronic procurement solution is to be decided by cross-organisational teams in enterprise-specific context. Below internet-based electronic procurement solutions are discussed in terms of support potential in strategic procurement and their respective fit in specific PRRs (see figure 2).

In the market research phase there are several opportunities for procurement support through sell-side (e.g. product catalogues, company information, product specifications and configuration) and neutral electronic marketplaces. (e.g. Global Sources, Thomas Register) [17, 18, 19]. Classic buy-side solutions provide search opportunities for procurement objects and procurement sources in tailored catalogues [20]. The demand owner within the enterprise is supported in the market research. However, since there is a limited access to the tailored procurement catalogues, and the content is centrally regulated, there will only be a limited possibility for researching new suppliers. Hence, the predefined supplier pool will not be extended. Classic buy-side solutions are well suited for general support of procurement in usage-driven procurement reference relationships [18, 20] as well as for specified procurement objects in market-oriented procurement reference relationships [6, 18] as defined in the planning framework. An opportunity for passive supplier search is provided by call for tenders / offer submissions of the demand on buyer-controlled electronic marketplaces or procurement web site of the buying enterprise [21]. This type of procurement market research can be used by all types of procurement reference relationships. However, when it comes to strongly standardised procurement objects (e.g. attached to market-oriented or usage-driven PRR), a more active market research is recommended.

In the supplier choice phase, the procurement is supported by sell-side and buy-side solutions in addition to neutral electronic marketplaces. Whereas sell-side solutions and classic buy-side solutions provide information through product and supplier catalogues, call for tenders and auctions are placed on buyer-controlled and neutral electronic marketplaces. These solutions are generally well suited for all types of procurement reference relationships. However, for supplier-dominated procurement reference relationships, market position of suppliers could limit the motivation to respond to call for tenders and auctions [18]. Therefore the deployment of such solutions seems only interesting if among suppliers there are so-called "price breakers" willing to compete on the basis of price. The decision must be based on the specific characteristics of the situation; generally it has to be assumed that the involvement of a "price breaker" especially in conservative markets will be challenging. The contracting phase is directly attached to and defined by the auction. Further on the possibility of contractual preparations like general terms and conditions or shipping and handling instructions are relevant.

4. CONCLUSION

Until now the deployment of ICT to support direct materials' procurement (e.g. classic EDI) was mainly performed at large enterprises and networks stable over a long time due to high initial investments in terms of organisation and technology. With internet-based electronic procurement solutions a wide variety of opportunities emerge which also SMEs can benefit from. The economic impact and industrial importance of internet-based procurement solutions is therefore expected to continue its growth.

Our research indicates that enterprises currently are seriously challenged by planning, design and deployment of internet-based electronic procurement solutions. Design concepts for integration of internet-technologies in the procurement are therefore crucial. With this background, an analysis- and design framework for the procurement of direct materials was developed and the support potential through internet-based electronic procurement solutions showed. The analysis framework is used for procurement objects and procurement groups analysis. In our research, based on proven procurement strategies in the machine and plant industry five characteristic procurement reference relationships were identified. The presented procurement reference relationships serve as test of support potential in the procurement as well as development of ideal concepts. Underlined by examples from the strategic procurement we show in which procurement process phase innovative ICT and internet-based electronic procurement solutions provide support. We also show in which procurement reference relationships this type of support appears in terms of a differentiated view on ICT. For the procurement the planning framework provides a navigation help in the integration of organisation and information technology and enables the strategic evaluation of the deployment of internet-based procurement solutions.

REFERENCES

1. Pfohl, H.-C. (2000). Supply Chain Management: Konzept, Trends, Strategien. In: Pfohl, H.-C. (Hrsg.): Supply Chain Management: Logistik plus? Logistikkette – Marketingkette – Finanzkette. Fachtagung der Deutschen Gesellschaft für Logistik e.V.; 15; Unternehmensführung und Logistik; Bd. 18. Berlin: Erich Schmidt; pp. 1-42.

2. Alard, R. (2002). Internetbasiertes Beschaffungsmanagement direkter Güter- Konzept zur Gestaltung der Beschaffung durch Nutzung internetbasierter Technologien. Dissertation ETH Zürich.
3. Alard, R.; Hieber, R. (2000). Lösungen für unternehmensübergreifende Kooperationen - Supply Chain Management und Business-to-Business Commerce. PPS Management 5 (2000) 2, pp. 10-14.
4. Arnold, U. (1997). Beschaffungsmanagement. Stuttgart: Schäffer-Poeschel.
5. de Boer, L., Harink, J., Heijboer, G. (2001). A Model for Assessing the Impact of Electronic Procurement Forms. Proceedings of the 10th International Annual IPSERA Conference, Jönköping.
6. Neef, D. (2001). e-Procurement - From Strategy to Implementation. Upper Saddle River: Financial Times Prentice-Hall.
7. Nellore, R. (2001). Managing buyer-supplier relations: the winning edge through specification management. London, New York: Routledge, Taylor & Francis Group.
8. Gelderman, K., van Weele, A. (2001). Advancements in the Use of a Purchasing Portfolio Approach: A Case Study. Proceedings of the 10th International Annual IPSERA Conference, Jönköping.
9. Syson, R. (1992). Improving Purchase Performance. London: Pitman Publishing.
10. Baumgarten, H.; Bodelschwingh, K. von (1996). Kostenreduzierung durch gestraffte Abläufe. Beschaffung aktuell 43 (1996) 2, pp. 35-38.
11. Hamm, V. (1997). Informationstechnikbasierte Referenzprozesse: prozeßorientierte Gestaltung des industriellen Einkaufs. Wiesbaden: dt. Univ.-Verl.; Wiesbaden: Gabler.
12. Gustafsson, M. (2001). Entwicklung eines Konzepts zum Materialgruppenmanagement im Bereich des Customer Service bei ALSTOM (Schweiz) AG. Diplomarbeit; Linköping Institute of Technology / ETH Center for Enterprise Sciences (BWI), Domain of Logistics and Information Management.
13. Wildemann, H. (1999). Das Konzept der Einkaufspotentialanalyse - Bausteine und Umsetzungsstrategien. In: Hahn, D.; Kaufmann, L. (Hrsg.): Handbuch industrielles Beschaffungsmanagement: internationale Konzepte - innovative Instrumente - aktuelle Praxisbeispiele. Wiesbaden: Gabler; pp. 435-452.
14. Schönsleben, P. (2000). Integrales Logistikmanagement - Planung und Steuerung von umfassenden Geschäftsprozessen. Berlin u.a.: Springer.
15. Fieten, R. (1994). Integrierte Materialwirtschaft. Beschaffung aktuell - Konradin Verlag: Leinfelden-Echterdingen.
16. Hunziker, D. (2001). Entwicklung eines Konzeptes zur Geschäftsprozess- und Organisationsgestaltung bei der Einführung von Electronic Procurement. Nachdiplomarbeit; ETH Center for Enterprise Sciences (BWI), Domain of Logistics and Information Management.
17. Harink, J.H. (1999). Excelling with e-procurement. Samson: Alphen aan den Rijn.
18. van Weele, A. (2002). Purchasing and Supply Chain Management: Analysis, Planning and Practice. London: Business Press, Thomson Learning.
19. Weber, S. (2001). Information Technology in Supplier Networks: A Theoretical Approach to Decisions about Information Technology and Supplier Relationship. Heidelberg; New York: Physica-Verlag.
20. Dolmetsch, R. (1999). Desktop Purchasing - IP-Netzwerkapplikationen in der Beschaffung. Dissertation Universität St. Gallen. Bamberg: Difo-Druck.
21. van Weele, A., Rozemeijer, F. (1999). Getting Organised for Purchasing and Supply Management in the Information Age. In: Hahn, D.; Kaufmann, L. (Hrsg.): Handbuch industrielles Beschaffungsmanagement: internationale Konzepte - innovative Instrumente – aktuelle Praxisbeispiele. Wiesbaden: Gabler, 1999, pp. 625-637.

24

SUPREME: SUPPLY CHAIN INTEGRATION BY RECONFIGURABLE MODULES

Y. Nishioka[1], F. Kasai[2] and Y. Kamio[2]
[1]*Hosei University,*
[2]*Toyo Engineering Corporation, Japan*
Em: nishioka@k.hosei.ac.jp

Abstract: This paper deals with dynamic supply chain configuration problems for one-of-a-kind production (OKP) environment. We propose an architecture named SUPREME, which supports collaborative planning and scheduling for web-based virtual enterprises. In this architecture, each system is located on the site of the network members as well as on the site of the supply chain coordinator, who performs as an agent configuring and managing the supply chain dynamically. A prototype system of SUPREME is developed and illustrated in order to evaluate effectiveness of the proposed architecture.

Key words: SCM, APS, OKP, Collaborative Planning, Scheduling, Web Services, Business Model, PSLX

1. INTRODUCTION

Manufacturing is changing drastically according to the improvement of the information and communication technologies (ICT) infrastructure. Since a manufacturer takes more partial roll in the whole manufacturing process, global supply chain management becomes an important topic. Moreover, to catch up customers' requirement just on the right time at the right place, the supply chain itself should be changed dynamically[1,2]. This paper proposes an architecture in which dynamic supply chains can be managed.

The proposed architecture is named SUPREME: SUPply chain integration by REconfigurable ModulEs. The architecture especially focuses on the APS (Advanced Planning and Scheduling) framework and XML

(Extended Mark-up Language) based communication protocols on the Internet. In this case, the scope of the APS systems is not only inside a factory, but also outside, e.g., logistic firms and supply chain coordinators offices. The interoperability of this distributed systems and their application integration are conducted by PSLX (Planning and Scheduling Language on XML specification) technologies[3].

The organization of this paper is as follows. First, in the section 2, the proposed system architecture is illustrated. Then some applications of the architecture to practical business scenarios are introduced in the section 3. Some of the scenarios are more elaborated and discussed in an experimental study in section 4. Finally, the section 5 describes some concluding remarks.

2. SUPREME ARCHITECTURE

The proposed architecture named SUPREME is designed to apply in a dynamic supply chain business models in an OKP environment. In this environment, each enterprise has to have a function to reconfigure the relationship of partners, because the products ordered by customers are one-of-a-kind so that they need a new supply chain every time. In order to do this, they should be able to select suppliers quickly, and also manage them consistently until the production will be completed. The proposed system architecture supports these business processes using ICT infrastructure and APS technologies.

In this framework, APS performs in planning and scheduling of each partner's activities, so that the throughput of the supply chain will be maximized. Moreover, APS deals with unexpected changes in the reality, adjusting their old schedule to the new environment immediately. Some times, the changes directly affect to the schedules in different enterprises through the supply chain network. In such a case, the system can maintain the consistency of the schedules by using APS systems and Internet technologies.

Features of the proposed architecture are as follows. First, the target of the system is OKP environment. There are many supply chain management systems for repetitive production. However, supply chain management for OKP needs different methods and technologies. Since OKP has to make a supply chain network for each customer's order, a partner selection process is required frequently. To shorten production lead time, the system should flexibly support the configuration and re-configuration of the network.

Second feature is that the system can support collaboration of planning and scheduling processes among different enterprises. The selection of partners is evaluated not only by the quality and cost of the products, but also by the total delivery date estimated by the APS module. Temporal relationship of delivery dates dynamically affects the lead time of the whole process of the supply chain. Therefore, the supply chain coordinator should manage the delivery date in order to synchronize the network. The APS modules in this framework provide re-configuration or re-scheduling functions. These functions will take a very important roll of the OKP environment.

The final feature is addressed on the communication protocols used in this framework. We use HTTP for data transportation on the Internet. Using this protocol, two different ways of communication are involved in this architecture. The first is a typical communication using a web server and an application server combination. This has an advantage that client side does not need any particular system except for a web browser. The other is a communication using a PSLX interface module, in which XML-SOAP technologies are used for implementing web-services. The detail specification of the PSLX interface is published in recommendations of the PSLX consortium[4].

In the SUPREME architecture, members candidate for the supply chain have a distributed system. Each distributed system is connected to the production planning system in the manufacturer. Operations in the production processes are managed by APSTOMIZER, which is an APS system developed in HOSEI University. On the other hand, the supply chain coordinator has another system, which deals with synchronization among the distributed system located in each manufacturer. This system concerns a virtual supply chain, calculating some recommendations of supply chain networks.

Communications among supply chain members and the coordinator are executed in two ways. One is web-based client-server communications, in which an enterprise gets another enterprise's information via a server-side system located in the coordinator. The second type is peer-to-peer communications that the supply chain members directly send their information each other. When communication processes are critical in terms of time constraint, the second type might be necessary in practice in order to establish synchronization of the business process.

3. BUSINESS SCENARIOS

Considering an OKP environment, there are various benefits if different enterprises can share their production information and make collaborative decisions on the SUPREME architecture. In those cases, planning and scheduling problems are important to catch up with the changing market environment. This section shows some business scenarios in which the SUPREME architecture will play an important role. Actors of the scenarios include customers, manufacturing enterprises and third-party-logistic (3PL) companies. Since these actors perform on the different decision-making systems, a supply chain coordinator, which is another actor proposed within the architecture, is due to establish synchronization for a win-win partnership.

3.1 Supply chain agent

The first scenario is that a customer, who requires an OKP from scratch, accesses the SUPREME coordinator in order to make a quotation and choose the best supply chain recommended by the system. In a quotation process, the customer puts his/her request in a form of process plan sheets. Then the coordinator asks the corresponding partners' manufacturing capability and also the availability to meet the schedule. Once the suppliers have registered their manufacturing capability to carry on some of the predefined manufacturing process candidates, the coordinator can select an appropriate supplier for each process, which is part of the input for the customer's request. Finally, the APS module calculates feasible plans and schedules, in which all temporal constraints such as precedence relations and lead time are satisfactory.

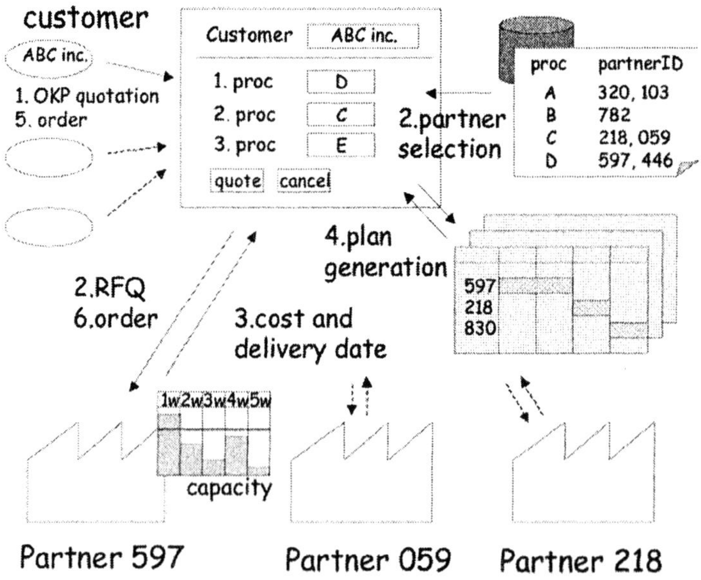

Figure 1. Supply chain configuration agent

In this case, the SUPREME coordinator performs as an agent who arranges the best plan to fulfill the customer's order of OKP. One of the differences between the agent and a supply chain planning staff is that former can be automatically executed on the Internet. Furthermore, after a supply chain is established, the members of the supply chain can be cooperated each other on the Internet until the supply chain network is dismissed. The coordination is managed in the SUPREME architecture. Figure 1 demonstrates this scenario.

3.2 Delivery process integration

The second scenario focuses on integration of a manufacturing process and a delivery process. While a 3PL partner manages delivery processes ordered by several manufacturers, connecting the customer's manufacturing processes with the supplier's manufacturing processes plays a very important role. This will be achieved by process integration between logistic enterprises and manufacturing enterprises. For instance, a truck driver has an accident and the estimated arrival time would be behind the schedule. Or a supplier's shipping process was delayed and excepted production volume would be shortage. In those situations, the scenario shows that a revised logistic schedule is sent to the coordinator, and manufactures relative to the change are informed to rearrange their current schedules.

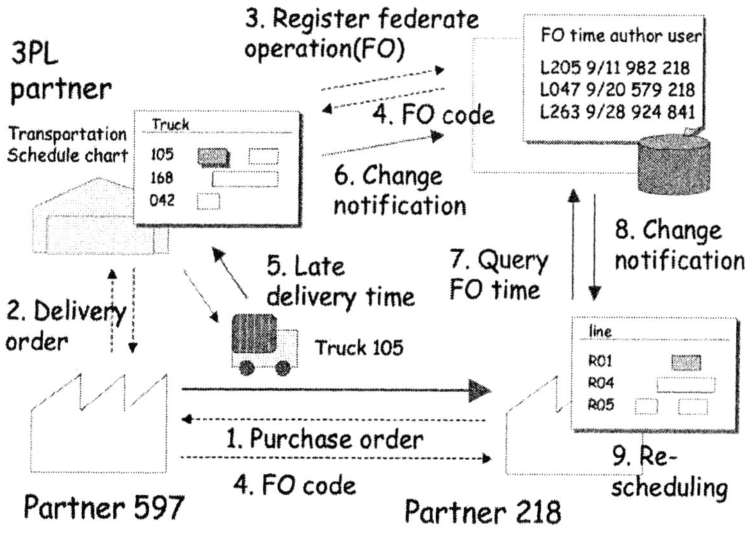

Figure 2. Delivery process integration

Detail of the collaboration is shown in Figure 2. A manufacturing enterprise sends purchase order to a supplier, who then asks a 3PL company to deliver the products. Regarding that both the customer and the supplier have their production schedule, the delivery schedule is uploaded to the SUPREME server in order to synchronize the production schedules. The customer frequently checks the delivery process as well as the supplier's production process. When some serious changes are occurred, the SUPREME suggests making reschedule the customer's plan.

3.3 Outsourcing management

The final scenario is addressed in outsourcing processes. When a manufacture makes production plan by loading and leveling the resource capacities, a part of the orders would be asked to some suppliers that have free capacity of resources for the corresponding production processes. In OKP, however, the selection of suitable suppliers takes a long time, because most of cases are the first time in its experiments. Like the first scenario, this partner selection for outsourcing can be supported by the SUPREME architecture.

The SUPREME also copes with the collaboration processes after the supplier selection is finished. As shown in the figure 3, the customer's production processes and supplier's production processes can be synchronized by peer-to-peer communication between them. This

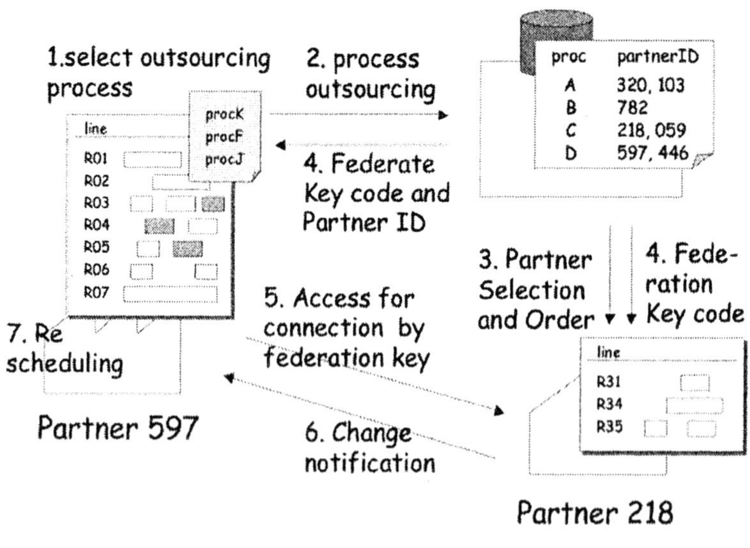

Figure 3. Outsourcing management

communication process has an advantage if changes in one partner's manufacturing processes immediately make a negative effect to the other. In this case, the SUPREME is a kind of broker, who only initiates a secure communication by "federation keys" for identifying their partner. By means of this peer-to-peer communication and APS modules, distributed processes in different enterprises can be synchronized.

4. DISCUSSION

Since SUPREME has defined as a system architecture for a supply chain management of OKP, particular systems implemented for each business environment would have different forms. In this study, a prototype system of SUPREME is developed for further discussions. The system performs pretty well on the first scenario and partially on the third. Some graphical user interfaces of the system are briefly described.

First of all, the *partner registration* form is used for new members who want to join the virtual enterprise network. In this form, each one registers its information of the company profiles and also describes its production capability, which will be used by the network members as a partial production processes in a future supply chain. Then the *order entry form* can be shown for the customer who is also a member of the network. In the order entry form, the customer can design by input any manufacturing processes in order to produce the appropriate products.

This information is sent to the SUPREME agent, who then gathers corresponding partner's manufacturing capacity. After the APS module calculates some supply chain configuration plans, the *plan recommendation form* shows them for the customer as selection candidates. If there is more than one plan, each plan is recommended with its cost, delivery date, and quality information. Furthermore, the *Gantt chart* that represents the whole supply chain production schedule can be shown for each candidate.

The user interface forms descried above are in the front end of the communication processes between an OKP customer and a supply chain coordinator. On the other hand, back end of the system is implemented on web-applications, web-services client/server modules, a RDB system, APS modules, and so on. The prototype system also has a function of peer-to-peer communication.

5. CONCLUSION

This paper deals with a dynamic supply chain for OKP environment. ICT support is necessary for current economical situation, because the supply chain setup time should be managed shorter and shorter. In the emerging supply chain, each collaboration scheme also has to change in accordance with the dynamics of the real environment.

SUPREME is a system architecture that has web-based collaborative planning and scheduling system for virtual enterprises. Therefore, the architecture has an APS modules and PSLX interfaces modules as well as the other technologies. These components are addressed both on a supply chain coordinator site and each member's local site in the supply chain network. Then the coordinator performs as an agent that dynamically configures and maintains the suitable supply chain.

A prototype system is developed and demonstrated for evaluation of the architecture. We show three business scenarios, both of which represent typical scenes of a strategic supply chain management in OKP. After the prototype system is briefly demonstrated, the experimental results shows that APS modules and PSLX interface modules play a very important role in the proposed system.

In conclusion, the SUPREME architecture, which has capability to re-configure and manage supply chains dynamically, is one of the most suitable solutions to achieve the goal of the supply chain management in OKP. In addition, the communication and collaboration methods in such an open network platform are required, so that PSLX interface and collaborative APS will be the important research topics in the future.

This research has been carried out under the IMS GLOBEMEN project. The authors would like to acknowledge all the members of the GLOBEMEN project for their valuable discussion and technical support.

REFERENCES

1. Kamio Y, Yamamoto E, Morita K, Fukuda Y and Nishioka Y (2001). Virtual Enterprise Architecture and its Supporting Methods/Tools for Managing Supply Chain System Life Cycle, *Global Engineering, Manufacturing and Enterprise Networks*, Kluwer Academic Publishers, 18-25
2. Nishioka Y, Kamio Y, Kawashima K. and Fukuda Y (2001). A Booking Type Production System as a Collaboration Method for Virtual Enterprises, *Global Engineering, Manufacturing and Enterprise Networks*, Kluwer Academic Publishers, 164-172
3. Nishioka, Y (2001). XML-based Standardization of Planning and Scheduling Problems for Virtual Enterprise Collaboration, *Proceedings of ICPR-16*, Czech Prague
4. PSLX Consortium Japan (2003). Planning and Scheduling Language on XML Specification Technical Documents, http://www.pslx.org/

25

TOOLS AND METHODS FOR RISK MANAGEMENT IN MULTI-SITE ENGINEERING PROJECTS

Mingwei Zhou[1], Laszlo Nemes[1],
Carl Reidsema[2], Ammar Ahmed[2], and Berman Kayis[2]
[1]CSIRO Manufacturing & Infrastructure Technology, Australia
[2]School of Mechanical and Manufacturing Eng., University of New South Wales, Australia
e-mail: Mingwei.Zhou@csiro.au

Abstract: In today's highly global business environment, engineering and manufacturing projects often involve two or more geographically dispersed units or departments, research centers or companies. This paper attempts to identify the requirements for risk management in a multi-site engineering project environment, and presents a review of the state-of-the-art tools and methods that can be used to manage risks in multi-site engineering projects. This leads to the development of a risk management roadmap, which will underpin the design and implementation of an intelligent risk mapping system.

Key words: Risk Management, Enterprise Engineering, Tools and Methods

1. INTRODUCTION

In today's highly global business environment, engineering and manufacturing projects often involve two or more geographically dispersed units or departments, research centres or companies. To enable the collaboration between these multi-site units across space and time, three essential elements are required:

1) ICT platform to support information sharing and collaboration;
2) Appropriate organizational structures and engineering processes to support collaborative work involving different functional groups; and

3) Identification, evaluation and treatment of the risks associated with multi-site engineering projects.

Research on Virtual Enterprises (VE) for global manufacturing [1,2,3] has extensively addressed the first two issues. This paper will focus on the third essential element, i.e. the risk management.

Recent research on project risk management has been mainly conducted in the software, finance, and service sectors. Fewer projects have extensively addressed the requirements of risk management for engineering projects, especially for those multi-site geographically distributed projects.

The IRMAS (Intelligent Risk Mapping and Assessment System) Project, an Australian research project described in this paper, aims to develop a risk mapping and assessment system for multi-site engineering projects, and therefore enhance the local manufacturers' capability of collaborating and competing in the global market.

This paper will firstly discuss the requirements for risk management in a Multi-Site Engineering Project (MSEP) environment. The Purdue Enterprise Reference Architecture (PERA) has been used to model the project lifecycle. This provides a guideline to establish the criteria for the review of the risk management tools and methods. This leads to the development of a roadmap for risk management, and enables the design and implementation of an intelligent risk mapping system.

2. RISK MANAGEMENT IN MSEP

2.1 MSEP as a Virtual Enterprise

With the advent of globalization of economies, enterprises are facing more and more rapid business changes. To seize and maintain its competitive advantage, an agile engineering and manufacturing enterprise must be able to react quickly to the changes in their business. As a consequence, a new business paradigm of "Virtual Enterprise" has been evolved to address this need.

A multi-site engineering project (MSEP) is a typical example of such a Virtual Enterprise, where two or more geographically dispersed business units or departments, research centers or different companies need to work together and exploit the fast-changing worldwide product engineering or manufacturing opportunities.

The MSEP is a competency-based temporary organization that uses the distributed intellectual strengths of its members, suppliers and customers.

The scope of its management is global, looking for the integration of the skills and contributions of those belonging to the value network. It explores the synergy necessary to satisfy the diversity demanded by customers, and to innovate, not just in product, but also in management practices.

However, because all MSEP members in each site need not only to work locally "business as usual" according to their own best practices, they also need to work globally by following the protocols and the plans agreed on in the MSEP. There may be gaps and conflicts in guiding project members' "local" and "global" behaviors. Therefore a set of effective risk management tools and methods is essential for the success of the MSEP.

2.2 Enterprise Integration and Risk Management

In order to manage the dynamic global manufacturing business of a MSEP more effectively, a clear understanding of the concept of virtual enterprise is needed. Therefore, enterprise models, which describe what an enterprise is conceptually composed of and how it works, as well as the Enterprise Integration methodologies become essential.

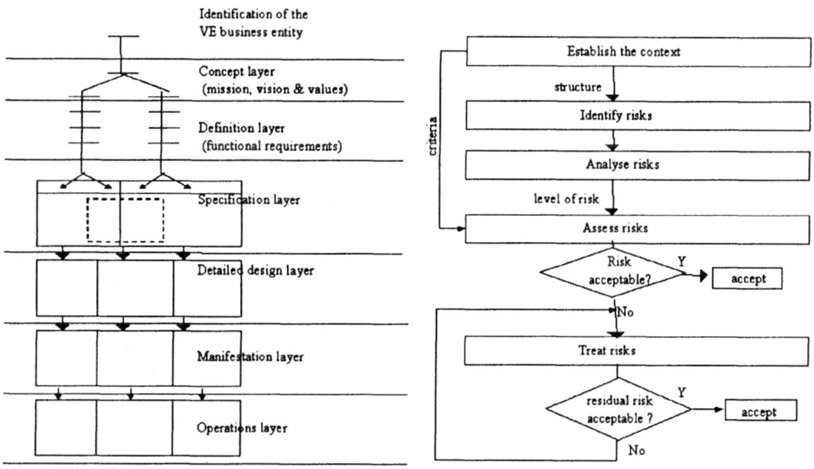

Figure 1. The PERA Methodology (left) and the Framework for Risk Management (right)

Zhou and Neff [4] proposed a virtual enterprise design methodology that utilize both the Enterprise Integration and Risk Manufacturing methodologies. It used the Purdue Enterprise Reference Architecture (PERA) [5] to depict the phases of the lifecycle of a virtual enterprise, and used the Australian/New Zealand Standard on Risk Management [6] as a framework to identify and analyze the risks involved in the building-up and the organization of a virtual enterprise.

As the MSEP is a typical virtual enterprise, this methodology can also serve as framework to help identify the requirements for risk management in MSEP environment.

2.3 Requirements for Risk Management in MSEP

The different phases of the PERA methodology can be used to guide our definition of the requirements for risk management in a MSEP. The first three phases are Identification, Concept, and Definition. The rest of phases represent an implementation view of a MSEP.

The Identification phase deals with the description of the external environment, and the criteria for creating or participating in a MSEP. At this phase, the risk management tools and methods should help the stakeholders to make strategic decisions such as why the MSEP is a feasible approach for the identified business opportunity.

The Concept phase aims to identify the motivation for creating a MSEP. At this phase, the risk management tools and methods should help the stakeholders and the potential project managers to identify and analyze the risks involved in the setting up and operation of the MSEP. Here the emphases will be the cross-site and cross-company boundary interactions within the MSEP.

The Definition phase is mainly focused on the explicit requirements concerning the planning and operation of the MSEP. It indicates a functional view of the lifecycle of a MSEP. The risk management tools and methods should be able to accommodate the specific function of the engineering projects. Here the emphases will be the ability to manage the risks involved in a specific engineering project, for example, an innovative product development project or a routine product improvement project.

The rest of the phases (Specification, Detailed Design, Manifestation, and Operation) are the Implementation phases of the MSEP, in which the MSEP is created, operated and finally dissolved after project completion. The risk management tools and methods for these phases are mainly focused on the implementation of risk management plans, monitoring, assessing and treating the identified risks as planned.

In the following section, we will present a survey and evaluation of the state-of-the-art tools and methods for risk management according to the phases of the MSEP lifecycle.

3. REVIEW OF TOOLS AND METHODS

3.1 The Identification Phase

Multi-site engineering projects, like other types of virtual enterprises, belong to a new manufacturing paradigm. Although many tools and methods have been developed for facilitating the design and operation of virtual enterprises, there are very few tools and methods on risk management for the identification phase.

The virtual enterprise design methodology [4] integrated with the Risk Management Framework can be considered as one of the early attempts to address this issue using enterprise engineering technologies.

Other methods have been largely aimed to achieve a better understanding of the organizational behavior in the "virtual" or the ICT connected multi-site environment. The establishment of Global Engineering and Manufacturing Networks [3], as well as the creation of Virtual Enterprise Communities [8], is aimed to stabilize the organizational structure of this virtual environment. The another term often comes with "risk management" in this phase is the term "trust", some people decide to establish trust *for* the purpose of minimizing risk, while others may prefer to establish trust *through* managing acceptable level of risk. IPR, liability, business intelligence, and many other legal issues are also active topics [8].

The COTS (commercial on the shelf) tools, such as FirstStep <http://www.interfacing.com/>, an enterprise modeling tool, can be used to do some scenario-based analysis and simulation for risk identification and analysis.

3.2 The Concept Phase

There are many tools and methods available for general project management, although none of them specifically developed for the needs of the multi-site engineering projects. Many of these methods are very generic and easy to understand, such as the PMBOK Guide [7]. Some of them have become national or international standards, such as [6].

Most of existing COTS tools for project risk management are for risk analysis and assessment. These tools include:

- Uncertainty modeling tools, such as RI$K <http://www.aceit.com/Products/risk.htm>, and Crystal Ball <http://www.decisioneering.com/>;

- Sensitivity and scenario analysis tools, such as @Risk <http://www.palisade.com/html/risk.html>;
- The "expert interview" tools, such as RiskTrak <http://www.risktrak.com/> and Trims <http://www.bmpcoe.org/pmws/>; and
- The Add-ons to the project planning tools, such as PertMaster <http://www.pertmaster.com/products/index.htm>.

These tools are useful when the identified risks are well understood and their probability and impact are known. It is unclear how effective these tools are in calculating propagated risks as the network of activities conducted in a multi-site engineering project, usually have a high degree of interdependencies.

The disadvantage of those "expert interview" tools is its highly domain specific nature, and seems to be suitable only to those highly structured processes. Those expert systems will need to be tailored for each specific domain.

3.3 The Definition Phase

For the definition phase, the tools and methods used for risk management are becoming more function oriented and domain specific. There are a number of risk management methods and tools already developed for many well-understood processes and domains, such as insurance, financial services, health service, security services, hazard management, and increasingly, the software development and information service domain.

There are many active research efforts looking at various types of engineering projects, such as product design, engineering development, assembly, and tooling. For example, our research team is currently investigating the risk assessment issues for a concurrent engineering project in a multi-site environment [9].

3.4 The Implementation Phases

There is a lack of the methods or guidelines that a project manager can use to develop a risk management plan, although many established companies have already developed their best practices in risk management into a template document for project risk management. However, there is also a lack of the COTS tools that can help a project manager to implement or execute the risk management plan, and manage the whole risk cycle, from risk identification, assessment, monitoring, to its treatment. There are no

COTS tools available that can capture and reuse the lessons learned from previous projects.

There is also a lack of the tools and methods to help a project manager to identify the gaps and conflicts during the monitoring and controlling of risks when different sites of the engineering project are using different risk management policies or at a different level of maturity.

4. TOWARDS A ROADMAP FOR RISK MANAGEMENT

Based on the review of the tools and methods for risk management in MSEP, we sensed an urgent need for a roadmap that can provide necessary guidelines for project managers to manage risks in multi-site engineering projects.

Such a risk roadmap is expected to provide guidance at an appropriate level of detail during the different phases of the lifecycle of a MSEP. Such a risk roadmap should also serve as the interface for project managers to identify and assess risk factors, as well as monitor the risk triggers and mitigate the risks through invoking corresponding risk management tools.

For the Identification phase, the risk roadmap should provide a "bird's eye view" map of the risk issues involved in the MSEP. This can be achieved by "interviewing" the project managers and major stakeholders to establish a risk profile for the new MSEP.

For the Concept phase, the risk roadmap should provide a "key" map of risk factors with a clear mark of the boundary of responsibilities for site to site, and site to MSEP interactions.

For the Definition phase, the risk roadmap should provide a detailed map that provides proactive advices on risk probability and impacts, helps to identify the critical path for the MSEP, and recommends the best practices for risk avoidance and risk minimization.

For the Implementation phases, the risk roadmap should provide an "on road" map that indicates the current project status, and current project positions in terms of the overall project progress against the project plan.

The development of such a roadmap for risk management will underpin the implementation of an intelligent risk-mapping tool for managing risks in multi-site engineering projects.

5. CONCLUSION

This paper discussed the risk management issues in a multi-site engineering project environment. The requirements for the risk management tools and methods have been discussed according to the lifecycle phases of the PERA model. The state-of-the-art tools and methods are reviewed against the identified requirements. Finally, the concept of a risk management roadmap is presented which will enable the development of an intelligent risk-mapping tool for multi-site engineering projects.

REFERENCES

1. Bernus, P.; Nemes, L.; Williams, T.J. (1996). Architectures for Enterprise Integration, Chapman & Hall
2. Zhou M. et al. (1998). A Framework for Design a Virtual Manufacturing Enterprise and its Implementation as a Workbench, in *Information Infrastructure System for Manufacturing*, edited by Miles J.J. and Kimura F., Kluwer Academic Publishers
3. GLOBEMEN. (2000 – 2003). Global Engineering and Manufacturing in Enterprise Networks. IMS 99004/ IST-1999-60002, http://globemen.vtt.fi
4. Zhou M., Neff P. (2000). Virtual Manufacturing Enterprises Design Using Enterprise Integration and Risk Management Methodologies, ICME 2000, Sydney
5. Williams, T.J. (1994). A Guide for Master Planning and Implementation based on PERA, Purdue University, Report No. 157
6. Australian/New Zealand Standard (1999). Risk Management, AS/NZS 4360
7. Project Management Institute (2000). A Guide to the Project Management Body of Knowledge (PMBOK® Guide) – 2000 Edition. Newtown Square, PA
8. Mo J.P.T., Nemes L. (2001). Global Engineering, Manufacturing and Enterprise Networks, Kluwer Academic Publishers
9. Amornsawadwatana S., Ahmed A., Kayis B., Kaebernick H. (2002). A Risk Assessment Approach for Concurrent Engineering Project in Multi Sites, Third Asia-Pacific Conference on Systems Integrity and Maintenance (ACSIM), Cairns

26

DEVELOPMENT OF AN AFTER-SALES SUPPORT INTER-ENTERPRISE COLLABORATION SYSTEM USING INFORMATION TECHNOLOGIES

Toshiaki Kimura[1],Fumio Kasai[2],Yoichi Kamio[2], and Yuichi Kanda[3]
[1]*Japan Scociety for the Promotion of Machine Industry, Japan*
[2]*Toyo Engineering Corporation., Japan*
[3]*Toyo University, Japan*
e-mail: kimura@tri.jspmi.or.jp

Abstract: This research paper discusses a manufacturing support system which supports not only maintenance services but also consulting services for manufacturing systems consisting of multi-vendor machine tools. In order to do this system enables inter-enterprise collaboration between engineering companies and machine tool vendors. The system is called "After-Sales Support Inter-enterprise collaboration System using information Technologies" (ASSIST). This paper describes the concept behind the planned ASSIST, the development of a prototype of the system, and discusses test operation results of the system.

Key words: After-Sales Support System, Maintenance, Inter-enterprise Collaboration

1. INTRODUCTION

Manufacturing support systems using IT (Information technologies) such as remote maintenance and remote monitoring systems are available today.

However, such support systems are usually machine tool-specific and depend on individual machine tool vendors. A mechanism does not exist that supports an entire manufacturing system consisting of multi-vendor machine tools implying cooperation between different machine tool vendors. Moreover, the usual scope of applications covering manufacturing support systems is often limited to the maintenance of machine tools and/or controllers, and does not include after-sales business consulting services.

This research paper discusses a manufacturing support system for a manufacturing system consisting of multi-vendor machine tools. The system not only supports maintenance services but also consulting services in both cases based on inter-enterprise collaboration between engineering companies and machine tool vendors. A hosting service environment is proposed for sharing data, information, and application systems among the collaborating engineering companies, machine tool vendors, and factories. A secure access control environment at factories for the exchange of maintenance information is also discussed. In this way, the manufacturing support system enables inter-enterprise collaboration for maintenance and consulting services using these environments. The manufacturing support system is called "After Sales Support Inter-enterprise collaboration System using information Technologies" (ASSIST).

Section 2 describes the concept for the planned ASSIST project. Section 3 describes the development of a prototype for the system and also evaluates the operation results of the system.

2. CONCEPT BEHIND ASSIST

2.1 Background and Objectives

Typically, maintenance of machine tools often depends on maintenance services for each individual machine tool vendor. Recently, however, machine tool vendors have developed remote maintenance systems using IT such as the Internet.

However, for maintenance of manufacturing systems consisting of multi-vendor machine tools, the overall condition of the manufacturing process has to be considered before specifying the maintenance areas. After assessment of maintenance areas, maintenance services by the machine tool vendors corresponding to the maintenance areas have to be procured. If engineering companies integrate multi-vendor manufacturing systems, the engineering companies are responsible for detecting maintenance areas. This means that it is the engineering companies that have to maintain the manufacturing system in collaboration with the machine tool vendors in question. For that reason, there is a need for a manufacturing support system that can handle maintenance of entire multi-vendor manufacturing facilities that can allow inter-enterprise collaboration between engineering companies and several different machine tool vendors.

On the other hand, in order to ensure improvement of manufacturing systems, it is necessary to collect the operational status information of these systems, analyse the data, track problems, and put together solutions for the problems including an evaluation of the solution plans [1]. For these activities, IT tools, such as TQC (Total Quality Control) tools and manufacturing system simulators, are useful. However, for smaller enterprises, it is difficult to deploy such IT tools because they are expensive and difficult to master. Therefore, ways are needed for engineering companies to contract these improvement activities as a consulting service and/or IT tools providing service by ASPs (Application Service Providers).

The objective of this research is to develop ASSIST as a manufacturing support system for multi-vendor manufacturing systems, which supports not only maintenance services but also consulting services carried out as an inter-enterprise collaboration between engineering companies and machine tool vendors.

2.2 Overview of ASSIST

The organization, which uses the system, consists of more than one multi-vendor manufacturing system, more than one machine tool vendor and an engineering company.

The system consists of these distributed enterprises and the hosting server integrated over the Internet as shown in Figure 1. The hosting server is a data-accumulating, data-providing, and application-providing server for inter-enterprise communication managed by the engineering company or a third party. Characteristics of the system are: managing information such as manufacturing system operational data in XML (eXtensible Mark-up Language) format, and providing data exchange among distributed enterprises using the hosting server. Possible usage scenarios for maintenance and/or consulting services are as follows.

a) Operational data from the manufacturing system is gathered and accumulated in the factory. The data is sent to the hosting server over the Internet when the factory needs maintenance and/or consulting service.
b) In the case of maintenance services, the engineering company provides support by finding trouble spots in the manufacturing system using the accumulated data from the hosting server.
c) Then, if the trouble area concerns a machine tool and therefore maintenance support performed by the corresponding machine tool vendor(s) is needed, the engineering company provides the accumulated data to the vendor(s) in question. After this, collaborative maintenance is

carried out through cooperation between the engineering company and the machine tool vendors.

d) In the other case of consulting services for improvement of the manufacturing system as ordered by the factory, the engineering company analyses the accumulated data from the hosting server using TQC tools and/or manufacturing system simulators. The engineering company then submits the results of the analysis, identifies problems, and proposes solutions for the problems to the factory.

e) Moreover, the engineering company can provide application systems such as TQC tools and manufacturing system simulators to the factory by using the ASP functionality of the hosting server.

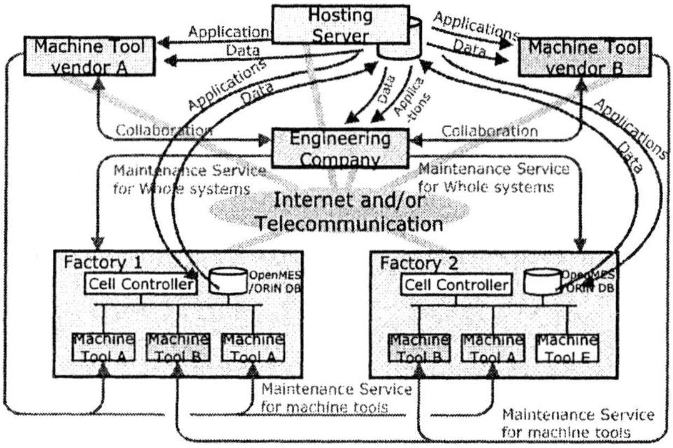

Figure 1. Concept of the ASSIST

3. DEVELOPMENT OF A PROTOTYPE SYSTEM

3.1 System Configuration and Functions of the System

A prototype system for ASSIST has been developed. This section describes the developed prototype system. An evaluation of the test results of the system is presented in the succeeding section.

The developed prototype system consists of a factory system, an engineering company system and a hosting server as shown in Figure 2. Functions and configuration of these systems are as follows.

a) Factory system

(Configuration)

- Multi-vendor machine tools and robots integrated using OpenMES [2], which is a middleware of MES (Manufacturing Execution System) base on CORBA.

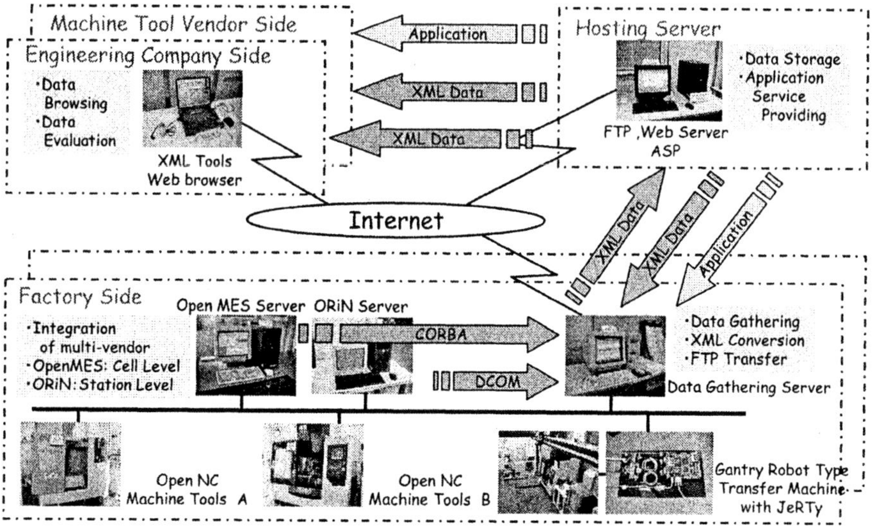

Figure 2. Configuration of the prototype system

- ORiN (Open Robot interface for Network [3]) server for condition monitoring of devices such as machine tools and robots.
- Data gathering server.
- Firewall with VPN (Virtual Private Network) server.

(Functions)

- Gathering of production log data as shown in the Table 1 for the manufacturing system from OpenMES server.
- Gathering of condition log data for devises as shown in the Table 2 from ORiN server.
- Conversion of the log data to an XML file.
- Periodic transfer of the XML file to the hosting server.
- The firewall with VPN (Virtual Private Network) server provides access permission control for machine tools of the factory for the machine tool vendors in question in order to allow maintenance of machine tools as shown in Figure 3.

Table 1. Production log data

Items	An example
Order information of the production	00000001
Product specification ID	MESPART1001
Scheduled production volume	1
Scheduled earliest start date and time	Fri Mar 29 13:00:00 JST 2002
Scheduled start date and time	Fri Mar 29 14:00:00 JST 2002
Actual start date and time	Fri Mar 29 13:52:00 JST 2002
Scheduled completion date and time	Fri Mar 29 14:12:00 JST 2002
Actual completion date and time	Null

Table 2. Device condition log data

Items	An example
Machine name	PV4-IIA
Controller name	MELDAS MAGIC 64
Controller ready status	True
Servo on/off status	True
Zero return status	True
Mode selection status	Auto
Operation status (Reset, stop, pause and start)	Stop
Error status	T02 0204
Error contents	Dry air pressure declined
Actual position X, Y, Z	0.001, 189.996, -0.021
Actual feed rate of the tool	0
Actual time of the controller	2002 Mar. 29 10:21:15

				Facility of the factory (VPN Server side)				
			Vendor name	Machine tool vendor A	Machine tool vendor C	Machine tool vendor B	Engineering company A	Engineering company A
			Facility name	Turning center P	Machining center Q	Machining center R	OpenMES server	ORiN server
Maintenance division of machine tool vendors and engineering companies (VPN Client side)	User name	User ID	IP Address / Password	192.168.xxx.uuu	192.168.xxx.vvv	192.168.xxx.www	192.168.xxx.xxx	192.168.xxx.yyy
	Factory User	AAAAAAAA	aaaaaaaa	permit	permit	permit	permit	permit
	Engineering Company A	BBBBBBBB	bbbbbbbb				permit	permit
	Machine tool vendor A	CCCCCCCC	cccccccc	permit				
	Machine tool vendor B	DDDDDDDD	dddddddd			permit		
	Machine tool vendor C	EEEEEEEE	eeeeeeee		permit			

Figure 3. Permission control of machine tools

b) Hosting server

(Configuration)

- Web service and FTP service using Windows NT and IIS (Internet Information Server) by Microsoft Corporation.

(Functions)

- Transmission and receipt of the XML file using FTP services.
- Web service of the XML files on a web server.

c) Engineering company system

(Configuration)

- Web browser.
- Statistics and analysis tools such as TQC tools that have an XML file interface.

(Functions)

- Receipt of the XML files from the hosting server.
- Analysis of production log data for the manufacturing system using the XML file.
- Preparation of graphs using the results of the analysis.

3.2 Evaluation of Test Operation Results

Test operation of the developed prototype system is performed. Consulting services by the engineering company through analysis of the operating conditions of the machine tools and usage time of the tools, using the production log data from the manufacturing system, are assumed for this test operation.

On the factory side, the manufacturing system worked for one week to manufacture five types of products: *MESPART1001, MESPART2001, MESPART3001, MESPART4001 and MESPART5001*. The production log data and device condition log data on the factory side are gathered and converted to XML files as shown in Figure 4 and sent to the hosting server once a day.

On the engineering company side, one week's worth of production log data as shown in the Table 1 and device condition log data as shown in the Table 2 for the manufacturing system are received from the hosting server. Machine tool operating conditions, including tool usage times, are analysed by use of statistics and, for instance TQC analysis tools. Graph charts for the results of the analysis are also produced. An example of a graph chart of tool usage times is shown in Figure 5.

A prototype test operation was carried out, and the performance and functions of the prototype system consisting of the factory system, the engineering company system and the hosting server integrated over the Internet were evaluated. All things considered, the test operations proved that the system performed quite well, showing promise for a future full implementation of the system.

However, the security system for the hosting server only uses a single user name and password. Therefore, the data sharing security system for the

Figure 4. An example of XML file

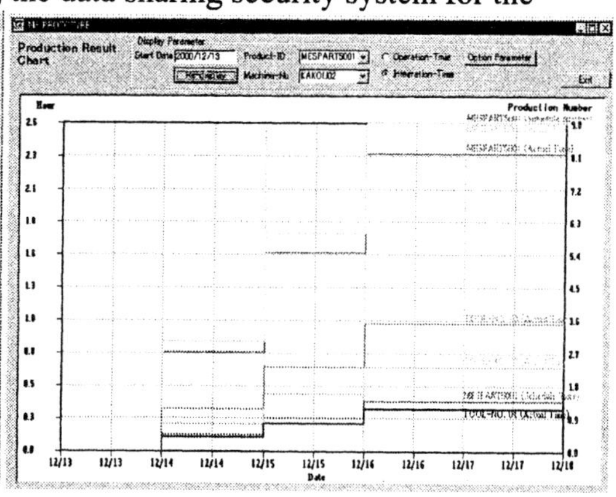

Figure 5. An example of graph chart

hosting server will have to be discussed in order to ensure dynamic sharing of data and information between several virtual enterprise partners. Also the ASP functionality of the hosting server will be discussed in this research. Moreover, test of operations that perform prevention, preservation, and maintenance of a multi-vendor manufacturing system are being considered.

4. CONCLUSION

The concept for a manufacturing support system named ASSIST has been proposed. ASSIST supports not only maintenance services but also consulting services for multi-vendor manufacturing systems carried out as an inter-enterprise collaboration between engineering companies and machine tool vendors.

A prototype system for ASSIST has been developed. A test operation using the prototype was successful. The test covered activities and functions of the prototype system consisting of a factory system, an engineering company system and the hosting server, all integrated over the Internet. According to the results of the test, the system performed quite well and the possibility that the ASSIST concept can be implemented has been confirmed.

The security system of the hosting server for dynamically sharing data among virtual enterprises consisting of more than one will be discussed. ASP functionality for the server will be also discussed in this research.

ACKNOWLEDGEMENTS

This research is a part of the IMS GLOBEMEN Project, and "Study on Manufacturing System using Open-oriented Information Technologies" which is a research project of Japan Society for the Promotion of Machine Industry. The latter project is supported by funding from the Keirin Association of Japan.

REFERENCES

1. Kimura, T. et al., A study of information processing system for the improvement of the manufacturing system, Annual meeting 1999 of IMS project, pp144-147
2. IBM Japan, Ltd., OpenMES Specification Ver.1.0 Draft Alpha, May 25. 1999
3. Mizukawa, M. et al., Matsuka, H., Koyama, T., Matsumoto, A., ORiN: Open Robot Interface for the Network, a proposed standard, Industrial Robots, MCB university press, Vol. 27, No.5, pp. 344-350

27

COLLABORATIVE SERVICE IN GLOBAL MANUFACTURING – A NEW PARADIGM

I. Hartel[1], Y. Kamio[2], M. Zhou[3]
[1]*Swiss Federal Institute of Technology Zurich – Center for Enterprise Sciences (ETHZ-BWI), Zürichbergstrasse 18, CH-8028 Zurich, Switzerland, Email: Ingo.Hartel@ethz.ch;* [2]*Toyo Engineering Corporation, Japan, Email: kamio@ims.toyo-eng.co.jp;* [3]*CSIRO Manufacturing & Infrastructure Technology, Australia, Email: Mingwei.Zhou@csiro.au*

Abstract: Small and medium-sized enterprises have to deal with increasing requirements on customer services. Furthermore these enterprises are not able to decrease their service delivery times and costs on their own and so, as a consequence, they have to team up with external partners to build up service cooperations. To support this development a model will be described to enable these enterprises fulfilling services collaboratively by using the basic concept of the virtual organization.

Key words: Collaborative Service, Service Management, Virtual Enterprise

1. INTRODUCTION

Global manufacturers are facing an increasing challenge in offering adequate and timely fulfilment of customer services (see [1]). This is particularly urgent for those small and medium-sized enterprises (SMEs) serving the one-of-a-kind industry, as they find it difficult, if not impossible, to maintain their own worldwide service networks at reasonable costs. One approach is to collaborate by selecting cooperating network partners based on their individual core competencies during the fulfilment of various after-sales services.

However, setting up and maintaining such a network organization, as well as developing and implementing a successful business process for global collaboration, are often beyond the knowledge and capacity of a SME.

Therefore, a reference model including guidelines to form inter-enterprise service collaboration is essential and can be critical for its success.

2. VIRTUAL ORGANIZATION

In the search for organization forms for the twenty-first century, the virtual organization concept is beginning to make headway as a dynamic structural pattern. Under this model, organizational units (also called as virtual enterprises) are created, restricted to the primary business purposes and thus, this structural simplicity allows maximum economic efficiency.

A virtual enterprise can briefly be characterized as a short-term inter-enterprise cooperation where individual enterprises join core competencies in order to establish a value chain configured exactly to meet a specific customer demand. When the customer demand has been fulfilled, the virtual enterprise is decommissioned.

For the one-of-a-kind production, different types of virtual enterprises can be set up based on their deliverables (see [2]). For example, in order to support the operation of a one-of-a-kind product such as a chemical production facility, a virtual service enterprise could be established to offer the plant owner with services such as maintenance or operation support.

This kind of virtual organization usually consists of three main elements: The service network of all potential service members, the virtual service enterprises (VSEs) of selected members, and the services as a product offered by the VSEs. To provide a structural arrangement and to capture the characteristics of these entities, the Virtual Enterprise Reference Architecture (VERA, see Figure 1) is currently being developed in the IMS GLOBEMEN project [3]. VERA is based upon GERAM [4].

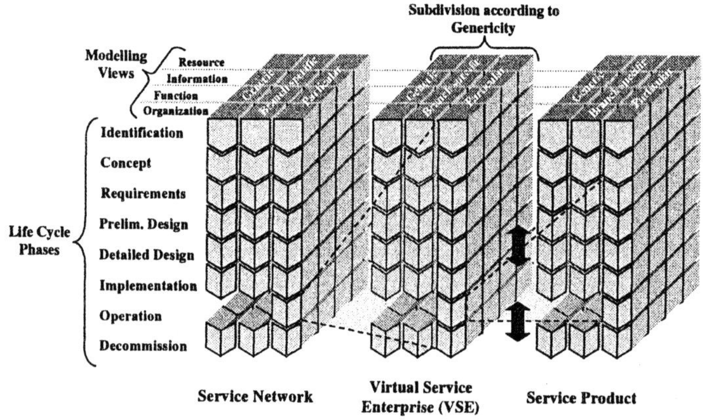

Figure 1. Virtual Enterprise Reference Architecture (VERA)

A **service network** in the operation phase has an array of service products, which they can offer to customers. A **service product** consists of one or many service modules, which targets different customer needs through its single or combined characteristics. A more elaborate description of service products and their configuration out of service modules can be found in [2]. If a customer reports a problem suitable network members will be selected and a **virtual service enterprise** will be set up that can best deliver the service product as defined by the given customer requirements.

3. COLLABORATIVE SERVICE

The reference model for collaborative service introduced in this paper describes the activities that are pertinent during the life-cycle phases of the three different entities described above. These life-cycle phases encompass all activities from identification to decommissioning of the entity (see chapter 3.1). To decrease the apparent complexity the reference model uses the view concept from GERAM that allows the operational process to be described as an integrated model, but to be presented to the user in different sub-sets, so called modelling views (see chapter 3.2).

3.1 Life-cycle

This chapter describes briefly the entire life-cycle of setting up a service network organization and its virtual service enterprises. In the reference model the life-cycle is modelled in various level of detail.

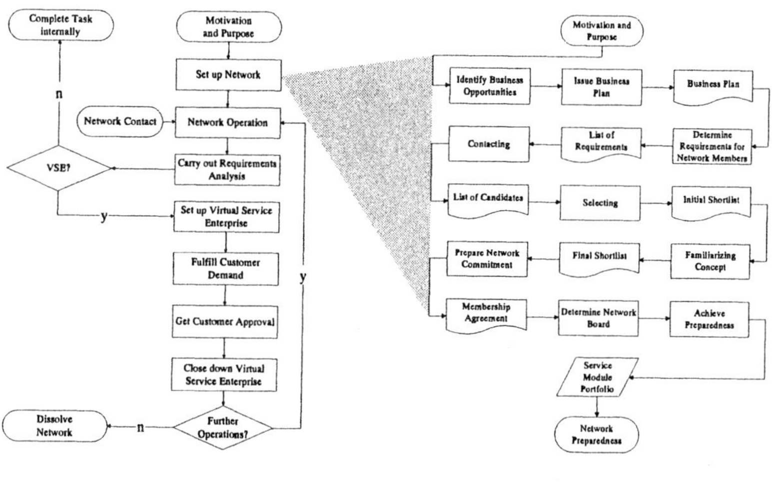

Figure 2. Top-level overview and set up network process [5]

The top-level overview helps to grasp the entire content and illustrates the process of setting up a network organization and its VSEs and finally, fulfilling the customer demand by providing the service product. Based on this the following level shows in detail, what has to be done in the different life-cycle phases. As an example Figure 2 illustrates on the left side, the overview of the business process and on the right side the depiction of the set up network process in detail.

3.2 Modelling views

Modelling views contain a subset of facts present in an integrated reference model allowing the user to concentrate on relevant questions during modelling of its service organization. The following subchapters aim to give an overview of the four different modelling views (organisation, function, information and resource), which are used in the reference model.

3.2.1 Organizational view

This view describes the organizational structure of the relevant entities and the roles that exist in theses entities with their responsibilities and tasks. The main organizational entities of a virtual service organization are the production facilities delivered from the one-of-a-kind producers, the service network, the resulting virtual service enterprises and the delivered service products.

Production facilities can be characterized by their locations, type of installed machines and installations, production processes, history of services, and modifications since production start up. The size of the owners may range from SMEs with one production location to large corporations with a number of globally distributed locations. The production facility can be made up of machines or installations purchased from one or more one-of-the-kind producers.

The **service network** consists of independent enterprises that are working together to exploit a particular service opportunity by offering **service products** jointly to the market, based on common interests and partnership-oriented business relations. Network members can be one-of-a-kind producers and a multitude of independent service companies, suppliers, vendors or ICT infrastructure providers. The distinctive features between them are competencies, technical aids, and available ICT, locations, and capacities. In addition the one-of-a-kind producers are characterized by size, branch, and position in the value chain. Thus, networks can be comprised of one-of-a-kind producers of the same or different sizes and in the same or different branches. As to position in the value chain, company activities can be at the same level (so-called horizontal) or downstream or upstream levels (so-called vertical). Collaboration among companies within the same branch, and thus potential competitors, is necessary when a critical mass is required in global markets in order to gain an advantage over local competitors.

The **virtual service enterprise** is formed of selected network members. Together these network members can fulfil the specified service. The service is divided into different tasks. Each network member in the VSE is responsible for performing a part of these tasks in accordance with its competencies and available technical aids and ICT. Sometimes collaboration is necessary in order to perform a task, such as, for example, when a local service company repairs a machine by installing new spare parts but does not have the know-how to reset machine controls after completion of the repairs. Under the remote direction of the one-of-a-kind producer, however, the service company can do the resetting. Here the network members depend upon modern ICT for coordination. Communication can be very efficient using video conferencing and wearable computers, for example.

The reference model distinguishes seven **roles**, six of them are active and one is auxiliary. The active roles include the customer, the network or project initiator, the network members, the project board, the network board and the audit team. The information system of the network organization plays a supporting role facilitating the information exchange between the active roles. More elaborate description of the roles can be found in [5].

If there is more than one one-of-a-kind producer in the service network, the question of **responsibility for tasks** like network operation, contacts with customers, evaluation of potential new network members, etc. must be clarified. Possible variants are described in [2]. Small and medium-sized one-of-a-kind producers have very oft not conditional on their size the financial and personal resources to manage a service network. If only this type of enterprises are part of the network then it can be useful to outsource management tasks to an external company and financial tasks to a bank or insurance company. Responsibilities might also be divided up according to market or customer segmentation.

3.2.2 Functional view

The functional view represents the hierarchical structure and the relations of functionalities (activities). In functional models activities related to the management and operation of the virtual service organization are represented, as well as support activities. An example of a function model for the VSE entity is illustrated in Figure 3. It is possible, that the same activity has to be fulfilled several times, e.g. in different life-cycle phases of one entity or in life-cycles of different entities.

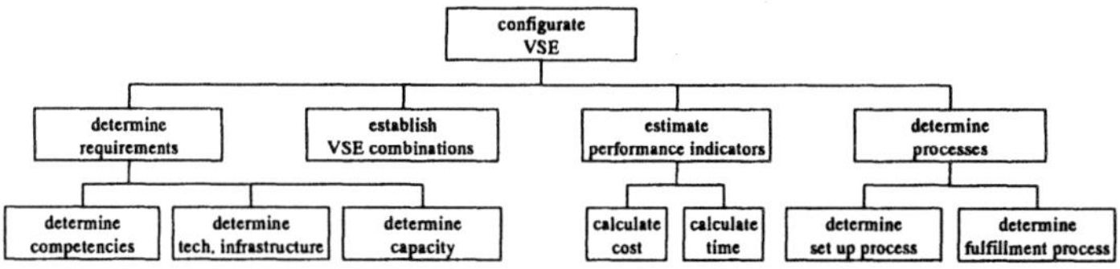

Figure 3. Function model for virtual service enterprise configuration

3.2.3 Information view

The information view is describing the content and flow of information, which is created, shared, used, modified and disposed along the life-cycle of the different entities of a virtual service organization. Junction points of information exchange among the defined network roles are indicated and key information occurring during the set up and operation of a service network and VSEs are structured and illustrated as different information objects. These information objects can be used to support the design of a data schema for an information system. Figure 4 presents exemplary the information exchange between different roles during the set up of the network and the information objects business plan and service module portfolio (**P** means information provider, **R** means information receiver).

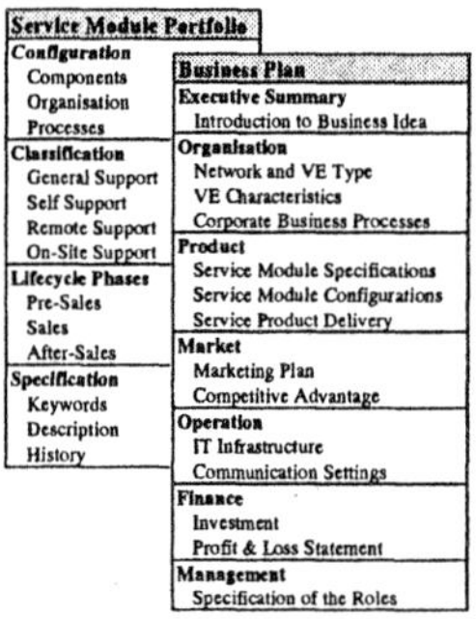

	Information Objects	Type of Information							Roles						
		Network	Project	Product	Production	Financial	Legal	Others	Customer	Network Initiator	Partner	Project Board	Network Board	Audit Team	Information System
1	Service Module Requirements			x					P	R					
2	Project Organisation Form		x						P	R					
3	Marketing	x	x	x		x		x		P/R					
4	Network Operation	x						x		P	R				
5	Financial Data					x				P	R				
6	Management	x	x					x		P	R				
7	Business Plan	x	x	x	x	x	x	x		P	R				
8	List of Requirements	x	x	x	x	x	x	x		P	R				
9	List of Candidates	x								P	R				
10	Preliminary Selecting Criteria	x	x	x	x	x	x	x		P	R				
11	Initial Shortlist	x								P	R				
12	Detailed Selecting Criteria	x	x	x	x	x	x	x		P	R				
13	Final Shortlist	x								P	R				
14	Network Principles	x	x	x	x	x	x	x			P/R				
15	Membership Agreement	x	x	x	x	x	x	x			P/R				
16	Members of the Network Board	x						x			P/R				
17	Service Module Portfolio			x							P/R				

Figure 4. Information exchange and information objects whilst setting up a service network

3.2.4 Resource view

The resource view represents the resources of the network members as they are used in the course of the network and VSE set up and operations. Resources can be for example competencies and know-how to fulfil service tasks, capacity of service technicians, material or machines needed to produce spare parts or specific diagnostic software and hardware. These resources have to be assigned using resource models to the defined activities and organizational entities.

4. INDUSTRIAL CASE STUDY

The proposed reference model is currently being established and exploited in a use case from the GLOBEMEN project. In this case a Japanese one-of-a-kind producer ("OKP1") develops and sells large chemical plants. For the manufacture of these plants, it has a large network of suppliers and vendors. Worldwide there are 100 of these plants in operation by customers. In the past, OKP1 developed service modules, such as a remote plant monitoring system and a training simulation system, to support after-sales services. However, OKP1 does not have its own team of service technicians. A customer ("C1") operates several plants in Asia and has its own teams for inspection and maintenance. A Japanese telecom company ("TC") maintains a communication network in the Asian region. TC offers also services to collect information from production equipment for safe control and use, provides network accounting, collection and authentication functions and a place to register a variety of applications to access remotely.

Below the first steps of modelling the organizational view will be briefly described:

Structure of the service network: The service network is established in the Asian region and made up of OKP1 with selected suppliers and vendors, C1's service teams and the network infrastructure from TC.

Collaborative service products: The operating service network will provide a wide range of service products, which are configured from existing service modules from the network members. Collaborative service products will be for example:

- *Hosting Service:* Long term data gathering of plant and process data and monitoring using the remote plant monitoring system, which is installed at TC. Suppliers and vendors can also access the remote plant monitoring system and are responsible for observing their specific parts of the plant.
- *Process or Equipment Performance Evaluation:* Based on gathered data from the hosting service an evaluation of the current status of the overall plant or specific equipment will be carried out. For the evaluation a process flow simulator from OKP1 and knowledge about plant operation from C1 are needed. The results of the evaluation will contain recommendations to change current operation condition (e.g. pressure, temperature or flow) or to establish plant services.
- *Plant Services:* These services are maintenance, repair, inspection, etc. and have to be carried out by service teams on-site.

Based on these products an enterprise configuration scenario could look as following: Due to the equipment performance evaluation the requirement for maintenance on a reactor and replacement of spare parts of the supply pipes is identified. The virtual service enterprise will then take the following configuration: A service team from C1, which has reactor training, goes on-site to the customer. At the same time, a supplier delivers to the customer the pipes it has manufactured or drawn from inventory. During the repair procedures, the member of the service team receives additional required information from engineers at OKP1 using mobile and internet-based ICT.

The benefits of the described type of cooperation are as follows: Through implementing various service modules, OKP1 can actively offer after-sales services and gather valuable experiences for new product design. C1 benefits from collaboration with OKP1, because they will have better utilization of their service team capacities. In addition, technicians from C1 will benefit from enlarged experience through service work at other companies.

5. CONCLUSION

The present contribution introduced a model of how future collaboration in after-sales services in the one-of-a-kind industry might look like. The proposed reference model for collaborative service is intended to provide a guideline to set up an inter-enterprise service organization based on the principles of the virtual organization. During the application of this model, in industrial use cases, it may be adjusted individually, but it still helps especially SMEs to increase the ability of setting up a network organization and of running its virtual service enterprises successfully.

REFERENCES

1. Hartel I, Burger G, Schnetzler M, Rüf C (2001). Company survey "Successful Service Management – Trends and Practices in the After-Sales", Zurich (CH).
2. Hartel I, Burger G, Billinger S, Kamio Y (2002). Virtual organization of after-sales service in the one-of-a-kind industry. In: Camarinha-Matos L M (Ed.). Collaborative Business Ecosystems and Virtual Enterprises, pp. 405–420, Kluwer Academic Publishers, Boston (USA), 631 pages.
3. GLOBEMEN (2000–2002). Global Engineering and Manufacturing in Enterprise Networks. IMS 99004/ IST-1999-60002, http://globemen.vtt.fi. The Swiss region is sponsored by the Swiss Commission for Technology & Innovation (CTI).
4. GERAM (1999). Generalized Enterprise Reference Architecture and Methodology. ISO/DIS15704, IFAC/IFIP Task Force on Architectures for Enterprise Integration.
5. Hartel I, Sonderegger S, Kamio Y, Zhou M (2002). The Information Roadmap of Virtual Enterprises. e2002 – eBusiness and eWork Conference, Prague (CS), 16–18 October.

28

REMOTE MAINTENANCE SUPPORT IN VIRTUAL SERVICE ENTERPRISES

Yoichi Kamio[1], Fumio Kasai[1], Toshiaki Kimura[2], Yoshiro Fukuda[3], Ingo Hartel[4], and Mingwei Zhou[5]

[1]Toyo Engineering Corporation, Japan, [2]Japan Society for the Promotion of Machine Industry, Japan, [3]Hosei University, Japan, [4]Swiss Federal Institute of Technology Zurich, Switzerland, [5]CSIRO Manufacturing & Infrastructure Technology, Australia, e-mail: kamio@ims.toyo-eng.co.jp

Abstract: This paper discusses a scheme that allows all parties involved in the maintenance of a chemical plant: the plant owner (customer), the engineering company, the equipment vendors, and the maintenance firms, to form a virtual service enterprise (VSE) whenever a maintenance service is necessary, in an attempt to provide required services more timely. This paper also discusses the requirements for knowledge management and risk management in such a VSE environment, and proposes a secure hosting service environment that allows all the parties to share information and applications, and to collaborate during plant operation and service fulfilment. To evaluate and demonstrate the feasibility of this scheme, a prototype of the remote maintenance system developed for a fertiliser plant in Indonesia is presented. Finally, a scenario of how to utilise such a remote maintenance system, the possibility of collaboration among VSE partners, and the advantages of utilising the proposed hosting services in VSE are also discussed.

Key words: Virtual Service Enterprise, Remote Maintenance, Hosting Service, After Sales Service, Inter-Enterprise Collaboration

1. INTRODUCTION

Modern chemical plants are getting more and more sophisticated, and the number of equipment vendors needed for each plant is also increasing. The provision of after-sales services, such as the maintenance of plants and keeping the plant at a good level of efficiency, safety and reliability, has become an increasingly challenging business opportunity for those engineering companies who constructed those plants. The customers who own and run those plants, on the other hand, are also trying to rationalize their plant operations by out-sourcing their maintenance services in order to concentrate themselves on their core competencies. All these demand for a new business scheme that can provide cooperative service by all parties involved, including the engineering company, the various equipment vendors, maintenance firms, and the plant owner (customer).

This paper attempts to address such issues. We are proposing herewith a remote supporting technology where engineering companies, equipment vendors and customers will be able to co-operate and solve problems quickly by exploiting the latest IT technologies. A case study in a large scale chemical plant has been taken as an example.

2. BUSINESS REQUIREMENTS

Plant owners are endeavoring day and night in order to operate their plants safely and enhance their production efficiency. However, the engineering companies who built the plant, and various vendors who provide equipment for the plant are getting more and more internationalized. It would be very hard, if not impossible, for any of these business partners to provide a win-win solution separately and independently. It would be necessary for them to work out a cooperative ways among themselves.

The business requirements from each party are described in more detail below.

2.1 Requirements from customers

Inherently, customers as plant owners usually maintain their own operation department or service department in an endeavor to operate their plants in an optimal manner. However, the systems have become much complicated due to the enhanced functions of plants and environmental considerations, which require the service personnel to have a higher level of analysis and judgment capability. In addition, while the service departments need a certain level of manpower for periodical inspection, such manpower may become redundant for daily routine inspection. Because it is unsustainable to keep a maximum required size of service personnel in house, the plant owners will have to find out ways to keep it minimum and to source out necessary work when required.

2.2 Requirements from engineering companies

Engineering companies have been traditionally concentrated on the so-called EPC (Engineering, Procurement and Construction) business sector, and minimize the involvement small and lingering after sales jobs whenever possible. However, due to the recent severe competition for new projects in the global market, and increasing demand from customers for complete solutions, the engineering companies are expected to extend their services to cover the entire life cycle of the plants and to keep a good relationship with customers.

However, customers are distributed throughout the world, it would not be cost effective for engineering companies to extend their support services world widely all by themselves. Therefore, engineering companies are looking for local partners to extend their support services. As the equipments used in the plants are procured from vendors scattered world widely, linkage with these equipment vendors is essential for servicing such components.

2.3 Requirements from equipment vendors

Vendors have recognized importance of maintenance service of their products and, therefore, have established maintenance service network systems by using manpower or telephone line systems. However, once their products are included in complicated systems such as plants, the effect of maintenance services by vendors alone would be limited and linkage with system analysis or simulation will be indispensable. Furthermore, it will be inefficient for each vendor to construct own service network. Hence, cooperative actions with engineering firms or other vendors will become necessary.

3. MODEL OF VSE [1],[2]

A virtual service enterprise is a temporary business entity established from a network of partners aimed to provide the after-sales services to a customer. Each partner is an independent entity that is equipped with its own unique capabilities and competencies, assuming responsibility to perform the allocated work. For example, when an engineering company, a group of equipment vendors and maintenance firms, and a customer (factory) organized a VSE using a hosting service (see Section 4.2), equipment in the plant made by various vendors can receive services by means of remote monitoring systems, some minor hitches could be rectified through remote maintenance. Furthermore, periodical repairing tasks can now be planned based on a more accurate judgment: when a material repair work (such as replacement of components) is necessary, the customer can order parts via Web system, and maintenance firms can schedule in advance to perform repairs and replacement of equipment on time. The engineering company can also carry out simulations of entire plant systems using actual data collected and will be able to provide the customer with valuable advices such as optimized operations. A conceptual image of a VSE is shown in Figure 1.

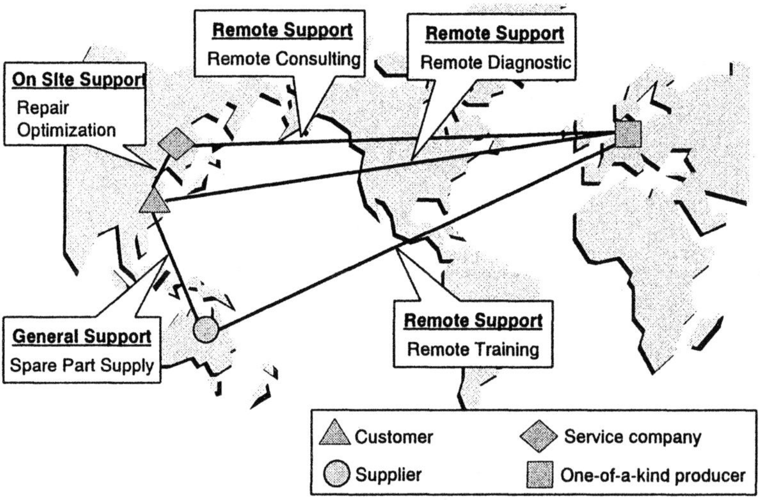

Figure 1. Example of a virtual service enterprise

4. ICT ENVIRONMENT FOR VSE

4.1 Remote Plant Monitoring System (RPMS)

Plant operations are usually controlled by the DCS (Distributed Control System), from which plant information, mainly the operational data, can be downloaded and forwarded to a data server through Internet connection.

Figure 2 illustrates our proposal on how relevant partners can share the information. As the plant operates continuously, its operational behavior does not change suddenly during normal operation. Therefore, the plant data can be transmitted on a daily basis at the least busy time zone of the network. Furthermore, only part of the information needs to be accessed by each partner. As shown on Figure 2, a list of data items, entries on P&ID diagrams and trend graphs have been made available. The time and frequency of data transmission may be adjusted according to the customer's requests or the engineering company's judgment.

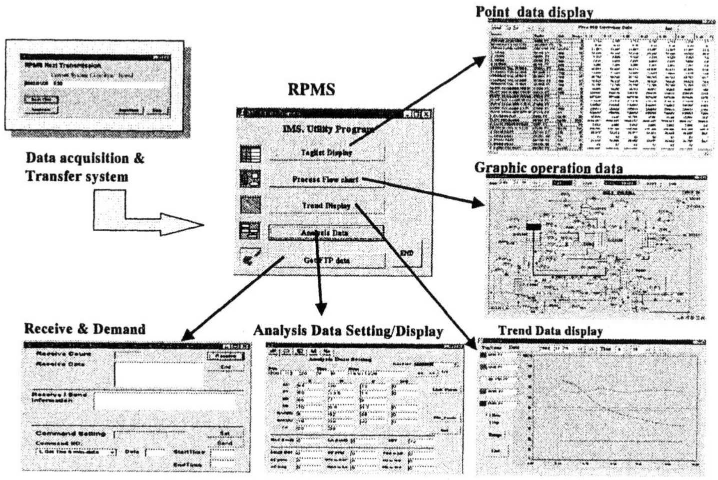

Figure 2. Remote Plant Monitoring System

4.2 Hosting Service

Normally, a virtual enterprise (VE) or a project team is organized when constructing a plant. An engineering company usually plays the role of a prime contractor, interacting with both the customer and the equipment vendors and coordinating the technical issues and work schedules.

This arrangement is convenient for the customer because the interaction between the equipment vendors is simplified and no day-to-day coordination is necessary. This is convenient for the equipment vendors too, because they no longer need to coordinate with each other on an individual basis.

After the construction of the plant was finished, and the manufacturing facilities were commissioned, the engineering company had successfully completed its commitment, and the responsibility for the plant was passed on to the customer (owner). Normally, a service department is setup by the customer to carry out all service and maintenance work. As we discussed above, this is a costly burden on the customer.

Naturally, we need to consider why not extending the VE or project team arrangement successfully used during the construction phase into the maintenance phase. The merit would be that all the parties including the customer and equipment vendors get benefits as in the construction phase.
We have adopted a hosting service, as shown in Figure 3, as an IT infrastructure to facilitate the VSE. The core of the system is a data center which links the customer (factory owner), the engineering company, the equipment vendors, servicing or maintenance firms. Through this service, the customer can execute the service and maintenance agreements with each

equipment vendors and the engineering firm. The management of data center security, services and computer maintenance will be carried out by an independent data center company. The data center company also manages all service applications for all registered companies. Conventionally, the customer needs to ask the engineering company or the equipment vendors to send in their engineers to do the repairs and maintenance of equipment installed at each plant. This has been a costly and time consuming task, and if not handled promptly, may interrupt the normal production in the plant. However, by exploiting the above mentioned hosting service, a more cost effective and speedy maintenance service could be expected. In addition, each vendor can provide more efficient after sales activities.

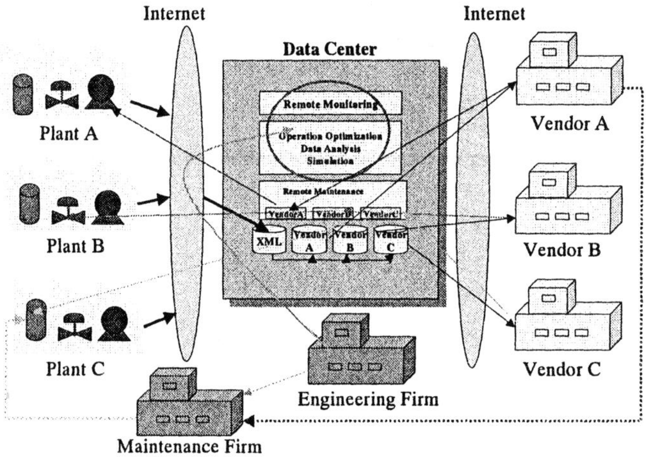

Figure 3. Hosting Service

5. INDUSTRIAL CASE STUDY

Toyo Engineering Corporation (TEC) as an engineering company has over 40 years history and has successfully constructed many chemical and petroleum plants all over the world. In particular, TEC has the license for urea processing, with which the company has built more than 100 plants and enjoys about 25% share of the world market. TEC has so far provided after sales services such as technical support, training simulator development and turn around (T/A) support. In order to expand the service base, the firm has decided to develop a business model for after sales services of urea processing plant, and invited one of the customers as a partner to jointly develop the new process. The first parts of the business model will be briefly described as follows:

Structure of the service network: After a review of techniques and tools that TEC and this customer can offer, it was found that we are in a reciprocal position. (For example, TEC owns design know-how and tools while the customer owns operation and maintenance know-how of plants.) It was also made clear that by combining the know-how, we would be able to offer more useful services. So we planed to construct a service network with TEC and the customer as the core, and establishing a VSE that best complies with requests of other urea plant owners and to provide best services. The other partners invited to the service network include the equipment vendors and maintenance firms.

Collaborative after sales service scenarios: A prototype developed is shown in figure 4. The hosting server is connected to Internet, and accessible from outside by using applications such as SecurID. Also it should be mentioned that the application developed can be operative on thin client PCs by using SBC (Server Based Computing) tool such as "Go Global". The data pertaining to the operation sent from plants are accumulated in the hosting servers through the data servers and utilized for various services such as plant monitoring, preventive maintenance, trouble shooting of complex units, total plant system simulation, equipment performance evaluation, system performance evaluation, training.

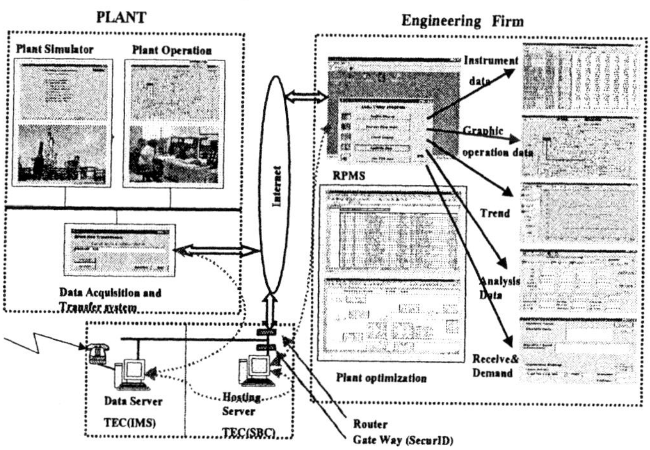

Figure 4. Prototype system

Conceivable examples of the services that use the above are: plant monitoring (basic), periodical training, performance evaluation of equipment and plant, and scenarios combining various trouble shooting measures.

By such set up as mentioned above the benefits are: Partners who participated in the service network can play the roles as VSE members without any additional need for ICT infrastructures. Also, by sharing data and IT infrastructures, and by reciprocating each other the technologies and tools, each partner will be able to provide services that were impossible by individual effort.

6. CONCLUSION

In this paper, we take a chemical plant as an example, and proposed a business model of collaborative after-sales services in "one-of-a-kind" industry and designed a prototype. Although we recognize that more specific aspects of business are yet to be studied in more detail, we believe that the methodology as introduced herein can be commonly applied to large-sized and complicated facilities where many business entities are involved.

ACKNOWLEDGEMENTS

This research has been carried out under the IMS GLOBEMEN project (see [3]). The authors would like to thank all other members of the GLOBEMEN project for their valuable discussion and technical support.

REFERENCES

1. Hartel I, Burger G, Billinger S, Kamio Y (2002). Virtual organization of after-sales service in the one-of-a-kind industry. In Camarinha-Matos L M (Ed.): Collaborative Business Ecosystems and Virtual Enterprises, pp. 405-420, Kluwer Academic Publishers, Boston (USA), 631 pages.
2. Hartel I, Sonderegger S, Kamio Y, Zhou M (2002). The Information Roadmap of Virtual Enterprises. e2002 - eBusiness and eWork Conference, Prague (CS), 16-18 October.
3. GLOBEMEN project web page: http://globemen.vtt.fi

PART III

Factory Floor Infrastructure

29

INTELLIGENT PROCESS PLANNING AND CONTROL FRAMEWORK FOR THE INTERNET

John P.T. Mo, Stuart Woodman
CSIRO Manufacturing and Infrastructure Technology, Australia
john.mo@csiro.au, stuart.woodman@csiro.au

Abstract: The use of computers has been growing rapidly in manufacturing since computer technology started to be used in machine controls. However, due to historical reasons, these technologies are isolated and their information cannot be shared. For example, to use the information from the CAD system for machining, a human expert engineer has to interpret the CAD meanings and then enters them into a CAM format. Integration of these manufacturing functions requires a technology break-through in automation of manufacturing technology. This paper describes the development of an integrated web-based CIM environment called J-MOPS which can intelligently transform CAD information into machine programs. The use of Java significantly enhanced the system with the advantages of robustness, object-oriented, network oriented, and platform independence.

Key words: Computer automated process planning (CAPP), Integrated computer aided manufacturing (ICAM), Object oriented programming

1. INTRODUCTION

The use of computers has been growing rapidly in manufacturing since computer technology started to be used in machine controls. In the beginning, each manufacturing system's component was developed separately resulting in a term known as *The Islands of Automation* [1,2]. For example, at the design stage in manufacturing, Computer-Aided Design (CAD) has been made feasible after the Cathode Ray Tube (CRT) was invented for generating graphic displays. At the same time, on the shop floor, Computer-Aided Manufacturing (CAM) has been developing, with technologies such as microprocessors for Numerical Control (NC) of machines. These technologies are isolated and their information cannot be shared. To use the information from the CAD system, a human expert engineer has to interpret the CAD meanings and then enters them into a CAM format. Likewise, to improve CAD by experiences in CAM, a human designer has to learn the essence of CAM programming and adjusts the design manually [3]. There is no direct computer path for these two activities to be linked automatically. Each *Island of Automation* was advanced separately until the need to use information from one source on another computer assisted application arose, i.e. Computer Integrated Manufacturing (CIM).

Integration of all manufacturing components requires a technology break-through in automation of manufacturing technology. One of CIM examples is CAPP systems [4]. CAPP helps manufacturing systems to integrate processes from the design to actual manufacturing. Typically, a Part-Program of a CNC machine is generated by an automated program in a CAPP system using data in the CAD model [5]. Although, there are so many CAPP systems nowadays, the development and improvement of CIM is still continuing.

CIM development inevitably involves substantial software development work. In the last decade, a new approach was introduced into computer software development, the object-oriented approach [6,7]. This approach speeds up the development of new programs, and improves the maintainability, reusability, and modifiability of software. Extending this philosophy to manufacturing, CIM system also adopts the object-oriented concept. Experience has shown that this concept brings significant improvement in manufacturing technology [8,9]. For example, Manufacturing Objects Processing System (MOPS), which is a framework for CAD/CAM system using object-oriented concepts, was developed in C++ programming language [10]. The framework is an intelligent manufacturing system for a complete integration from design (CAD) to actual manufacturing of a product (CAM) [11].

2. SYSTEM REQUIREMENTS

Recent development in computer networks has revealed that a computer is not just a single unit tool, but is also a part of a complex network with other computers. An incompatible system amongst computers is a common problem for an integration system. Every different computer system has its own platform of software and hardware. This prevents information being transferred across the system boundary. For example, a PC compatible application cannot run in a UNIX system and vice versa. This will be a great obstacle for now and in the future, when the Internet and network play an important role in computer communication linking different computer systems or platforms electronically [12].

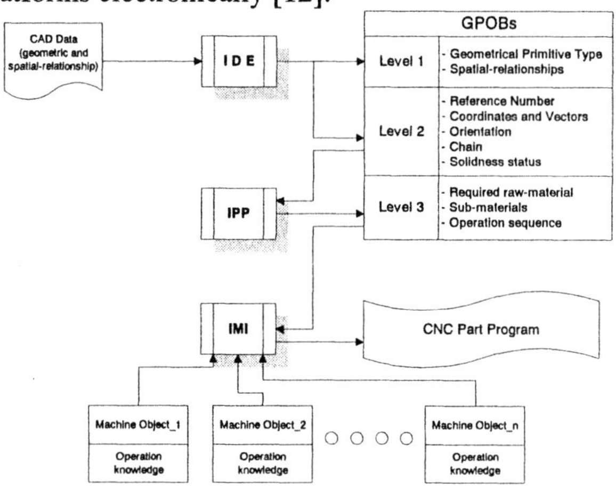

Figure 1: Structure of MOPS.

The widespread use of computer networks and especially the availability of the Internet has a significant impact on application accessibility [13,14]. Applications which are tied to a particular platform are therefore losing out on the general usability. Many applications were designed for single-user single-platform and this prohibits people on other computer environments from access. For example, Manufacturing Objects Processing System (Figure 1) depends on PC environment and cannot be moved onto the World Wide Web.

3. JAVA MANUFACURING OBJECT PROCESSING SYSTEM

This paper describes the development of an environment with significant enhancement in graphical and web-based delivery capabilities. The system, which is called J-MOPS (Java MOPS), is created using Java language. Java has advantages of robustness, object-oriented, network oriented, and platform independence [15]. J-MOPS integrates the design stage to the manufacturing activities in the NC machine, showing the examined functions and elements of the objects. J-MOPS is developed with an aim to be moved to the World Wide Web and hence it is independent of hardware or operating system, thereby providing the biggest possible usage in the modern global CIM era. J-MOPS incorporates the object-oriented philosophy of MOPS and extends its graphical functionality. Geometrical Primitive Objects (GPOBs) received from a CAD data system are extracted according to their geometrical characteristics and relationships. Then the extracted data is processed in an object-oriented approach into a detailed description of the next manufacturing processes and machine programs.

J-MOPS uses the latest Java3D [16] technology to implement its graphics modelling of the manufactured and raw material product in 3D display. The result is a completely new 3D network-enabled graphical computer aided process planning (CAPP) system, which can be accessed by any web browser. Using the interactive 3D viewing capability, users can rotate, zoom, move the objects on the screen as if they are held manually. The effect is a much better visualisation of the product being designed and hence eliminates as many design errors as possible.

It is obvious that the prototype and model of the actual real world plays an important role for helping decision making in industries. The approach of bringing the real objects in a three-dimensional modelling graphical display is one of the aims of the research. The other improvement is to move the framework into a more user-friendly application. Using the Java language, J-MOPS provides a platform independent system similar to MOPS which can be accessible from any place in the world using a web browser. The system is an expansion of an earlier version in FORTRAN and C [17]. In the last couple of years, latest software design technologies have been applied to improve the modularity and connectivity of the system [18,19].

The system uses the concept of three levels data abstraction, called Geometrical Primitive Object (GPOB). The GPOB is a universal set of objects that can interact with processing modules. The first level of GPOBs consists of objects with basic geometrical attributes. This information is retrieved from the CAD data model when it was loaded by the system. In the further processing, the CAD data models are extracted and assembled into the second level of GPOBs, consisting of attributes such as the information of coordinates, vectors, orientation, the relationships to other object, and the nature of solidness.

In actual implementation, the system consists of three main modules: Intelligent Data Extractor (IDE), Intelligent Process Planning (IPP), and Intelligent Machine Interface (IMI); and one additional module for three-dimensional graphical display. From the CAD system, which is the input to the system, until the CNC Part-Program result, which is the output of the system, the data are processed object by object. In the system, all of these modules are acting as objects themselves.

4. THE INTELLIGENT DATA EXTRACTOR

Intelligent Data Extractor (IDE) is the entry module to activate the system. Its aims are to extract CAD data models and to assemble them as GPOB information for the input of next module. The CAD data model, retrieved by IDE, supplies information for GPOBs level one: geometrical primitive type and spatial-relationships (Figure 2). Then, IDE extracts this information to produce GPOBs level two, such as the reference number, vertices and vectors, volume, orientation, spatial-relationships, and object's nature as solid or hollow object.

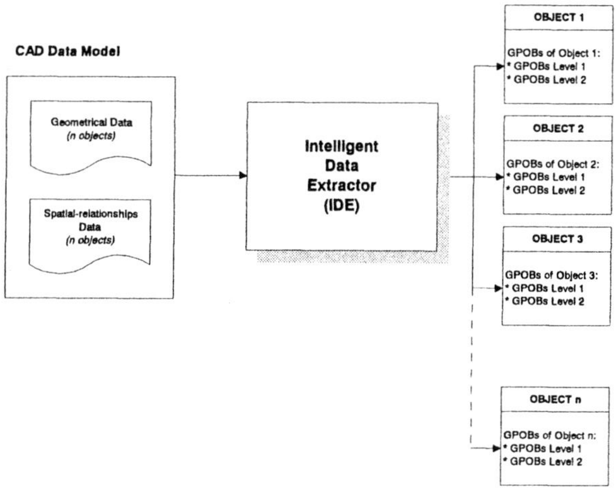

Figure 2: GPOBs in Intelligent Data Extractor.

The nature of the object, in term of its solidness, is determined by analysing chains among GPOBs. Using an algorithm developed by Liang et al [4], IDE determines the chains, which will be used to identify the solidness of the objects. The algorithm consists of iterations of chain list and utilises a series of postulates and reasoning to decide the characteristics of the objects in the product. All of the GPOBs, assembled by IDE, are used for further manipulation in the next modules. Thus, the information assembled by IDE has to present informative data for smooth transforming of the CAD model data to the actual manufacturing data.

5. THE INTELLIGENT PROCESS PLANNER

The second module of the system is called Intelligent Process Planner (IPP). This module receives and processes the GPOBs information from IDE in order to produce an operation sequence needed for the manufacturing activities. This sequence will be used for the guideline operation of CNC machine on the shop floor. IPP works in three stages of operation. The activities are to transform the information from IDE into material stock requirements and operation sequence in manufacturing.

IPP consists of several sub-modules in order to accomplish those three levels of operation. The first sub-module is the Information Interpreter. This sub-module is responsible for receiving any information coming as input for IPP and interpreting this information, so it is ready to be used and processed in other sub-modules. The second module, the Data Share Stock, is the central database for IPP. It is responsible for receiving and providing all the information from and to the sub-components in IPP. The third sub-module is the Material Generator. This sub-module works to accomplish the second stage of IPP operation. It consists of two components: Master Material Generator and Sub-Material Generator. These components have the role of generating base material and sub-materials, which are the required materials for manufacturing the part objects. There are algorithms and some rules in order to generate the base material and sub-materials.

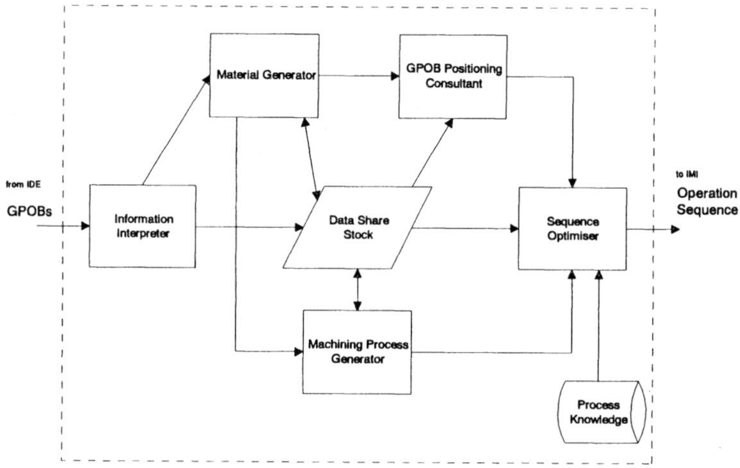

Figure 3: Sub-modules of IPP

Using the same philosophy of IDE, IPP also generates the base material and each of sub-materials in its own GPOB properties. The nature of the base material and sub-material is made as a hollow object. The reason is because the materials contain the manufactured part objects, which are solid objects.

The next two sub-modules, the Machining Process Generator and the Sequence Optimiser, are the important sub-modules for the last stage operation in IPP. The operation sequence for machining is generated by the

Machining Process Generator. In general, there are three considerations for generating the operation sequence.

The processes in the operation sequence are the removal action. It is possible because the GPOBs and sub-materials are objects. Then the sequence is passed to the Sequence Optimiser to be optimised. Based on the knowledge base and information from the GPOB Positioning Consultant sub-module, the sequence from the Machining Operation Identifier is updated.

6. INTELLIGENT MACHINE INTERFACE

Intelligent Machine Interface (IMI) is the third module in the system. This module generates part programs for machining processes on the shop floor. IMI consists of machine objects, which contain the knowledge base of controlling the manufacturing machine.

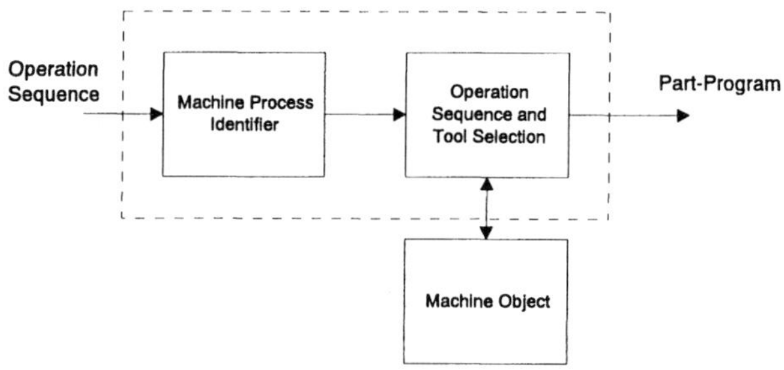

Figure 4: Sub-modules of IMI.

IMI has two sub-modules: Machine Process Identifier and Operation Sequence and Tool Selection. It also uses the knowledge base of controlling machines, which is located in the Machine Objects. The Machine Process Identifier identifies and groups each machining process information of the operation sequence, received from IPP, into some machining process categories, such as milling, drilling, and grinding, so that same category of machining processes can be sent to the right machine for manufacturing operation. However, in this system the machining process is still limited only to the CNC milling machine.

After that, the Operation Sequence and the Tool Selection will work to assign the machining process sequence and the tool selection for each of the machining process categories in every machine, using the knowledge base of machines provided by the machine object. Then, a part program is ready to be generated with the GPOBs as its referred information.

7. THE JAVA 3D GRAPHIC MODULE

Three-dimensional Graphic Display is the last module in the system. It presents the final manufactured parts and the raw materials as three-dimensional graphical objects and interacts with the user to give the best viewing. The module consists of three sub-modules: Object Recogniser, Graphical Object Generator, and Appearance Properties. The module draws the final manufactured parts using the GPOBs from the output of IDE, whereas the raw materials will be drawn according to their GPOBs received from IPP. The Object Recogniser receives the GPOBs and distinguishes the object to produce necessary information, such as: material or part, hollow or solid object, cube or cylinder, etc.

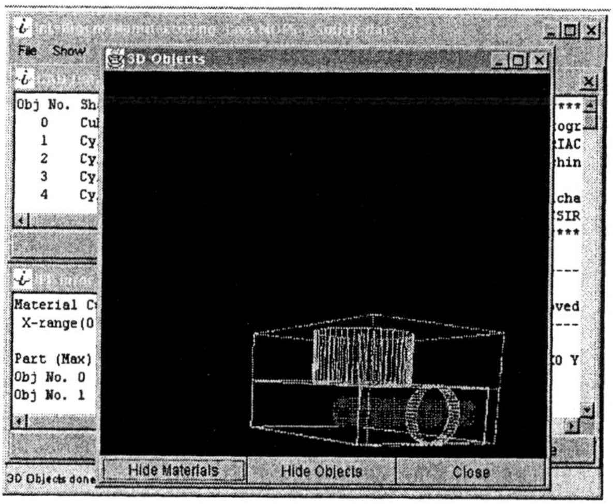

Figure 5: 3D Display in wire-frame mode.

The module works in object-oriented philosophy, which means that every object is drawn individually as an object. The object is created using a suitable class, chosen according to its characteristic from the Graphical Primitive Objects. The third sub-module is Appearance Properties. This sub-module supplies the appearance properties such as colour and drawing mode to the Graphical Object Generator, so it can generate the object in particular appearance according to the information received from Object Recogniser.

Using the inherent capability of Java 3D, the system can be easily programmed to display the objects in wire-frame mode (Figure 5) or rendered mode (Figure 6).

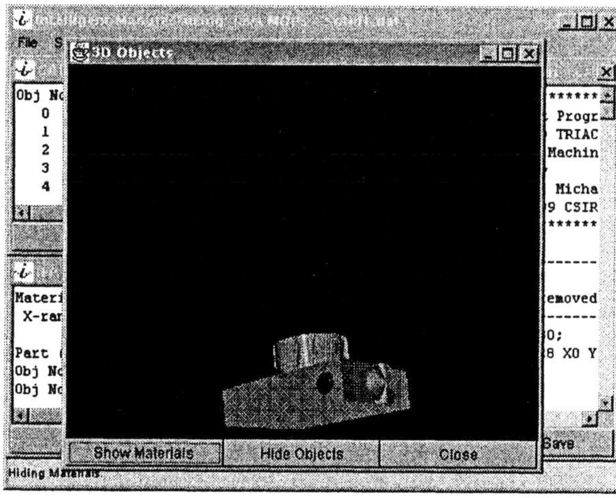

Figure 6: Display in rendered mode.

8. SUMMARY

J-MOPS is developed based on MOPS philosophy. It represents a major advancement of offering the complete CAPP and 3D simulation capabilities on the World Wide Web. It simplifies the system requirements for the user and removes the dependency on platforms. The results of this study provide the basis for supporting design to be done intelligently on a global basis.

REFERENCES

1. Chakravarty A.K. Dimensions of Manufacturing Automation, Int. J. Prod. Res., Vol.25, No. 9, 1987, pp.1339-1354
2. Harhalakis G., Lin C.P., Hillion H., Moy K.Y., Development of a Factory Level CIM Model, *The Journal of Manufacturing Systems*, Vol.9, No. 2, 1990, pp.116-128
3. Lin A.C., Wang H.P.B. The Development of a Computer-Integrated Design and Manufacture System for Plate Cams, *Manufacturing Review*, Vol.1, No.3, 1988, October, pp.198-213
4. Subrahmanyam S., Wozny M., An Overview of Automatic Feature Recognition Techniques for Computer-aided Process Planning, *Computers in Industry*, Vol.26, 1995, pp.1-21
5. Liu C.H., Perng D.B., Chen Z., Automatic Form Feature Recognition and 3D Part Reconstruction From 2D CAD Data, *Computers & Industrial Engineering*, Vol.26, No. 4, 1994, pp.689-707
6. Ainger A.W.S., Schmid F., The Helical Approach to Software Design, *Journal of Computer Integrated Manufacturing Systems*, Vol.8, No. 2, 1995, May, pp.105-115

7. Troll M., Brandon J.A.,Vollmer E., Object-oriented Methodology for Optimizing Information Flow in Manufacturing, *International Journal of Flexible Automation and Integrated Manufacturing*, Vol.4, No. 3&4, 1996, pp.137-153
8. Westfechtel B., Integrated Product and Process Management for Engineering Design Applications, *Integrated Computer-Aided Engineering*, Vol.3, No. 1,1996, pp.20-35
9. Usher J.M., A STEP-Based Object-Oriented Product Model for Process Planning, *Computers and Industrial Engineering*, Vol.31, No. 1/2, 1996, pp.185-188
10. Mo J.P.T., Fernandez G., Liang G., "An Object Oriented Framework for Intelligent Manufacturing from Computer Aided Design Features", ETCE '93, Computers in Engineering Symposium, ASME, 31 January, 1993, Houston, USA
11. Mo, J.P.T. and Ranganathan, D., From Design to Manufacture by Objects, *International Journal of Flexible Automation and Integrated Manufacturing*, 4(2): 95-110, 1996.
12. Nemes L., Bertok P., Cooperative Design on the Internet,IFIP World Congress on IT Tools, 2-6 September, 1996, IFIP - International Federation for Information Processing, Canberra, Australia, pp.419-428
13. Bertok P., Mo J.P.T., "Collaboration and Application Integration: Distributed Design with Virtual CAD", 4th International Conference on Design of Information Infrastructure Systems for Manufacturing, 15-17 November, 2000, Melbourne
14. Jiang H.C., Mo. J.P.T., "Computer Aided Design over the Internet", 26th International Conference on Computers and Industrial Engineering, 15-17 December, 1999, Melbourne, Australia, Vol.1, pp.497-501
15. Au, E. and Makower, D., Java Programming Basics, MIS Press, New York, 1996.
16. Sowizral, H., Rushforth, K. and Deering M., The Java 3D API Specification, Sun Microsystems, California, 1998.
17. Liang, G., Fernandez, G. and Mo, J.P.T., Intelligent CAD Data Recognition for an Object-Oriented CAD/CAM System, Proceedings of International Conference on Data and Knowledge Systems for Manufacturing and Engineering, Hong Kong, 2-4, 7-12 May 1994.
18. Ooi S.T., Bertok P., Zhao L.P., Jiang H.C., Mo J.P.T., "Patterns as Reengineering Front-End in Distributed Environments", *Proceedings of 2nd Australian Workshop on Software Architectures*, 23 November, 1999, Melbourne, Australia, pp.75-82
19. Bertok P., Mo J.P.T., "Software Technology for Design System Integration", 4th International Conference on Design of Information Infrastructure Systems for Manufacturing, 15-17 November, 2000, Melbourne

30

IMPLEMENTATION OF A DATA GATHERING SYSTEM WITH SCALABLE INTELLIGENT CONTROL ARCHITECTURE

Masayuki Takata[1] and Eiji Arai[2]
[1] *Information Processing Center, The Univ. of Electro-Communications, Japan*
[2]*Dept. of Manufacturing Science, Graduate School of Eng., Osaka Univ., Japan*
e-mail: takata@cc.uec.ac.jp

Abstract: This paper describes a system named "Glue Logic", which is a infrastructural system designed for factory automation control system, and a sample implementation of a layer structured control system architecture, which is named "Scalable Intelligent Control Architecture" reported in the DIISM2000. The "Glue Logic" supports the real-time controlling and monitoring system, by means of realizing communication and synchronizing among multiple agents. Using the active database technique, this system includes event notification message sending and condition monitoring features to eliminate data polling. This system also supports efficient programming environment, by increasing modularity and reusability of the software assets. Using the Glue Logic, we are now designing a real-time data gathering system in practical manufacturing lines according to the Scalable Intelligent Control Architecture, which permits expansion of control systems not only in spatial dimension but also in intelligence.

Key words: Factory Automation architecture, Manufacturing work-cell control system, Infrastructural Software System, Distributed Programming / Execution Environment

1. INTRODUCTION

The system named ``Glue Logic" is an infrastructural system which is designed to make building manufacturing work-cell control systems easy and flexible[1,2]. This system binds multiple application software modules, referred as ``agents", and coordinates those agents by means of inter-process massage passing. As the Glue Logic supports event notification and condition monitoring features based on active database scheme, users can easily build real-time event-driven application agents, waiting for notification messages from the Glue Logic.

As all of the data and agents in a system are abstracted, and are handled with symbolic names defined in the Glue Logic, agents can be built without any knowledge on implementation of others. As the result, the Glue Logic compliant agents are easy to re-use, and the users can build up large libraries of application agents. As some agents having rather general purpose may be shared among users, and the software development cost is greatly reduced.

This paper describes the Glue Logic designed as the infrastructural system for factory automation applications, and also a sample implementation of the real-time intelligent control system. In Section 2, description of the Glue Logic is discussed. Section 3 discusses ``Scalable Intelligent Control Architecture (SICA)," which is designed to fit multi-agent control systems, and required extension in order to implement single control system with multiple server processes of the Glue Logic. Lastly, in Section 4, the architecture of the coming multi-agent real-time control systems are discussed, and the extended Glue Logic is evaluated.

2. THE GLUE LOGIC

The Glue Logic is the minimal set of functionality to coordinate multiple real-time application agents, which consists of an active database and message passing mechanism. There are some other "middleware" systems alike, such as CORBA, but we think those systems are too complex to implement real-time embedded applications with compactness.

2.1 Architecture

The Glue Logic has been developed to support application programming by means of data sharing, event notification and condition monitoring. As the system uses inter-process communication internally over the network, the Glue Logic can play the roles of the infrastructure of the distributed manufacturing work-cell control systems. This makes development and maintenance of the event driven application easier.

The Glue Logic relays all inter-process communication among its agents, and manages all data shared by those. Because of this, the Glue Logic can send the change notification messages, when the values of the shared data are altered. As the virtualizing the counterpart of the communication can be achieved by relaying all of the inter-agent communication, each agent can be independent from adding, deleting and altering other agents.

2.2 Overall Implementation

In the first implementation of the Glue Logic, its design is based on the client-server model of transaction processing, as shown in Figure 1, though there is no need for the users to know about its implementation. All agents' processes communicate only with this server process over network, and there is no redundancy in this system.

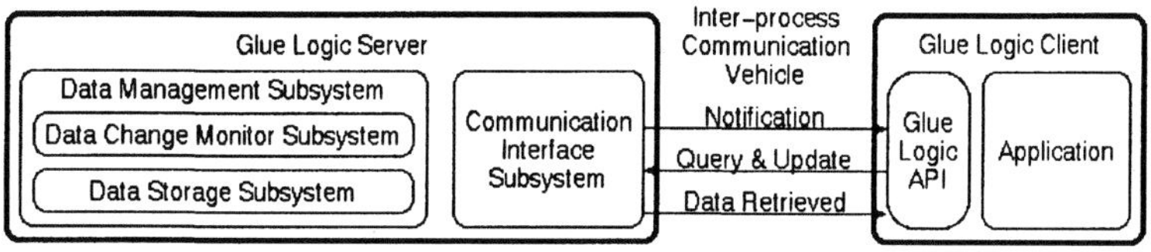

Figure 1. Configuration of the Glue Logic.

The Glue Logic consists of two major parts: the communication interface subsystem and the data management subsystem. The communication interface exchanges information with agents running in both the same work-cell controller and remote controllers connected with the network system.

The data management subsystem consists of also two parts: the data change monitor subsystem and the data storage subsystem. The data storage subsystem manages the association pair of the *name* and the *value* of the object. The data change monitor subsystem monitors the changes in the data storage subsystem and sends out the data change notification messages, and executes depending data evaluation.

2.3 Data Handling of the Glue Logic

The atomic data element of the Glue Logic is the tuple of a *name* and its *value*. The *name* resembles variable identifier, and can have a value. The name is a sequence of identifiers, separated by a period, such as abc.ijk.xyz. Using this format, the agent programmers can implement arbitrary data structures. For all elements of one structure, their names contain same identifier sequence in their leading part, but trailing part of names differs from each other. The common part is called a *stem* and the trailing part is called a *variant*.

As the value of the name, application programs may specify one of followings; integer, floating point real, character string, expression and link.

As the name itself is not typed, users may bind any types of data in turns. If an agent accesses the name bounded to an expression value, the expression is evaluated and the result is returned as the value. Using the link, users can point another name. Each name may have some *attributes*. The attributes denote optional characteristics of corresponding names, and the Glue Logic changes its behavior according to the values of attributes.

2.4 Interlocking Features

To share resources safely among multiple agents, the Glue Logic has the feature of interlocking operation. Single transaction from an agent to Glue Logic server process is processed indivisible way, and it may include multiple accesses, for which three access methods are prepared; read, write and compare. In general, the transaction of the Glue Logic may contain arbitrary number of accesses, and there is no restriction on order of accesses.

The transaction for interlocking compares the value of name with the value shown in the transaction, and assigns another value shown in the transaction to the name if the compared data are identical. This transaction provides means of updating shared data. When one transaction includes more than two sets of comparison and conditional assignment, users can maintain a linked list safely from insertion and deletion.

2.5 Active Database Features

In order to eliminate the needs of data polling and to decrease the network load, the feature of data change notification is introduced. The agents, which want to receive a change notification of a certain name's value, can register the ID of the agent itself to the name. The agent ID list of the notification destination is kept as the value of InformTo attribute. On the time when the Glue Logic server system receives data update request, the system searches for the agents registered as the notification destination, and then notify the fact of change to all the registered agents.

Some agents may need to know the value of name being a certain constant value, or the values of names satisfy a certain condition. The Glue Logic can be set to send a notification message only if a certain condition is met. As shown in Figure 2, each name in the Glue Logic can have dependence lists as the values of *Triggers* and *TriggeredBy* attributes. If one or more elements of the list in the some name's *TriggeredBy* attribute is updated, the value of the name itself is updated to have the result of an expression, which is also registered as the value of *IfTriggered* attribute. If this new value differs from the former, the data change notification is sent to its notification destinations.

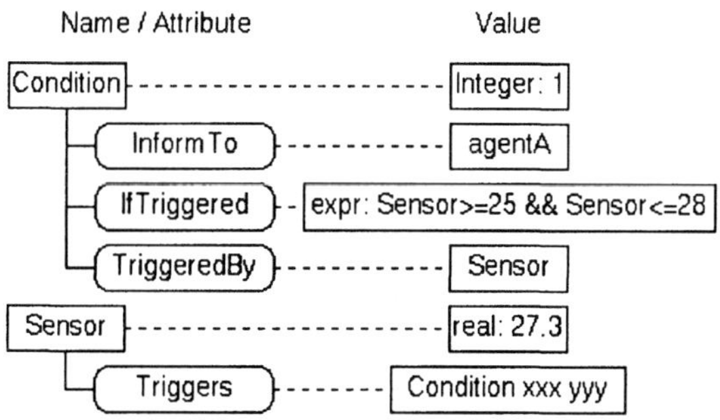

Figure 2. Data used by the condition monitoring.

3. SCALABLE INTELLIGENT CONTROL ARCHITECTURE

3.1 Overview

The control systems ever built with the Glue Logic are implemented in accordance with the ``Subsumption Architecture" proposed by Brooks[3]. With his original architecture, it is very easy to add-on much more intelligent controlling and planning abilities, by integrating much more higher level layer which monitors status of lower layer and emits directive information.

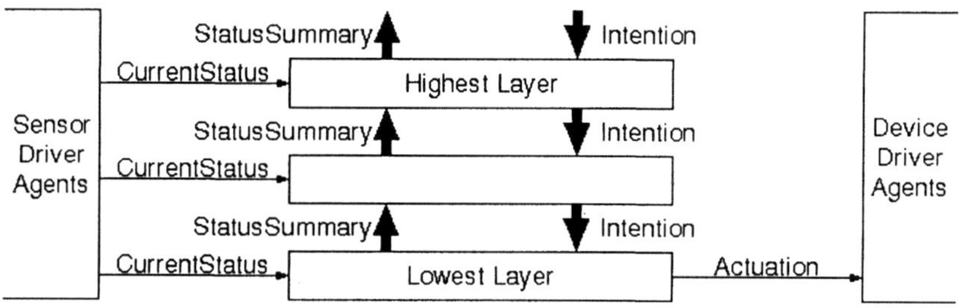

Figure 3. The Scalable Intelligent Control Architecture

The authors have modified this architecture to fit it to multi-agent control systems. In authors' ``Scalable Intelligent Control Architecture (SICA)" illustrated in Figure 3, which permits expandability not only on the breadth of system to be controlled but also on the abstractness of the controlling aim. Each layer consists of one or more agents, and their communications are supported by message exchange feature of the Glue Logic.

In our SICA paradigm, lower layers send messages to next higher ones to show the current ``summarized'' or ``abstracted'' status of the controlled system. On the other hand, higher layers send messages to next lower to show the ``intention'' of the control or ``targeting goal state,'' not to show how the lower layer to behave.

3.2 Adapting the Glue Logic to the Scalable Intelligent Control Architecture

In order to make the control systems scalable, it is necessary for the Glue Logic to have capability of data locality control. In the Scalable Intelligent Control Architecture, detail data are used frequently in lower layers, and such data are accessed from limited number of agents within those layers. This is the locality of data, and it is better for the Glue Logic to separate data into multiple Glue Logic server processes.

3.2.1 The Name Space

To achieve this aim, the naming convention of the Glue Logic should be extended to control locality of the data. Multiple Glue Logic server processes consisting the unique control system must share same name space, and the name of the data should imply the locality of the data, and also imply the place (i.e. process) the data is kept.

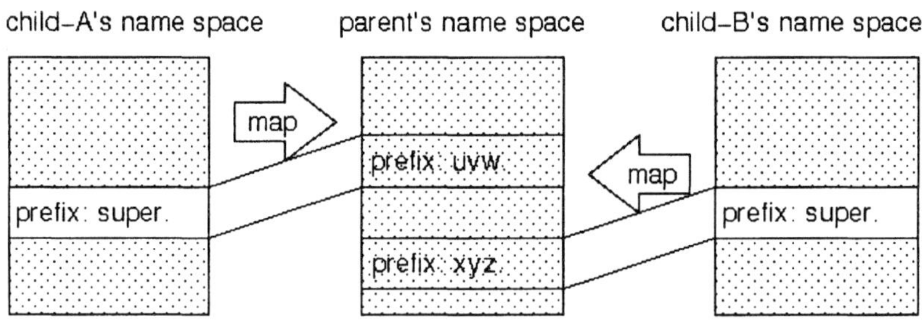

Figure 4. Name Spaces of Parent and Two Child Server Processes.

In the current implementation of the Glue Logic, we structured server processes into tree structure. The names begin with string super. show that the data with those names reside in the remote Glue Logic server process, and other names show they are not. This remote server process is called ``super'' server or ``parent'' server. In the name space of a parent server, each child server is associated to each name prefix, named ``anchor''. The data with those name prefixes are kept in the parent server itself, but are also accessible from one of child server process of the parent server, which is associated to the name prefix. This relation is illustrated in Figure 4.

3.2.2 Communication Among Glue Logic Server Processes

With this naming convention, no server process can communicate with its brother or sister server process directly. To achieve this, a relay agent attached to the parent server process must copy information between two data area resides in the parent server itself. Such an agent is overhead for this information system, but it frees other application agents from knowing of their brothers and sisters.

Alike this, a relay agent is also required to communicate between one server process and its grandparent server process. In this case, as there is two levels of difference, the characteristics of data are also different. The data to be handled by agents in the higher layer should be summarized, in order to reduce data volume and to make those data more abstracted.

So the relay agent which relays ``vertically upward" from one server process to its grandparent server process must have summarizing feature. In the reverse direction, relaying data ``vertically downward" from one to its grandchild, a relay agent should have some functionality which divides one intention into a set of multiple sub-intentions. In many cases, this process is called ``goal deduction," and needs intelligence to solve complex problem.

3.2.3 Inter-Layer Interlocking

As the implementation of interlocking operation is based on the fact that one transaction is indivisible, there may be some troubles if the names to be interlocked resides in two Glue Logic server processes.

To keep consistent operations, we introduced some restrictions on multiple ``comparison and assignment" sets within one transaction. Thought the restriction rules are now under evaluation, the current restriction set on the order of comparison and assignment for the names implementing interlocking includes; all assignment should come after all of comparison, all comparison of name with super. prefix should come after all comparison of name without super. prefix, and all assignment of name with super. prefix should come before all assignment of name without super. prefix.

3.3 Implementation Example

Accordance with the Scalable Intelligent Control Architecture, we are designing a data gathering system. This system gathers point-of-production data and process equipment condition data from multiple remote production facilities. Processed data can be monitored at multiple sites, and warning messages are given to the operators. This system consists of two layers, lower one is for collecting data and warning to manufacturing floor when unusual data is found, and upper one is for logging and warning to management office when abnormal data is found.

The upper layer has its own Glue Logic server process, while lower one has multiple Glue Logic server processes each of which are associated to physical manufacturing lines or cells. The lower layer gathers raw data continuously, and each data is checked against its own control limit.

4. CONCLUSION

In this paper, the new implementation of the Glue Logic and Scalable Intelligent Control Architecture are described. The authors believe in the effectiveness of the concept of infrastructural system for control agents, and the paradigm of the scalable intelligence, through some application implementation.

To develop flexible manufacturing system control software, it should take less cost, less time and more reliability, as one may have to develop a new manufacturing control software for producing only one instance. The library which consists of widely used agents is strongly required for this requirement, and the concept of our control architecture is one of the most efficient way to grow manufacturing systems up step-wisely.

The authors would like to emphasize that the smart mechanism of the Glue Logic is the very thing to make the control system powerful, intelligent, and easy to be programmed.

REFERENCES

1. Takata, M., Arai, E. (2000): ``Implementation of a Layer Structured Control System on the "Glue Logic"", *Global Engineering, Manufacturing and Enterprise Networks (Ed. Mo, J., Nemes, L.), Proc. of DIISM 2000*, pp.488 - 496, Kluwer Academic Publishers on behalf of IFIP, 2001.
2. Takata, M., Arai, E. (1997): ``The Glue Logic: An Information Infrastructure System for Distributed Manufacturing Control," *Proc. of the Int'l Conf. on Manufacturing Milestones toward the 21st Century*, pp.549 - 554, Tokyo, Japan, July 23-25, 1997.
3. Brooks, R. A. (1986): ``A Robust Layered Control System For A Mobile Robot," *IEEE Journal of Robotics and Automation*, Vol. RA-2, No. 1, 1986

31

CREATION OF FEATURE SETS FOR DEVELOPING INTEGRATED PROCESS PLANNING SYSTEM

Hendry Muljadi, Koichi Ando, and Makoto Ogawa
Graduate School of Eng., Shibaura Inst. of Technology, Japan
e-mail: m701102@sic.shibaura-it.ac.jp

Abstract: Our research puts its goal in the generation of a computer-aided process planning system for machining process in the dynamic manufacturing stage. For this goal, we develop a Feature Sets Creator that can lead to the generation of multiple process plans. The generation of multiple process plans of a part is intended to provide alternative plans in the determination of optimal sets of process plans for all the part types in the shop, and to provide flexibility in the shop floor scheduling. The Feature Sets Creator is divided into 2 phases. First, by using Modified Super Relation Graph Method, we recognize interacting and overlapping features in several different ways, corresponding with different kind of machining operations. The second phase is feature organization phase. Here, we organize recognized features into multiple feature sets. In the paper, we will present a case study to confirm the validity of the feature set creator.

Key words: Feature Recognition, Feature Organization, Process Planning System.

1. INTRODUCTION

Dynamic changes such as increased production, machine breakdowns etc are ordinary occurrence in the manufacturing stage. Skilled workers deal with these dynamic changes by changing the manufacturing schedules, or even by changing the process plans. Thus, we presupposed the need to integrate process planning, scheduling and manufacturing activities. Our research puts its goal in the generation of a CAPP system that can integrate process planning, scheduling and manufacturing activities [1]. Figure 1 shows the overview of our proposed process planning system. The system

consists of 3 steps.
Step 1: feature sets creation from the product design data (CAD data).
Step 2: generation of process plan of a part based on the created feature sets.
Step 3: determination of optimal set of process plans for product mix.

The optimal set of process plans obtained in Step 3 is used for the shop floor scheduling. During the shop floor monitoring, re-scheduling may be occurred to handle the dynamic changes in the manufacturing stage. In the re-scheduling stage, we can return to Step 3 to determine of the optimal set of process plans for the present shop floor or production planning condition.

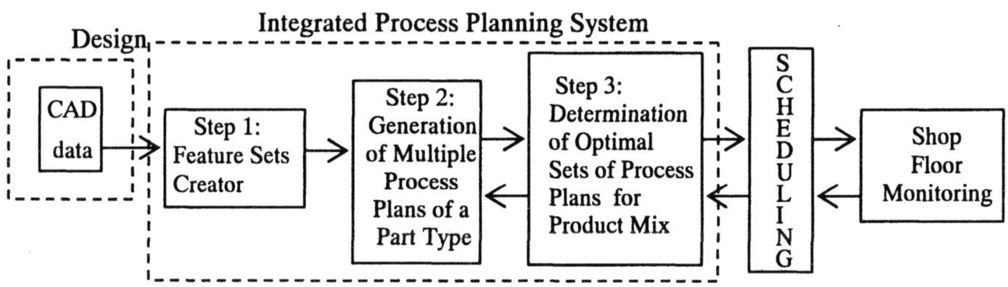

Figure 1. Overview of Integrated Computer-Aided Process Planning System

In order to bring this integrated process planning system to realization, we develop a Feature Sets Creator that can lead to the generation of multiple process plans. The Feature Sets Creator consists of 2 phases: Feature Recognition Phase and Feature Organization Phase. For the Feature Recognition Phase, we implemented Super Relation Graph method [2]. In SRG method, B-Rep product data structure is converted into graph structures, and by extracting feature representing graph patterns from the graph structures, features are extracted. In general, the recognition of interacting and overlapping features cannot be handled well with graph-based approach [3], but the SRG method is capable of recognizing interacting or overlapping features in several different ways, corresponding with different kind of machining operations. However, the proposed SRG method was developed only to deal with features constructed by plane faces. We proposed Modified SRG method to deal with curve faces. We will describe the Modified SRG in section 2. After features have been recognized by the Modified SRG method, features are organized into one or more features sets. The idea of organizing features into feature sets came from the fact that there are interacting features which are automatically machined by machining the other features. For this Feature Organization phase, we use our proposed Feature Spatial Relation Method. We will describe the Feature Organization phase in section 3. In section 4, we will use a case study to show the validity of our Feature Sets Creator.

2. MODIFIED SRG METHOD

2.1 Modifying the Super Relation Graph Method

In SRG Method [2], feature recognition is made possible by using two relations between faces, super-concavity relation and face-to-face relation. Super-concavity relation and face-to-face relation can be defined by Equation (1) and Equation (2) respectively.

$$n^+_{f_i}.n^+_{f_j} \neq -1;\ f_i \text{ n } S(f_j)^{|+|} \neq \varnothing \ \text{ and } f_j \text{ n } S(f_i)^{|+|} \neq \varnothing \quad (1)$$

$$n^+_{f_i}.n^+_{f_j} = -1;\ f_i \subset S(f_j)^{|+|} \text{ and } f_j \subset S(f_i)^{|+|} \quad (2)$$

where $n^+_{f_i}$ is the positive face normal of face f_i (figure 2(a)), and the strict positive half space of face f_i, $S(f_i)^{|+|} = \{x |\, n^{+\,T}_{f_i} x > k\}$ is the positive half space which exclude the embedding plane of face f_i, $P(f_i) = \{x |\, n^{+\,T}_{f_i} x = k\}$ (figure 2(b),(c)). $n^+_{f_j}$ and $S(f_j)^{|+|}$ are defined similarly as above.

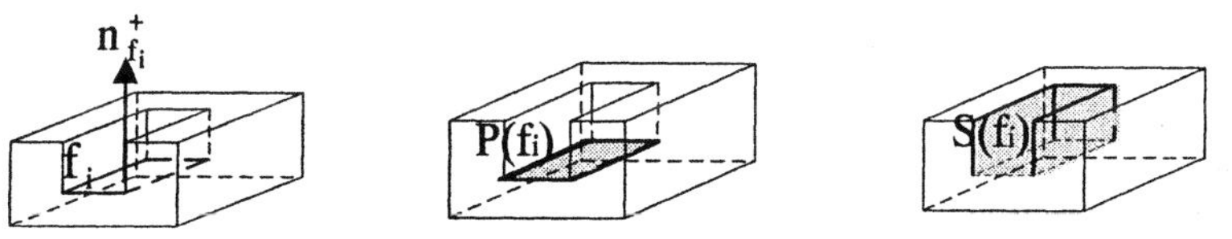

(a) Positive Face Normal of Face fi (b) The Embedding Plane of Face fi (c) Strict Positive Half Space of Face fi

Figure 2. Explanation of terms used in SRG

Figure 3 shows the example of the super-concavity relation and face-to-face relation. In general, the above two relations between faces are called super relations. Using these two relations, features can be defined using super relation graphs (SRGs). A SRG is an undirected graph, where the node in the SRG corresponds to a face in the feature, and two types of links are used to represent relations between two faces. Solid links are used to represent super-concavity relations and dotted links to represent face-to-face relations. But a problem with the SRG method is that geometrically and topologically different kind of features may have one same SRG, if the node is used to correspond to both plane and curve faces. Figure 4 shows that a step feature and a partial round notch feature are represented by same SRG.

In our research, we modified this SRG method. We divide the usage of node for plane faces and curve faces. In the Modified SRG method, a circle node in the SRG corresponds to a plane face in the feature, and a double circle node corresponds to a curve faces. Figure 5 shows the SRG of Partial Round Feature using the Modified SRG method.

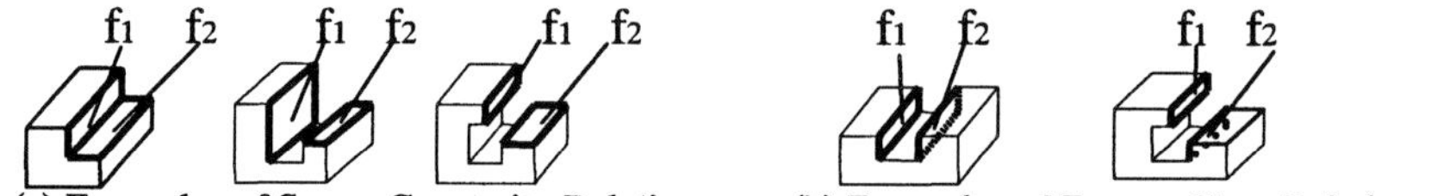

(a) Examples of Super-Concavity Relation (b) Examples of Face-to-Face Relations

Figure 3. Example of Super Relations

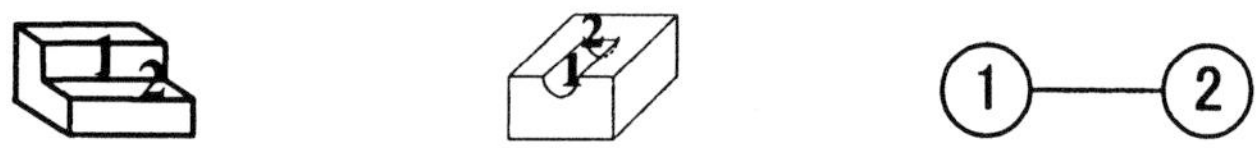

Step Feature Partial Round Feature (b) SRG of features in (a)

(a) Features

Figure 4. Example of features represented by same SRG

Figure 5. Partial Round Feature and its SRG in the Modified SRG Method

2.2 Procedure for Feature Recognition

The procedure for feature recognition using the Modified SRG method is as follows.

Step 1: Obtaining the Envelope Face Set. Envelope face set is the set of faces on the envelope of the object. An envelope is defined as the convex hull of an object and it should consist of primitive shapes such as cubes, parallelepipeds, cylinders, cones or prism. Figure 6(b) shows the envelope faces of the sample part in figure 6(a). Thus, the envelope face set is ENV= $\{f_8, f_9, f_{10}, f_{11}, f_{12}, f_{13}, f_{14}\}$.

Step 2: Constructing the Global Graph. A global graph is a face adjacency graph that shows the neighbourhood relationship of the faces of the part. Nodes of the graph represent the faces, and arcs between the nodes represent the neighbourhood relationship between the faces. Figure 6(c) shows the global graph of the part in figure 6(a).

Step 3: Obtaining Interacting Feature Face Set (IFFS). IFFS is a set of faces that surround the removal volume of the part. IFFS is obtained by constructing interacting feature face graphs, which are constructed by removing the nodes belonging to the ENV and the links incident to those nodes from the global graph. An IFFS can be obtained by grouping all the nodes in one IFFG. Figure 6(d) shows the IFFGs of the part in figure 6(a). Thus, the IFFSs of the part are $IFFS_1=\{f_1, f_2\}$ and $IFFS_2=\{f_3, f_4, f_5, f_6, f_7\}$.

Step 4: Generating SRGs. For each IFFS obtained in Step 3, its SRG consists of the nodes corresponding to the elements of the IFFS, and the links representing either super-concavity relations or face-to-face relations. Figure 7 shows the SRGs generated from each IFFS.

Step 5: Extracting Features. Feature extraction is done by extracting the feature representing SRG from SRG of each IFFS. Figure 8(a) shows that step feature representing SRG is extracted from SRG of $IFFS_1$. Extracted feature representing SRGs from SRG of $IFFS_2$ are shown in figure 8.

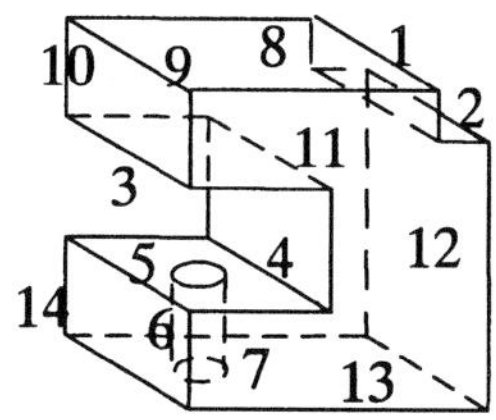

(a) A Sample Part

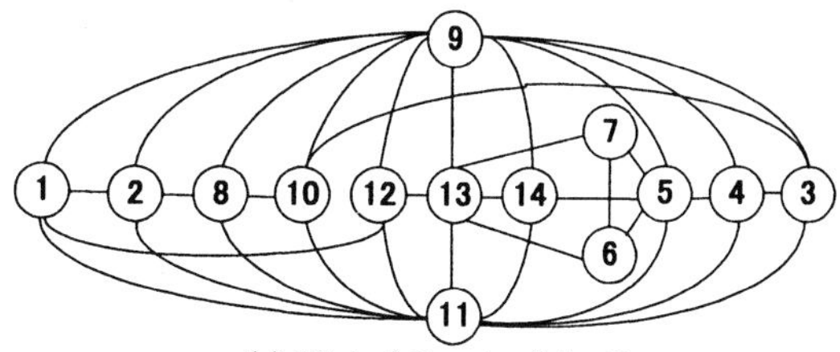

(c) Global Graph of the Part

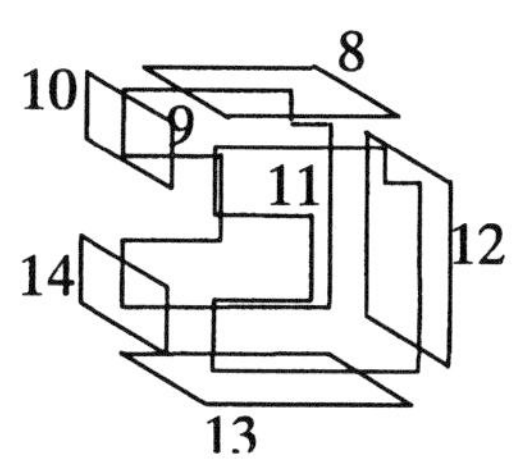

ENV: $\{f_8, f_9, f_{10}, f_{11}, f_{12}, f_{13}, f_{14}\}$

(b) Envelope Face Set of the Part

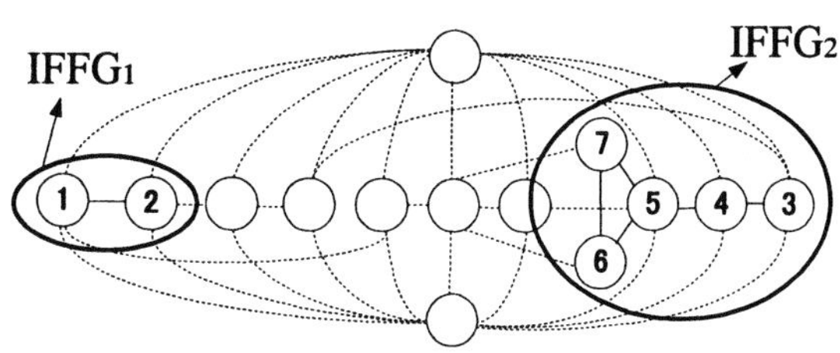

(d) the interacting feature face graphs of the part, $IFFG_1$ and $IFFG_2$, obtained from Global Graph by deleting the nodes of Envelope Face Set and links incident to those nodes

Figure 6. Example of Global Graph, Envelope Face Set and Interacting Feature Face Graph of a Part

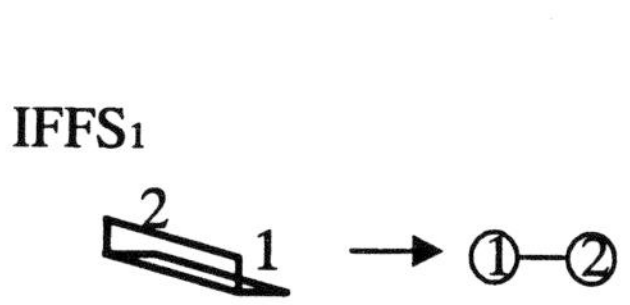

(a) Super Relation Graph of $IFFS_1$

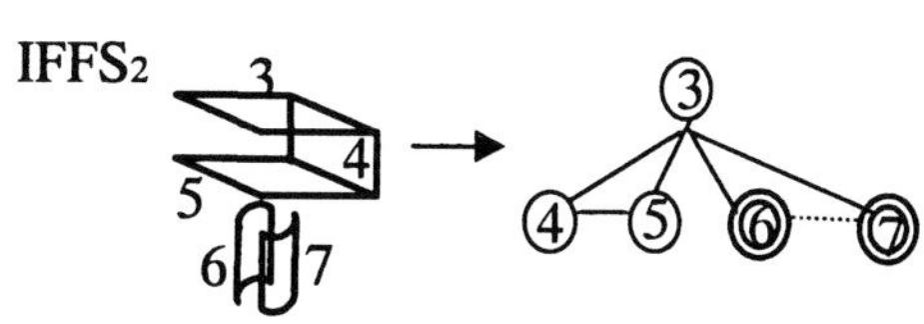

(b) Super Relation Graph of $IFFS_2$

Figure 7. Super Relation Graphs of the Sample Part using Modified SRG Method

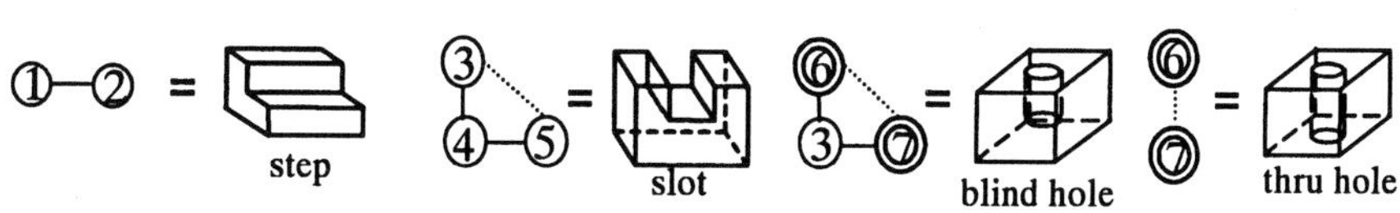

(a) Step feature representing SRG extracted from SRG of $IFFS_1$

(b) Feature (slot, blind hole, thru hole) representing SRGs extracted from SRG of $IFFS_2$

Figure 8. Extracted feature representing SRG from SRG of each IFFS of the part

3. FEATURE ORGANIZATION

3.1 Machining Feature

As we consider the interaction relation of features recognized by the Modified SRG method, there are 3 types of interaction need to be considered.
1. Features that have no interaction with other feature(s). Step feature shown in figure 9(a) has no interaction relation with other features. This feature can be called independent feature.
2. Features that interact other feature(s) partially. Slot feature shown in figure 9(b) and blind hole feature shown in figure 9(c) are interacting with each other partially. Both features can be called interacting features.
3. Features that are covered by other feature(s) completely. This type of interaction shows the existence of overlapped feature. Thru hole feature shown in figure 9(d) is covered by blind hole feature in figure 9(c). Also, blind hole feature in figure 9(c) is covered by slot feature in figure 9(b) and thru hole feature in figure 9(d). In each cases, thru hole feature and blind hole feature can be called overlapped feature.

In the machining point of view, when features that cover an overlapped feature being machined, the overlapped feature will be automatically machined. It means that overlapped feature should not be considered as a machining feature. Thus, only independent and interacting features that can be called machining features. We proposed Feature Spatial Relation (FSR) method to organize machining features into feature sets that have no overlapped features.

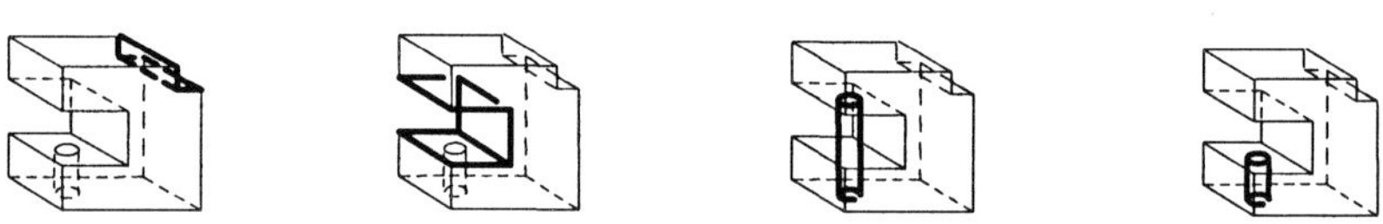

(a) Step Feature (b) Slot Feature (c) Blind Hole Feature (d) Thru Hole Feature

Figure 9. Extracted Feature of the Part in figure 6(a)

3.2 Feature Spatial Relation Method

The FSR method can be represented as follows. Let $t_i \in T_x(IFFS)$, where $T_x(IFFS)$ denotes the feature set in IFFS, t_i be the element of T. Let $S(t_i)$be the space for feature t_i to occupy. Feature set $T_x(IFFS)$ is derived from the equation below.

$$T_x(IFFS)=\{t_i \mid \bigcup_{i\in N} S(t_i) \supseteq S(IFFS),\ \bigcup_{i\in N} S(t_i)\cup S(t_{i+1}) \not\subset \bigcup_{i\in N} S(t_i),\ \bigcup_{i\in N} S(t_i)\cup S(t_{i+1}) \not\subset S(t_{i+1})\} \qquad (3)$$

where N={1,⋯, n}, and n is the total number of features in IFFS.

3.3 Algorithm of FSR Method

The algorithm of FSR method to obtain T_x(IFFS) is as follows:

Step 1. Form a list of features in a IFFS;
Let n = number of features in a IFFS; x =1;
If n > 2, then go to step 2; Else T_x(IFFS) = {features}; End

Step 2. Initialize k = 2

Step 3 Initialize T_x(IFFS) = {t_1}, i = 1

Step 4. While i + 1 < n {
For $S(t_i) \cap S(t_{i+1}) = \varnothing$ {
$S(t_i) \leftarrow S(t_i) \cup S(t_{i+1})$; Add t_{i+1} into T_x(IFFS); i + 1 ← i + 2; }
For $S(t_i) \cap S(t_{i+1}) \subset S(t_i)$ { i + 1 ← i + 2; }
For $S(t_i) \cap S(t_{i+1}) \subset S(t_{i+1})$ {
T_x(IFFS) ={t_{i+1}}; $S(t_i) \leftarrow S(t_{i+1})$; i + 1 ← i + 2; } }

Step 5: List all features in T_x(IFFS);
let y = number of features in T_x(IFFS); G ←T_x(IFFS); j =1;
While j < y {
For $S(t_j) \cap S(t_{j+1}) \subset S(t_{j+1})$ { Delete t_j from G; j ← j + 1 }
For $S(t_j) \cap S(t_{j+1}) \subset S(t_j)$ { Delete t_{j+1} from G;
If j + 1 < y , j + 1← j + 2; Else j ← j + 1 }
For $S(t_j) \cap S(t_{j+1}) = \varnothing$ {If j + 1 < y , j + 1← j + 2; Else j ← j + 1 } }
T_x(IFFS) ← G;

Step 6: If k = n, End;
Else {for i = 1 , do i←n+1; i←i-1; k=k+1; x=x+1; Repeat Step 3;}

As we apply the above algorithm to the recognized features in figure 9, two feature sets obtained from the IFFS: T_1(IFFS)={slot, blind hole} and T_2(IFFS)={slot, thru hole}. And as each feature set will lead to a different process plan, it shows that the combination of Modified SRG method and FSR method is useful for the generation of multiple process plans of a part.

4 CASE STUDY

Using another sample part, as shown in figure 10(a), we confirm the effectiveness of our feature set creator. Figure 10(b) shows the IFFS of the sample part. Using the Modified SRG method, 9 features are recognized: t_1 is open pocket feature. t_2, t_3 are blind holes. t_4, t_5, t_6, t_7 are thru-holes (see figure 10(c)). And using the Feature Spatial Relation Method, 5 feature sets are constructed: Feature Set T_1={t_1, t_2, t_3}, Feature Set T_2={t_1, t_3, t_4}, Feature

Set $T_3=\{t_1, t_5\}$, Feature Set $T_4=\{t_1, t_6\}$ and Feature Set $T_5 = \{t_1, t_2, t_7\}$. Each feature set will require a different kind of process plans. Thus, it shows that our Feature Set Creator is effective for making multiple process plans.

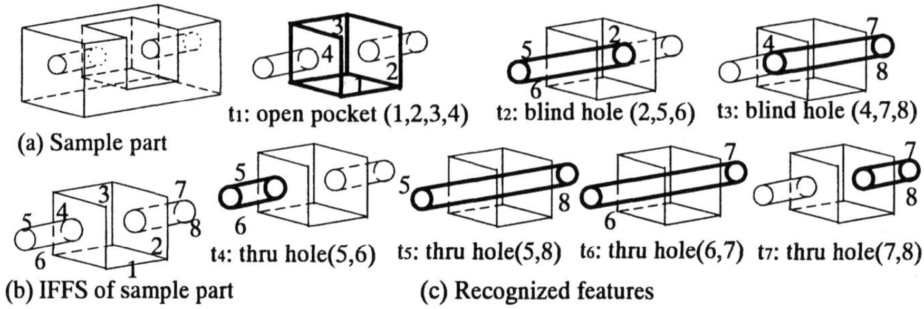

Figure 10. Sample Part, its IFFS and the recognized features

5 CONCLUSION

In this paper, we presented our Feature Sets Creator which consists of 2 phases: Feature Recognition Phase and Feature Organization Phase. In the Feature Recognition Phase, our proposed Modified SRG Method is effective to recognize interacting or overlapping features constructed by plane and curve faces from the boundary data in several different ways. In the Feature Organization Phase, our proposed Feature Spatial Relation Method is effective to organize features into one or more feature sets, which correspond to different kind of process plans. Thus it shows that our Feature Sets Creator is a useful basis for our aim integrated CAPP system.

REFERENCES

1. Sakurai T., Muljadi H., Ando K., Ogawa M. (2000). Research of Dynamic, Flexible and Integrated Process Planning System. Proceedings of the JSPE Spring Conference, 44 (in Japanese)
2. Kao C.Y., Kumara S.R.T., Kasturi R. (1995). Extraction of 3D Object Features from CAD Boundary Representation Using the Super Relation Graph Method, *IEEE Transaction on Pattern Analysis and Machine Intelligence*, **17**(12), December, 1228-1233
3. Gao S., Shah J.J. (1998). Automatic Recognition of Interacting Machining Features based on Minimal Condition Subgraph, Computer-Aided Design, **30** (9), 727-739

32

PROPOSAL OF MODIFICATION STRATEGY OF NC PROGRAM IN THE VIRTUAL MANUFACTURING ENVIRONMENT

Hirohisa Narita[1], Lian-yi Chen[2], Hideo Fujimoto[2]
Keiichi Shirase[3], and Eiji Arai[4]
[1]*Tsukuri College, Graduate School of Eng., Nagoya Institute of Technology, Japan*
[2]*Omohi College, Graduate School of Eng., Nagoya Institute of Technology, Japan*
[3]*Dept. of Mechanical Engineering, Kobe Univ., Japan*
[4]*Dept. of Manufacturing Science, Graduate School of Eng., Osaka Univ., Japan*
e-mail: narita@vier.mech.nitech.ac.jp

Abstract: Virtual manufacturing will be a key technology in process planning, because there are no evaluation tools for cutting conditions. Therefore, virtual machining simulator (VMSim), which can predict end milling processes, has been developed. The modification strategy of NC program using VMSim is proposed in this paper.

Key words: virtual machining, machining strategy, cutting force, machining error

1. INTRODUCTION

Recent computer systems have been enhancing the manufacturing performance. In the design stage, the concept of concurrent engineering has been introduced and it has changed the conventional sequential process to overlapped process for the design process efficiency. The concept of CAD embedded CAE also has been realized, so evaluation of product shape can be carried out easily and quickly in the computer environment. In the machining stage, some advanced CAM systems can achieve the effective process planning, NC program generation and so on. But there are no evaluation tools for cutting conditions, so the test cutting is carried out for the evaluation and modification of cutting conditions after CAM generates

NC programs. It is not a feasible solution, however, especially for the small amount product such as mould and die, because it is based on trial-and-error.

Virtual manufacturing technique is regarded as the innovative solution for the reason that machining status can be verified effectively. So, the CAM embedded virtual manufacturing system is proposed in this research.

Some machining simulator related to virtual manufacturing has been studied and developed. The various approaches of geometric verification for NC machining have been studied based on solid modeling technique[1]. The cutting simulation system using a physical model has been developed to evaluate machinability and adjust the cutting conditions before the real machining operation[2]. In addition, the radial depth of cut is controlled by modifying a tool path distance at the circle path segment and adding tool path segments at the corner in terms of evaluating the physical models[3].

The objective of this research is to show the feasibility of CAM embedded VMS by proposing the modification strategy using the cutting process simulator termed VMSim (Virtual Machining Simulator).

2. OVERVIEW OF A PROPOSED SYSTEM

At first, the definition of virtual manufacturing is introduced. Generally to say, virtual manufacturing is the manufacturing in the computer environment. But this definition is too abstract to develop the system, hence some advanced definitions are proposed so far. The definition of "The use of computer models and simulations of manufacturing processes to aid in the design and production of manufactured products"[4] are selected from these definitions, because this definition is suitable for this research.

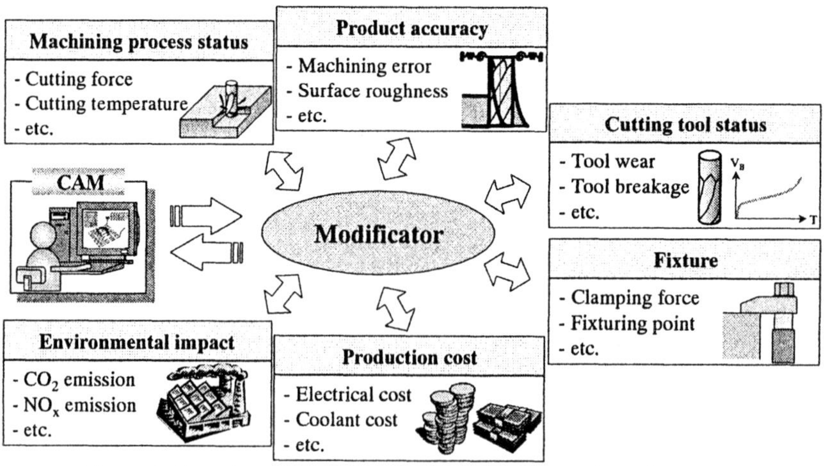

Figure 1. Overview of a CAM embedded virtual manufacturing system

Figure 1 shows the overview of a proposed system in this research. This is a CAM embedded virtual manufacturing system and consists of a CAM, a modificator and 6 evaluation blocks which are machining process status, product accuracy, cutting tool status, fixture, production cost and environmental impact. The modificator decides the machining method automatically or interactively by referring the results of 6 evaluation blocks. The contents of this paper correspond to modificator, machining process status and product accuracy evaluation blocks.

3. MODIFICATION PROCESS OF NC PROGRAM

Figure 2 shows the process flow of NC program modification in this paper. The modification system of NC program consists of 4 process blocks: NC program interpreter, NC program re-generation, modificator and cutting process simulator. The input data to modification system are NC program, tool list, workpiece and tool geometry, simulation parameters, requirement to the machining operation. Requirement to the machining operation is threshold of cutting force, machining error and so on, and simulation parameters are used to predict the machining status in cutting process simulator (VMSim). Modification of NC program is also divided to 2 more modules, which are feed rate modification and tool path modification. When an NC program is modified, feed rate is modified at first. If the requirement can not be satisfied by the feed rate modification, tool path is modified. The modification processes, which are feed rate modification and tool path modification, are described in next chapter.

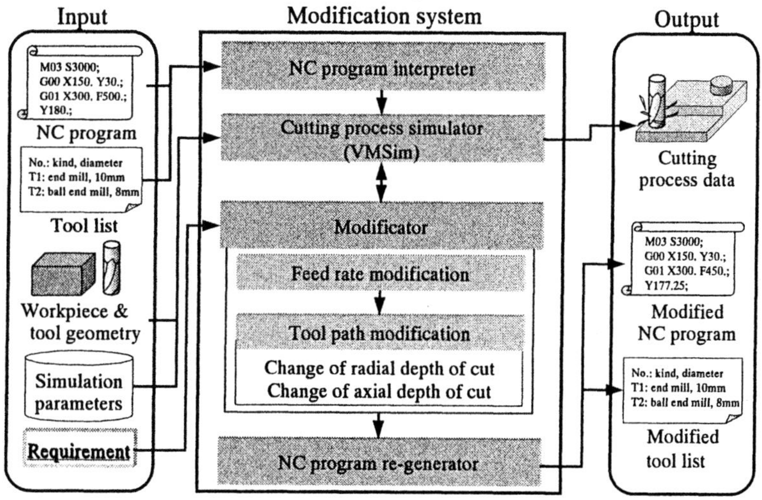

Figure 2. Process overview of NC program modification

4. CUTTING PROCESS SIMULATOR

In this research, a cutting process simulator termed VMSim (Virtual Machining Simulator) for end milling operation shown in Figure 3 has been developed to evaluate the machining process and to adjust the cutting conditions automatically by changing the feed rate[5]. The VMSim consists of a geometric simulator and a physical simulator.

The geometric simulator calculates change geometries of the product machined and extracts cutting conditions like a radial depth of cut, axial depth of cut, spindle speed and feed rate from a workpiece geometry, a tool geometry and NC data. The physical simulator estimates instantaneous cutting force and machining error based on the cutting models from the extracted cutting conditions, cutting coefficients and rigidity. The estimated result of the cutting force and the machining error has already been verified to coincide with the measured result of them.

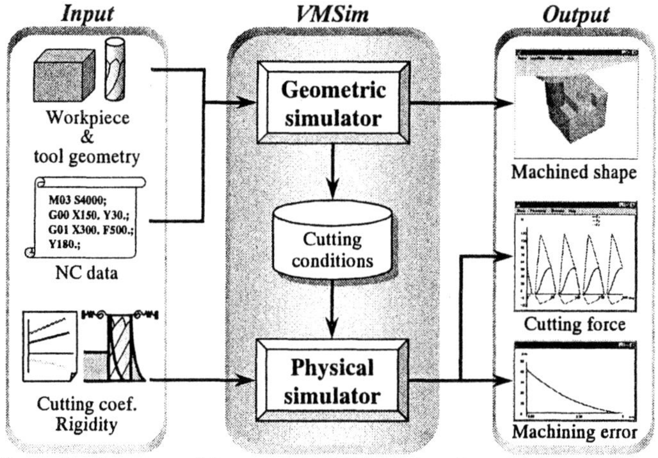

Figure 3. System architecture of VMSim (virtual machining simulator)

5. MODIFICATION OF CUTTING CONDITIONS

5.1 Modification of feed rate

A procedure of the feed rate modification with considering the cutting force is described at first. The feed rate value is calculated by equation (1).

$$F = F_c \cdot \frac{CF_o}{CF_{\max}} \qquad (1)$$

where F_c is a current feed rate value, CF_{max} is absolute value of maximum or minimum cutting force predicted from F_c, and CF_o is a threshold of cutting force. After that, the modificator iterates to change the feed rate value in small steps until satisfying the requirement (threshold) value by referring to VMSim. The procedure with considering the machining error is the same, too. Also, this calculation process is adapted not only over the thresholds but also under them to shorten the total machining time.

Modification method of NC program is the following. Cutting force data, machining error and cutting conditions are recorded with NC block number by VMSim verification. The cutting conditions may be difference at each tool position in same NC block, so one NC block is divided into some blocks according to need and feed rate is calculated with satisfying the requirement for divided NC blocks.

Experimental verification of this function is carried out. The test shape, machined shape and tool path pattern are shown in Figure 4 (a), (b) and (c), respectively. Work piece is cast iron (FC250), cutting tool is carbide square end mill with 10 mm diameter, 2 flute and 30° helical angle in this experiment. The requirement to the machining operation is that cutting force is between -600 to 600 N and the machining error of product surface is not over 30 μm.

(a) test shape *(b) machined shape* *(c) tool path pattern*

Figure 4. Test shape and tool path pattern of case study of feed rate modification

Figure 5 (a), (b) and (c) shows the relationship of changes of cutting forces and machining time of original NC program in X, Y and Z axes. Estimated cutting forces are indicated maximum and minimum cutting force, and measured cutting force are indicated all data in the figures. Figure 6 shows the comparison of measured and estimated machining errors of a finished surface pointed by red circle. Machining error caused by tool deflection is presented and biggest value means the one at the bottom of pocket. As shown in the figures, estimated cutting force and machining error coincide with measured them very well. Both of the cutting force and the machining error, however, exceed the requirement obviously. So, this NC program is modified by changing the feed rate value by referring to the VMSim.

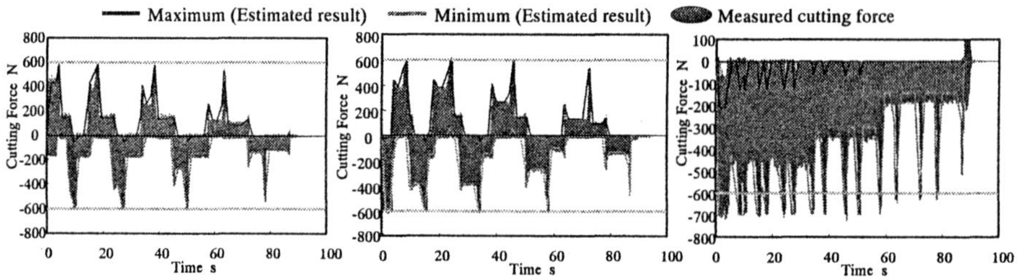

(a) cutting force along X axis (b) cutting force along Y axis (c)cutting force along Z axis

Figure 5. Comparison of predicted and measured cutting force (original NC)

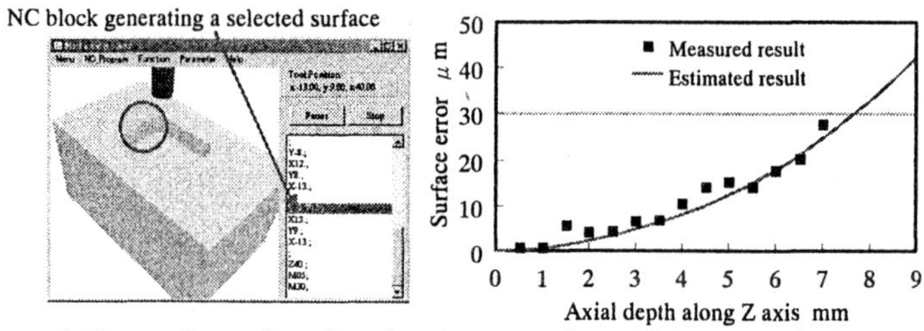

Figure 6. Comparison of predicted and measured machining error (original NC)

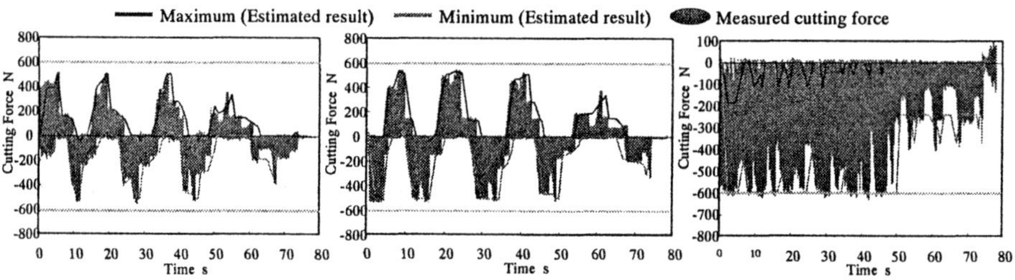

(a) cutting force along X axis (b) cutting force along Y axis (c)cutting force along Z axis

Figure 7. Comparison of predicted and measured cutting force (modified NC)

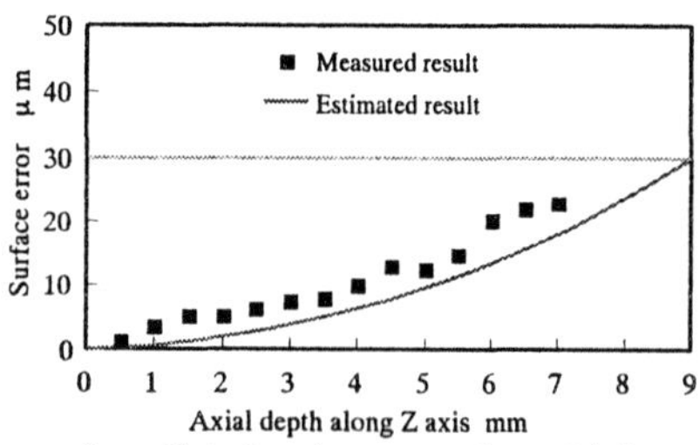

Figure 8. Comparison of predicted and measured machining error (modified NC)

In the modification process, rough machining of NC program is considered only cutting force and finish machining of one is considered cutting force and machining error. Rough and finish machining are distinguished automatically by using recorded cutting force and machining error data. Figures 7 and 8 show the cutting force results and a machining error result of modified NC program. It is clear that the cutting force and the machining error are limited within the required value. The modification system can modify cutting conditions to satisfy the requirement in this case. But, in case that the feed rate modification can not provide a feasible solution, the tool path has to be changed.

5.2 Modification of tool path

The procedure of the tool path modification is the following. First, tool path pattern is recognized based on generation process of tool swept volume and tool motion vector. In this research, tool motions are classified into 3 types, i.e., hole making motion, linear motion and arc motion. Tool swept volumes are generated based on 3 types of tool motion, and directions of these generation process are recognized like figure 9 (large arrow). For example, if the direction of generation process is one way, tool path pattern is unidirectional or zigzag pattern. In this case, these are recognized based on tool motion vector (small arrow). If the direction of generation process is in all direction, the tool path pattern is contour. The fault of machining process also may also be recognized. For example, if the cross feed is larger than tool diameter in zigzag pattern, cross feed and tool path have to be modified.

The change of axial depth of cut is the following. The modification is simple, because the tool path information is shifted to the upper Z value face and the feed rate is also calculated to adjust the machining status mentioned already. In some case, product shape of cross section of the upper Z value may be different from one of lower Z value. The system requires the tool path re-generation to CAM system with indicate the problematic NC blocks.

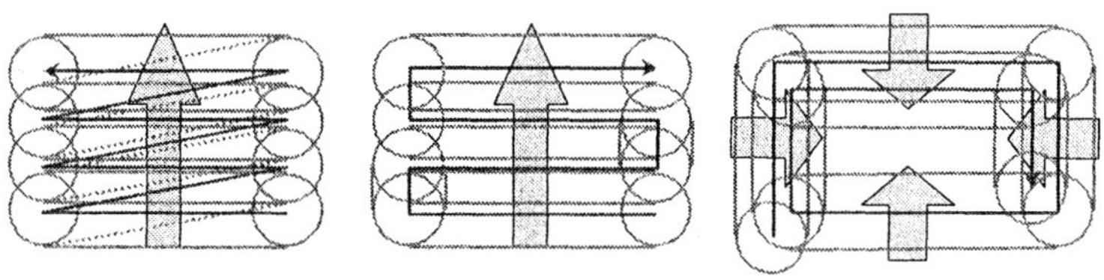

(a) unidirectional pattern (b) zigzag pattern (c) contour pattern

Figure 9. Recognition method of tool path pattern

The change of radial depth of cut is the following. If an initial one decided by CAM operator is large, the system requires the tool path re-generation to CAM system with indicate the problematic NC blocks. If there are some problems in some parts of the machining process, radial depth of cuts are changed in order to realize the constant cutting force machining automatically. Especially for the corner parts, 2 types modification are prepared for automatic modification shown in figure 10. In first type, the loop tool path[6] is generated, and if necessary, loop paths are generated until residual parts disappear. In second type, cutting tool is changed to smaller diameter, and small arc path is inserted to generate the required corner shape. These algorithms will be constructed to the modificator.

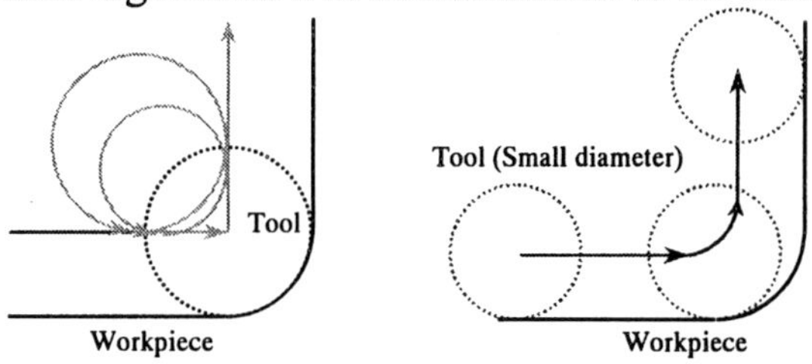

(a) first type modification (b) second type modification

Figure 10. Tool path modification for corner parts

6. CONCLUSIONS

1) The CAM embedded virtual manufacturing system is proposed to realize effective decisions of machining strategies.
2) The strategies for automatic and interactive modification of NC program using VMSim has been proposed. NC program modification is carried out in two stages, feed rate modification and tool path modification.
3) An NC program, which machines simple cavity shape, has been modified to satisfy the requirements. That is to say the feasibility of feed rate modification has been verified by experimental milling.

REFERENCES

1. Leu M. C., Lu F., Blackmore D. (1998). Simulation of NC Machining with Cutter Deflection by Modeling Deformed Swept Volumes, *Annals of the CIRP*, **47**(1), 441-446.
2. Takata S. (1993). Generation of a Machining Scenario and Its Applications to Intelligent Machining Operations, *Annals of the CIRP*, **42**(1), 531- 534.
3. Tsai M.D., Takata S., Inui M., Kimura F., Sata T. (1991). Operation Planning Based on Cutting Process Models, *Annals of the CIRP*, **40**(1), 95- 98.
4. Lin E., Minis I., Naus D. S., Regli W. C. (1997). The Institute for Systems Research (ISR) CIM Lab. Virtual Manufacturing Home Page, March. <http://www.isr.umd.edu/Labs/CIM/virtual.html>.
5. Narita H., Shirase K., Wakamatsu H., Tsumaya A., Arai E. (2002). Real-Time Cutting Simulation System of a Milling Operation for Autonomous and Intelligent Machine Tools, *International Journal of Production Research*, **40** (15), pp.3791-3805.
6. Iwabe H., Fujii Y., Saito K., Kishinami T. (1989). Study on Corner Cut by End Mill – Analysis of Cutting Mechanism and New Cutting Method at Inside Corner-, *Journal of the Japan Society for Precision Engineering*, **55**(5), 841- 846 (in Japanese).

33

DYNAMIC CO-OPERATIVE SCHEDULING BASED ON HLA

Keiichi Shirase[1], Hidefumi Wakamatsu[2], Akira Tsumaya[2], and Eiji Arai[2]
[1]*Dept. of Mechanical Eng., Kobe Univ., Japan*
[2]*Dept. of Manufacturing Science, Graduate School of Eng., Osaka Univ., Japan*
e-mail: shirase@mech.kobe-u.ac.jp

Abstract: In an advanced factory, higher and higher flexibility is required to meet a great variety of customers' requirements. In this environment, a dynamic management architecture is required for distributed production system. In this paper, a newly distributed simulation architecture called HLA (High Level Architecture) is utilized to achieve a distributed scheduling simulation for dynamic work assignment and flexible working group configuration in a distributed production system. It is verified that the distributed scheduling simulation based on HLA is effective to achieve higher flexibility for a distributed production system, by some case studies.

Key words: HLA, dynamic scheduling, distributed simulation, intelligent and distributed production system

1. INTRODUCTION

Because of the drastic change in computers and information technology, production systems have to be changed. Some mass production systems or transfer line production systems have become old fashioned, and YATAI production systems or cell production systems have been introduced to satisfy a great variety of customers' requirements[1]. In this environment, a dynamic management architecture is required for such distributed production systems to have and to maintain higher flexibility and agility.

In distributed systems, information processing in each subsystem can be reduced as compared with the centralized module in conventional systems.

However, the heavier load is imposed on the network because frequent information exchange among subsystems is needed. Furthermore, synchronization among subsystems is also required because information processing speed of subsystems are different from each other.

In this study, HLA (High Level Architecture)[2], which can realize effective communication and time management among subsystems, is applied to the distributed production system. Furthermore, cooperation methods among subsystems which do not need any centralized module are proposed in order to realize flexibility and agility of the whole system. Finally, the effectiveness of our developed system based on HLA is demonstrated with some case studies.

2. DISTRIBUTED PRODUCTION SYSTEM BASED ON HLA

HLA is an architecture in order to realize distributed simulation, which has been developed by the U.S. Department of Defense[2]. It has the following characteristics ;

- Reusability and interoperability of simulators
- Easy realizability of complex distributed/parallel processing

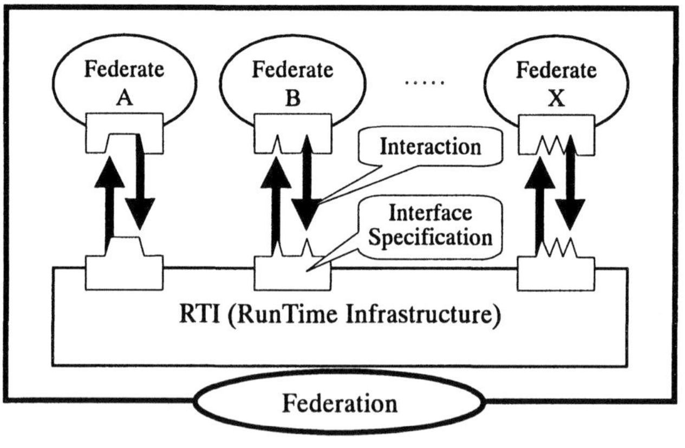

Figure 1. Conceptual diagram of HLA

Figure 1 shows the conceptual diagram of HLA. A federation is a simulation system which consists of several subsystem simulators referred to as federates. RTI (Run Time Infrastructure) is a software supporting communication among federates. Each federate can exchange only data which are defined by FOM (Federate Object Model) with others through RTI.

A new federate can be added to the federation at any time if it has only an adapter which satisfies interface specification in order to connect RTI.

Consequently, we apply HLA to the distributed production system architecture. Each cell/AGV then corresponds to a federate. They have the capability to estimate their processing/transporting time with respect to works and exchange only input/output of the estimation with each other through RTI. Furthermore, a cell/AGV can be attached/detached to the system at any time. This implies that the structural flexibility of the system can be realized.

We also provide a scheduler as one of federates. As a production system becomes large and complex, it takes a longer time to optimize a production schedule. However, the generated schedule may become null immediately if disturbances such as change of production requirements or breakdown of facilities occur frequently during production. Moreover, all processes for production must be stopped until rescheduling is finished. This inhibits the agility of the system as a result. Therefore, in our study, the scheduler has a quick scheduling method based on a simple rule without stop of the whole system. It is referred as to dynamic scheduling in this paper. Note that its result can be not optimal totally.

Our developed system consists of some cell groups, a storage, a scheduler, and RTI. Each cell group consists of some cells. Each cell has the capability to determine processability and to estimate operating time with respect to works. They can exchange such information with others in the same cell group. Cell groups exchange another information such as request of assistance with each other. A production schedule in a cell group is determined by a scheduler, but cooperation among cell groups is executed without any centralized function/module.

This system is similar to a centralized and distributed production system based on active database which we developed[3]. In this system, the active database of the previous system is replaced by RTI. Then, cell groups can be synchronized by RTI. A procedure to generate a master schedule in a cell group is equivalent to that in the active database system. Let us explain it briefly. Production requirements are transmitted to each cell group through RTI at first. Each cell in each cell group determines processability with respect to each work and estimates operating time of works if it can be processed by them. Estimation results from each cell are sent to a scheduler through RTI, then, the scheduler generates a master schedule. Note that scheduling is executed after the scheduler receives replies from all cells in this phase.

We explain the rescheduling flow of our developed system based on HLA in the next section.

3. DYNAMIC RESCHEDULING

In this paper, as disturbances of the system, the following events are considered ;

- Change of production requirements
- Change of properties of subsystems

The former includes change of the kind/number of parts to be manufactured, the latter includes change of functions/abilities of subsystems. Not only facilities but also human workers/craftsmen can be regarded as one of subsystems. When a disturbance occurs, our developed system tries dealing with it by rescheduling in each cell group. If it is impossible, that is, the final result of rescheduling can not satisfy due date, each cell group tries cooperating with others by reassigning works or facilities. Let us explain this procedure.

3.1 Rescheduling in One Cell Group

In this section, let us consider rescheduling in one cell group. When a disturbance occurs, only manufacturing operations which are being executed and can be continued are assigned to running cells until they are finished. The rest of operations should be cancelled and rescheduled. Information with respect to operations to be rescheduled is delivered to each cell and processability determination and operating time estimation are performed by each cell. Note that the current operation in each cell never be stopped during its processability determination and operating time estimation. After that, each cell replies a result of them to the scheduler.

As processability and operating time with respect to each operation, which are determined/estimated by each cell when a master schedule was generated, are stored in the scheduler, they are used for rescheduling without asking each cell about them again. Therefore, needless inquiries to cells can be omitted.

However, time needed for processability determination and operating time estimation by each cell is different because function, ability, and/or information processing capability of each cell is different. So, the scheduler can not receive replies from all cells at a time.

Moreover, when the number of operations increases, it takes much time for the scheduler to receive replies from cells because much time is needed for determination /estimation of each cell.

In our system, the scheduler repeats rescheduling whenever it receives a reply from a cell. This process flow is illustrated in Figure 2. In order to finish rescheduling with a short time, SPT (Shortest Processing Time) rule is adopted as a dispatching rule in the scheduler. Consequently, a new result by one rescheduling does not reflect all processability determination and operating time estimation with respect to operations which have not been executed yet. Furthermore, this system does not guarantee that the finishing time of all operations becomes earlier by repeating rescheduling. Therefore, the system repeats rescheduling with adopting only an improved rescheduling result on the previous one until the result reflects all processability determination and operating time estimation with respect to un-started operations.

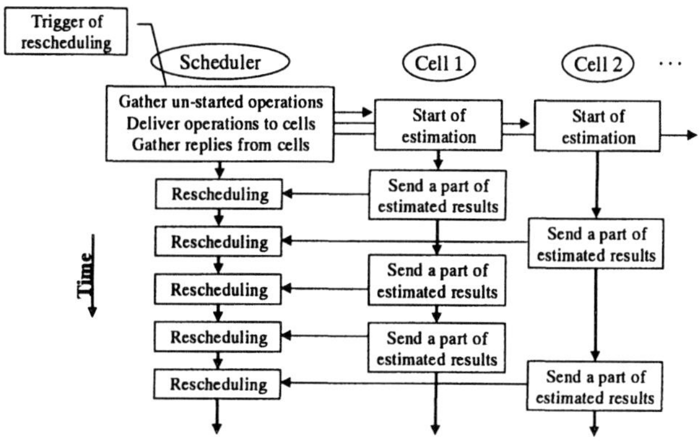

Figure 2. Rescheduling procedure in one cell group

3.2 Rescheduling among Cell Groups

In the previous section, we explained rescheduling in one cell group of our developed system. However, the result of rescheduling in the group may not satisfy its due date. Then, rescheduling among cell groups is needed so that all cell groups can satisfy their due dates. In this study, for such cooperation among cell groups, the following means are prepared ;

- Dynamic work assignment among cell groups
- Dynamic operator assignment among cell groups

First, let us explain dynamic work assignment. When disturbance such as breakdown of facilities occurs in one cell group, rescheduling mentioned in the previous section is executed at first. If the due date of this cell group is not satisfied in spite of rescheduling, the group asks for assistance, that is, taking on operations with respect to one lot to other groups. Other groups simulate rescheduling when they accept it, with executing their current operations. If their results satisfy their due dates, they inform that they can accept that request. Then, the lot is reassigned to the cell replied earliest. The asking cell reschedules again in order to check whether its due date is satisfied by assistance of another cell or not. If it is not satisfied yet, the cell continues to request assistance of another lot. Other cells also continue to determine its request and accept if possible. When the due date of all cell groups is satisfied, rescheduling of the whole system completes.

When we assume that cells are not fully automated facilities but machines controlled by human operators or human workers/craftsmen themselves, they can move around in a factory. This implies that dynamic reconfiguration of cell groups is possible. In such environment, dynamic operator assignment, one more cooperative scheduling method can be introduced.

A cell which becomes unable to satisfy its due date asks for sending of one operator to others. Other groups simulate rescheduling when one operator in them is sent to another group without stop of their operations. Note that which operator should be sent depends on each cell group. If their due dates are satisfied in spite of decrease of their operators, one operator is actually sent from the earliest responding cell group. After that, requests and acceptances/rejects are exchanged among cell groups until all due dates are satisfied.

Thus, by using two types of dynamic cooperative scheduling methods, our proposed system can deal with disturbances without any centralized module.

4. CASE STUDIES

In this section, the effectiveness of our developed system is demonstrated with some case studies. First, rescheduling in a cell group is shown. Next, two kinds of cooperative rescheduling among cell groups are shown.

4.1 Rescheduling among Cell Groups

Figure 3 shows an experimental rescheduling result of one cell group, which is a part of our developed system. This cell group consists of five machining cells. The first Gantt Chart in Figure 3 shows a master schedule. The finishing time of all operations is 21:08. If the ability of cell 1 descends at a certain time, the system generates the rescheduling result when cell 2 through cell 5 are used as shown the second chart in Figure 3. Because this result does not reflect an operation assigned to cell 1 and been executed and operations following it, the chart includes many blanks. The finishing time becomes 22:22. The scheduler reschedules whenever processability and operating time with respect to an operation been executed and not finished by cell 1 are sent from each cell. Incidentally, those with respect to other operations are stored in the scheduler. The finishing time reaches 24:33 temporally when processability and operating time of one cell are reflected. Whenever those of a cell are transmitted to the scheduler, rescheduling is executed and the more improved result, that is, the shorter one between a new one and the previous one is adopted. After replies from all cells are received, the finishing time becomes 22:13 as the result of final rescheduling.

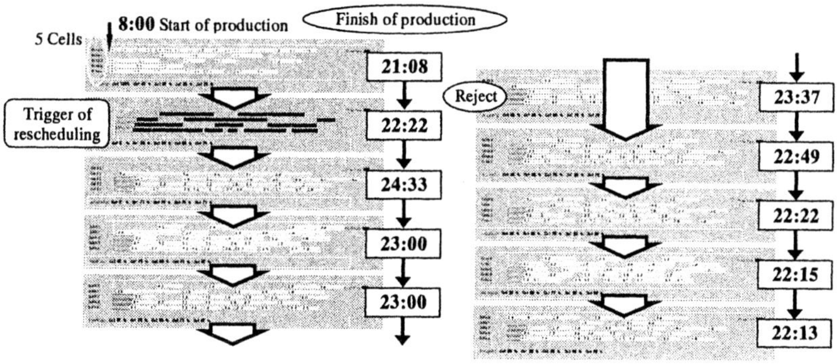

Figure 3. Experimental rescheduling result in one cell group

4.2 Rescheduling Among Three Cell Groups

Next, let us show examples of cooperation among cell groups. Our developed system consists of three cell groups. Cell group 1, 2, and 3 also consist of five, seven, and three machining cells, respectively.

Figure 4 shows an example of dynamic work assignment. At a certain time, cell group 1 becomes unable to satisfy its due date because the ability of one cell descends. Then, cooperative rescheduling mentioned in the previous section is started. After three lots and two lots are reassigned into cell group 2 and 3, respectively, the due date of cell group 1 is satisfied.

Figure 5 shows an example of dynamic operator assignment. When the due date of cell group 1 becomes unable to be satisfied, the group asks for sending of one operator to cell group 2 and 3. In this case, cell group 2 can accept this request but cell group 3 can not accept it as a result of rescheduling simulation. Therefore, cell group 2 sends one operator to cell group 1. By this assistance, cell group 1 can satisfy its due date.

Thus, our developed system can deal with some disturbances by rescheduling in each cell group in the first step, and by rescheduling among cell groups with two kinds of cooperation methods in the second step.

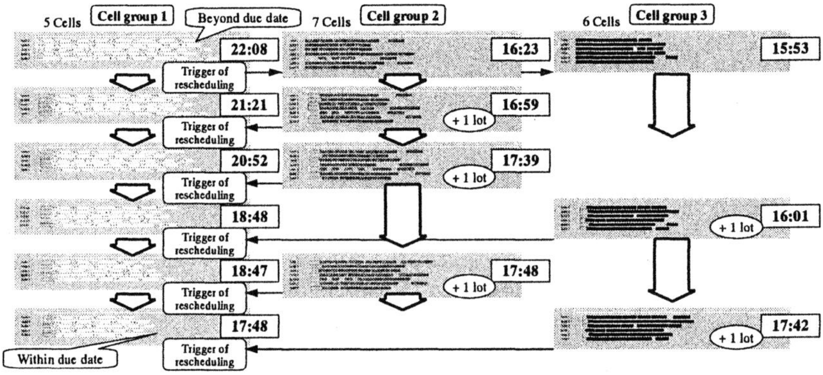

Figure 4. Rescheduling result of dynamic work assignment

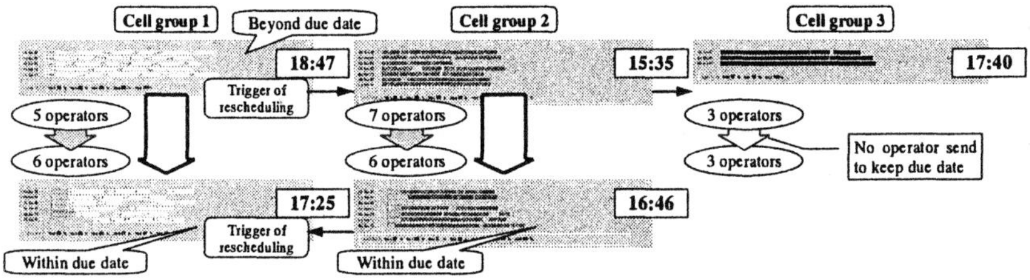

Figure 5. Rescheduling result of dynamic operator assignment

5. CONCLUSION

In this paper, a distributed production system based on HLA was proposed. It has a function of dynamic scheduling for flexibility and agility of the whole system. Furthermore, two kinds of cooperation methods, dynamic work assignment and dynamic operator assignment, were introduced in this system. Then, by cooperation among subsystems, the system can deal with disturbances without any centralized module.

REFERENCES

1. Fujii S., Kaihara T., Morita H., Tanaka M. (1999). A Distributed Virtual Factory in Agile Manufacturing Environment, *Proceedings of 15th Conference of the International Foundation for Production Research*, II, pp.1551-1554.
2. Kuhl F., Weatherly R., Dahmann J. (2000). Creating Computer Simulation Systems – An Introduction to the High Level Architecture, Prentice Hall.
3. Arai E., Wakamatsu H., Takagi S., Uchiyama N., Takata M. (1996). Auction Based Production System through Active Database System, *Proceedings of the pacific conference on manufacturing '96*, pp.371-376.

34

A STUDY ON DATA HANDLING MECHANISM OF A DISTRIBUTED VIRTUAL FACTORY

Kentarou Sashio[1], Susumu Fujii[2], and Toshiya Kaihara[1]
[1]*Graduate School of Science and Technology, Kobe University, Japan*
[2]*Faculty of Engineering, Kobe University, Japan*
e-mail: sashio@kaede.cs.kobe-u.ac.jp

Abstract: To cope with diversified consumers' needs, recent manufacturing systems are required to accommodate the agility in manufacturing by shortening the lead time, reducing the indirect costs and so on. To evaluate the performance of the total manufacturing system taking the complicated information flow as well as the material flow among areas into consideration, Distributed Virtual Factory (DVF) has been proposed. To construct DVF by integrating area level simulators, Communication, Synchronization and Data Handling Mechanism are required. The Data Handling Mechanism supplies external information required by area level simulators. In this paper, a Database Interface is developed to integrate databases into a DVF and the Data Handling Mechanism is implemented by utilizing database.

Key words: simulation, Distributed Virtual Factory, manufacturing, Database

1. INTRODUCTION

To deal with the diversification of demands in the market, manufacturing systems are strongly required to accommodate agility in manufacturing by shortening the lead time and reducing the indirect costs. Through utilizing rapidly developing information technology (IT), computer networks connect separate sections of a manufacturing industry as well as related industries and the time lags in the communication among them are reduced considerably. The effectiveness of simulations for evaluating manufacturing systems has been widely recognized and simulations at the area level, such as processing area, assembly area and so on, have been performed independently. The reduction of the time lags in communication, however, made the relationship among areas or sections closer and the mutual effect in the material flow among areas became difficult to neglect. This suggests the importance to model the total manufacturing system for the simulation, in addition to the independent simulations of areas.

To solve the difficulty to model a large manufacturing system, a Distributed Virtual Factory (DVF) has been proposed, which integrates existing area level simulators via computer networks [1,2]. Each simulator of an area level is executed on the computer of that area and all simulators cooperate with each other based on the concept of distributed simulation[3]. A large scale simulation system can be constructed efficiently by reusing existing area level simulators, and efficient execution is obtained by distributing loads of computing. To deal with the changes in manufacturing systems, only the reconstruction of the corresponding area level simulator in the DVF is needed, and thus the high flexibility for the changes in manufacturing systems can be realized by the DVF.

To develop DVF of real manufacturing systems, it is necessary to consider the difference of objectives and the difference of developing environments of area level simulators. In the former research work[4], the communication system among the simulators has been reported enabling to integrate the simulators developed under various OSs and by various simulation languages at the communication level. In this study, we focus to the difference of accuracy and information description (e.g. material name) among area level simulators and a data handling mechanism which supplies information appropriately is developed.

2. DISTRIBUTED VIRTUAL FACTORY

2.1 The Concept of the DVF

The DVF is developed by integrating existing area level simulators via computer networks and utilizing the concept of distributed simulation. In addition to area level simulators, a simulator of transportation system is required to represent the material flow among areas and a factory management system is required for the decision making for the whole manufacturing system, such as for the production control.

To utilize area level simulators as subsystems of the DVF, a mechanism for communicating with other simulators via computer network is required. Since the difference of simulation time among area level simulators arises during the process of simulation, a synchronization mechanism[5] which copes with the difference of simulation time is also necessary. In addition, to supply the external information which required by area level simulators for simulation of DVF, data handling mechanism is required. In this study, a set of these three functions is named as the communication interface.

In this study, we employ the Time Bucket Method for the synchronization[6] and the communication system by utilizing TCP/IP[4] which realizes data exchange between simulators developed on various OSs and simulation languages.

2.2 Problems on Integrating Area Level Simulators

Area level simulators are developed independently and the description of products and the detailness in the modelling are different. Each simulator assumes that the material arrival and the production orders to the area are externally supplied or are randomly generated at a source node based on the prespecified distribution and that the completed products are simply terminated at a sink node in the simulation after recording the production statistics as shown in Figure 1. In reality, the products terminated at an area A represented by the simulator A are transported to an area B and become source material there.

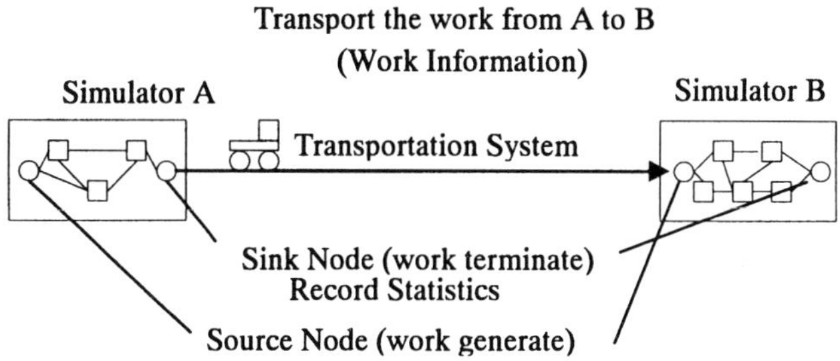

Figure 1. Construction of Area Level Simulators and Concept of DVF

In a DVF, this implies that the completed products need to have the information of the destination and the name in the destination area level simulator and to request the transportation to the transportation system. To supply these information, we develop a data handling mechanism to utilize the databases already developed in a real manufacturing system, such as those in MES, POP, PDM and so on. For simplicity, this study concentrates on only two kinds of information required for area level simulators in DVF. One is the destination area of material after all processes are completed in an area and its name in the simulator at the destination area. Other information includes the processing sequence and time, and so on, if they are not prepared in the simulation at the succeeding area, i.e., the destination area.

Data handling mechanism supplies the information by the following procedures.

For the preparation of simulation, a master database which includes information required for the data handling mechanism is developed by utilizing the existing database of the real manufacturing system. The master database consists of six tables i.e., the Production Instruction Table, the Process Information Table, Process Sequence Table, The Bill Of Materials Table, the Transportation O-D Table and the Rename table, as shown in Figure 2 including the following information.

- production schedules
- process information
- transportation information
- Bill Of Materials (BOM)

Since the local names of materials in each area level simulator may not be included in the databases of real manufacturing system, the Rename Table is required to be carefully developed by checking the material flow from simulator to simulator.

When the production instruction is received by an area level simulator, following procedures are required to obtain the process information from the master database.

1. By the instruction ID, identify a record of instruction from the Production Instruction Table.
2. Identify the process information from the Process Information Table with the Process ID (Arrow ① in Figure 2).
3. Repeat the arrows of ②,③ and ④ for number of processes in an area.

The destination area of the transportation is obtained from the Process Sequence Table by following procedures when a product becomes ready to transport after completing all processes in the area.

1. Identify the Transportation ID from the Process Sequence Table with the Material Name and the Area shown as arrow ⑦ in Figure 2.
2. The destination area is identified from the Transportation O-D Table with the Transportation ID.

The material name in the destination area level simulator is obtained by following procedures.

1. When all processes of a material finished in an area, the material changes into a part and its name also changes into a part name. To obtain the part name, the Bill Of Materials table is utilized shown as arrow ⑤ and ⑥ in Figure 2.
2. From the Rename Table, the local name of the part in the destination area is identified by procedures discussed above and the part name shown as arrow ⑧ and ⑨ in Figure 2.

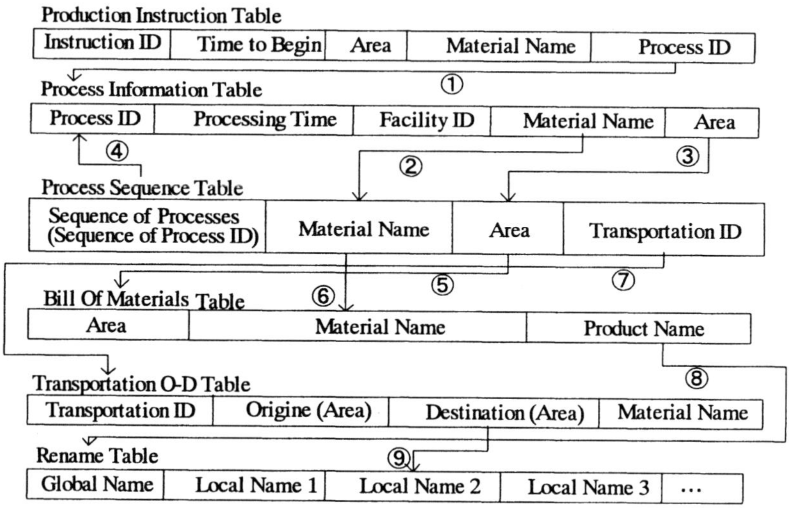

Figure 2. Structure of the Master Database for the DVF

3. INTEGRATION OF DATABASES

In this study, various databases are integrated into a DVF by developing an interface for Relational Database Management Systems (RDBMS) of widely used databases such as, Oracle, SQL Server and ACCESS.

3.1 Database Interface

The storing schemas of different RDBMSs are different from each other. However, interfaces of input and output are based on a common specification called ODBC (Open Data Base Connectivity). Thus in this study, databases are integrated through ODBC to utilize a variety of RDBMSs. To integrate databases into a DVF, both the interfaces for RDBMSs and the interface for the DVF are required. The module, which has these functions, is called a database interface in this study. The database interface is developed by the following processes and the concept of database interface is shown in Figure 3.

- **Interface for RDBMS**

  ActiveX Data Object (ADO) and Visual Basic are employed to develop an interface for RDBMS. ADO provides a unified procedure to connect RDBMS. Visual Basic enables to construct graphical user interface easily.

- **Interface for the DVF**

For integrating databases into the DVF, several configurations are considered (detail is discussed in section 3.2). To implement them, following two methods of mounting databases are required.

(1) Database is mounted on a computer apart from simulator.

(2) Database is mounted as a subsystem of area level simulator.

By utilizing communication interface, method (1) can be realized. Method (2) can be realized by connecting an area level simulator and the database interface on a computer with the call back IP address of the local host and is generally specified as 127.0.0.1. To fit the objective system, these methods can be combined.

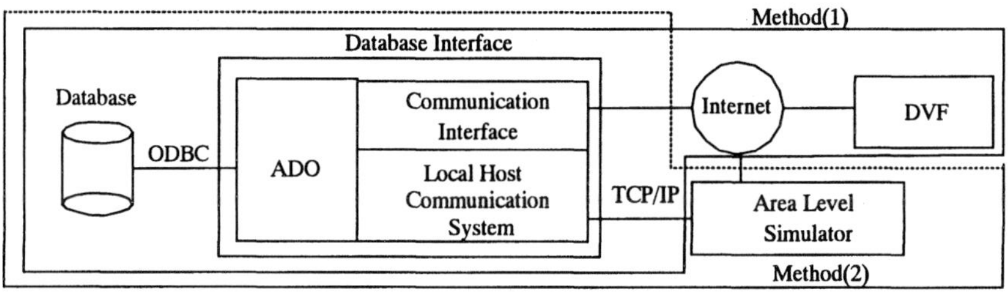

Figure 3. Database Interface

3.2 Configurations of the Database

On integrating the database, following three different configurations of the database are considered in this study.

Type 1. Only one master database is developed in the DVF for integration.

All area level simulators access the master database to obtain information. In this configuration, each area level simulator needs to access the master database every time when a product is completed, even if the required information has been obtained before.

Type 2. Each area level simulator provides a full copy of the master database.

When the information required by an area level simulator, the area level simulator refers to its copied database. To implement this configuration, copies of the master database for area level simulators are required at the preparation stage.

Type 3. Master database is developed in the factory management system and empty databases are prepared in area level simulators.

When the information required by an area level simulator, the area level simulator refers to its own database at first. If the required information is not stored in the database, the area level simulator accesses to the master database to obtain the necessary information. The obtained information is stored in the database of the area level simulator.

When two or more databases are provided in DVF as configurations of Type 2 and Type 3 above, the contents of the databases are required to be the same and consistent. In this study, databases are updated at a certain time span by informing the changes of contents among the databases.

4. EXPERIMENTAL SYSTEM

Based on the concepts discussed above, experimental implementation of the data handling mechanism is performed. The configuration of the objective DVF is shown in Figure 4. In this paper, we explain the system based on the configuration Type 2 although Type 1 and 3 were also implemented. By updating databases, production instructions are distributed among area level simulators. By receiving production instructions from the Factory Management System at storage 1, materials which are stored in the Storage 1 is transported to the Processing A or B by the transportation system. According to the instructions, materials are processed in processing areas and transported between processing areas. After completing the all processes in processing areas, materials are transported to the Storage 2 and terminated.

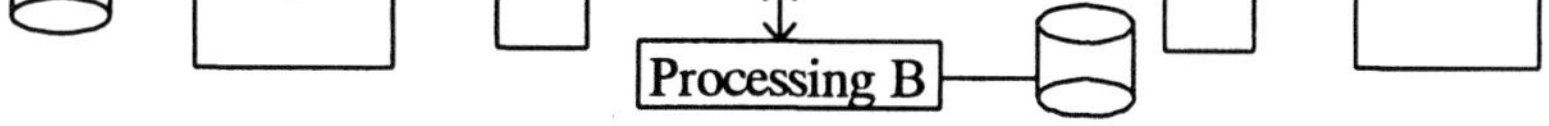

Figure 4. Configuration of the Objective DVF

Communications among the area level simulators are realized by exchanging messages in the DVF. Thus, the messages of transportation request and database updating are defined before execution of simulation. Message format of transportation request is shown in Figure 5(a) as serial arrangement of a set of time, area name, message type, material name, destination of the transportation, instruction ID and number of finished areas. Message format of the updating message is shown in Figure 5(b).

Time : Area Name / MessageType / MaterialName / Destination / InstructionID / Number of Finished Areas

(a) Transportation Request

Time : Area Name / MessageType / MaterialName / InstructionID / Sequence of Areas / Process ID / Time to Begin

(b) Database Updating

Figure 5. Message Formats

5. CONCLUSION

In this paper, the data handling mechanism integrating databases is discussed on constructing a DVF by integrating area level simulators. To realize the data handling mechanism, the database interface is developed. The data handling mechanism is implemented in a simplified DVF for verifying its performance. In this study, we assumed that the information required for area level simulators is only the material name at the destination area and the destination of transportation. However, the same procedure is considered to be applicable for other information which resembles in characteristics, such as facility name, process information and so on.

In this study, the database interface for the WINDOWS system is developed, where one for UNIX system is left for the future research. Further study is also necessary to clarify the characteristics of information which can be supplied by the data handler.

REFERENCES

1. Susumu FUJII, Haruhisa TSUNODA, Atsusi OGITA, Yasusi KIDANI (1994). DISTRIBUTED SIMULATION MODEL FOR COMPUTER INTEGRATED MANUFACTURING,*Proceedings of the 1994 Winter Simulation Conference*,pp.946-953
2. Susumu FUJII, Yoko HIRASHIMA, Haruhisa TSUNODA, Mahiro YAMANE, Masaru HIRANO (1994). A BASIC STUDY ON A DISTRIBUTED SIMULATION FOR VIRTUAL MANUFACTURING, *New Directions in Simulation for Manufacturing and Communications*, pp.277-281
3. RICHARD M. FUJIMOTO. (1990). PARALLEL DISCRETE EVENT SIMULATION, *COMMUNICATIONS OF THE ACM.*,33,10,pp.30-53
4. Susumu FUJII, Toshiya KAIHARA, Kentarou Sashio. (2001).A Development of Distributed Virtual Factory –Integration of Area Simulators Modeled by Various Simulation Softwares-, proceedings of the 3rd Manufacturing & Machine Tool Conference, pp.37-38 , in Japanese
5. DAVID R. JEFFERSON (1985). Virtual Time, *ACM Transactions on Programming Languages and Systems* , 7,3, pp.404-425
6. Susumu FUJII, Atsushi OGITA, Yasushi KIDANI, Toshiya KAIHARA. (1999). Synchronization Mechanisms for Integration of Distributed Manufacturing Simulation Systems, *SIMULATION*, MARCH, 72, 3, pp.187-197

35

A STUDY ON REAL-TIME SCHEDULING METHODS IN HOLONIC MANUFACTURING SYSTEMS

Koji Iwamura, Yoshitaka Taimizu, and Nobuhiro Sugimura
College of Eng., Osaka Prefecture Univ., Japan
e-mail:iwamura@mecha.osakafu-u.ac.jp

Abstract: Recently, new architectures of manufacturing systems have been proposed to realize flexible control structures of the manufacturing systems, which can cope with the dynamic changes in the volume and the variety of the products and also the unforeseen disruptions, such as failures of manufacturing resources and interruptions by high priority jobs. They are so called as the autonomous distributed manufacturing system, the biological manufacturing system and the holonic manufacturing system. Rule-based scheduling methods were proposed and applied to the real-time production scheduling problems of the HMS (Holonic Manufacturing System) in the previous report. However, there are still remaining problems from the viewpoint of the optimization of the whole production schedules. New procedures are proposed, in the present paper, to select the production schedules, aimed at generating effective production schedules in real-time. The proposed methods enable the individual holons to select suitable machining operations to be carried out in the next time period. Coordination process among the holons is also proposed to carry out the coordination based on the effectiveness values of the individual holons.

Key words: Manufacturing system, Holonic Manufacturing system, Real-time scheduling system, coordination

1. INTRODUCTION

Recently, automation of manufacturing systems in batch productions has been much developed aimed at realizing flexible small volume batch productions. The control structures of the manufacturing systems developed, such as FMS (Flexible Manufacturing System) and FMC (Flexible Manufacturing

Cell), are generally hierarchical. The hierarchical control structure is suitable for economical and efficient batch productions in steady state, but not adaptable to very small batch productions with dynamic changes in the volumes and the varieties of the products.

Computer systems and manufacturing cell controllers have recently made much progress, and individual computers and controllers are now able to share the decision-making capabilities in the manufacturing systems. The network architectures are widely utilized for the information exchange in the design and the manufacturing, and some standardized models, such as STEP [1] and CNC data model [2], have been developed for the information exchange through the information networks for the design and the manufacturing.

New distributed architectures of manufacturing systems are therefore proposed, in order to realize more flexible control structures of the manufacturing systems, in order to cope with the dynamic changes in the volume and the variety of the products and also the unforeseen disruptions, such as failure of manufacturing equipment and interruptions by high priority jobs. They are so called as autonomous distributed manufacturing system, biological manufacturing system, and holonic manufacturing system [3]-[13]

The objective of the research is to develop holonic control architecture of the manufacturing systems. This research is now being carried out as a part of an international cooperative research organized as a HMS (Holonic Manufacturing System) consortium. HMS consortium has defined holon as an autonomous and cooperative building block of a manufacturing system for transforming, transporting, storing and/or validating information and physical objects. The present paper deals mainly with the following issues.

(1) Procedures for the real-time scheduling in HMS,
(2) Real-time selection of suitable machining operations to be carried out in the next time period,
(3) Coordination methods among the selected machining operations by the individual holons, and
(4) Case studies for the real-time scheduling by the proposed methods.

2. REAL-TIME SCHEDULING PROCEDURE IN HMS

2.1 Scheduling Procedure of Existing System

One of the important objectives of the HMS is to provide the system components with the flexible and robust capability against the unforeseen disturbances of the manufacturing systems, such as failure of machining equipment and interruptions by high priority jobs. A real-time production scheduling system has therefore been proposed to control of the components of the HMS. The real-time scheduling means that the production schedules of the workpieces and the machining equipment are determined dynamically only when the status of the manufacturing system and its components are changed due to some events occur in the manufacturing system. Therefore, the scheduling system only determines the schedules of the workpieces and machining equipment in the next time period. The time period means the period between the time when one event occurs and one when another successive event occurs.

The scheduling system consists of a set of holonic components named job holons and resource holons, which represent the workpieces to be manufactured and the machining equipment, respectively. A distributed real-time scheduling method has been proposed, in the previous paper [13], to determine suitable production schedules dynamically, based on the decision-makings of the individual holons. The procedure to determine the schedule is summarized in the followings.

The individual holons in the HMS firstly modify their status, if one of the following events occurs. The status of the resource holons and the job holons are represented by 'operating' or 'idling'.

(1) a machining process of a job holon is finished,
(2) a new job holon is inputted to the HMS,
(3) a resource holon is broken down, or is recovered, and
(4) a status of a job holon is changed from normal one to high priority one.

In the second step, all the job holons which are 'idling' at that time select suitable resource holons, which are 'idle' and can carry out their machining processes in the next time period. Some collisions may occur among the selections of the job holons. For example, more than one job holons select a same resource holon for their next machining processes, as shown in Fig. 1 (a). If a resource holon is selected by more than one job holons, the resource holon selects a most suitable job holon, as shown in Fig. 1 (b), in order to avoid the collisions, in the third step. The job holons and the resource holons select most suitable ones by applying their own decision rules.

The procedure mentioned above provide the holons with an effective distributed scheduling system, however, the individual holons make decisions based on only their own decision-rules, which do not reflect the objectives of the whole HMS. Therefore, a new scheduling procedure is proposed, in the present paper, to determine a suitable combination between the job

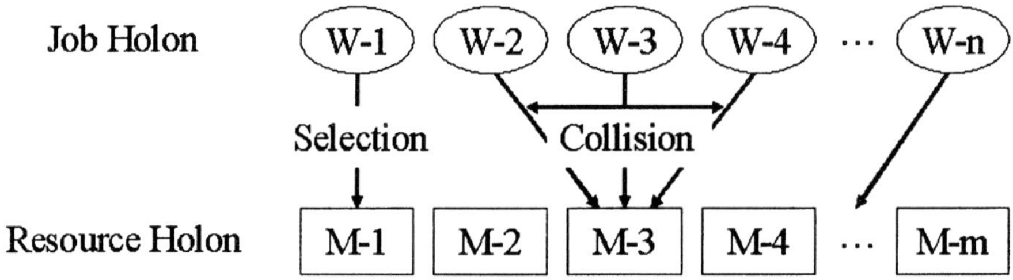

(a) Selection by job holons

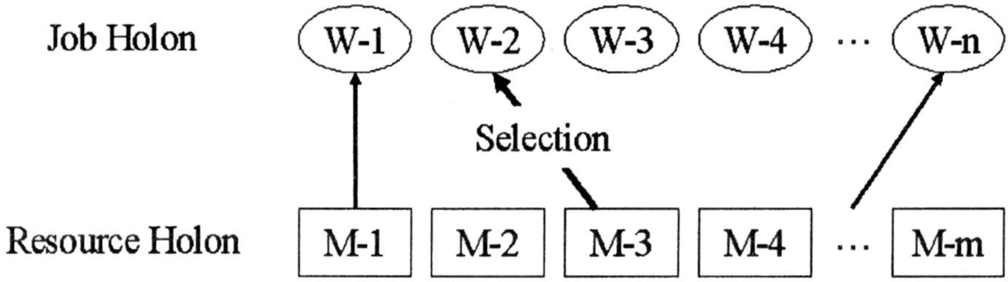

(b) Selection by resource holons

Figure 1. Rule-based real-time scheduling process [13]

holons and the resource holons from the viewpoint of coordination among all the holons.

2.2 New Scheduling Procedure

It is assumed here that the individual job holons have the following technological information representing the machining process of the jobs.

M_{ik}: k-th machining process of the job holon i. (i = 1,..,a), (k = 1,..,ß).
A_{ik}: Required machining accuracy of machining process M_{ik}. It is assumed that the machining accuracy is represented by the levels of accuracy indicated by 1, 2, and 3, which mean rough, medium high, and high accuracy, individually.
R_{ikm}: m-th candidate of resource holon, which can carry out the machining process M_{ik}. (m = 1,...,γ).
T_{ikm}: Machining time in the case where the resource holon R_{ikm} carries out the machining process M_{ik}.

The individual resource holons have the following technological information representing the machining capability of the resources for the machining process M_{ik}.

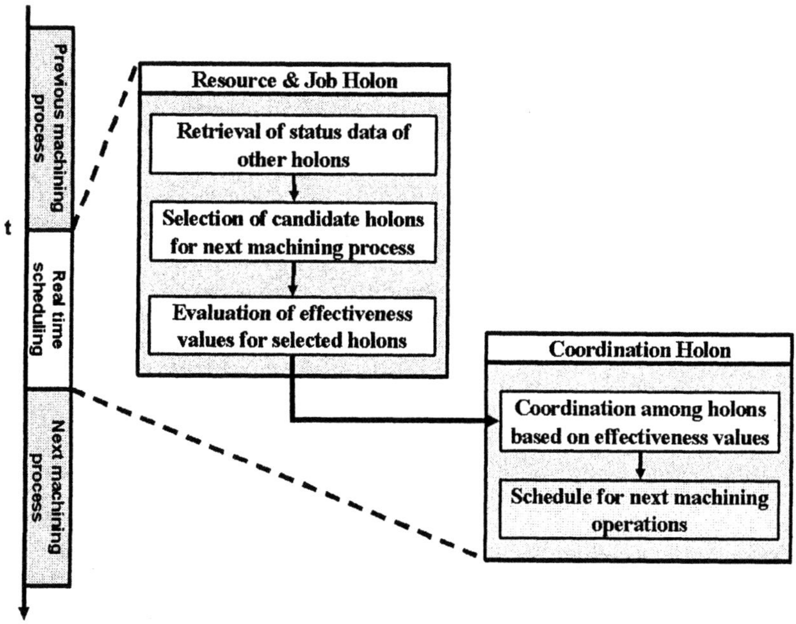

Figure 2. Real-time scheduling process proposed

MA_{ikm}: Machining accuracy in the case where the resource holon R_{ikm} carries out the machining process M_{ik}. MA_{ikm} is also represented by the level of 1, 2 and 3.
MC_{ikm}: Machining cost /unit time in the case where the resource holon R_{ikm} carries out the machining process M_{ik}.

Based on the information above mentioned, a procedure shown in Fig. 2 is proposed, in the present research, to select a suitable combination between the job holons and the resource holons which carries out the machining process in the next time period.

In the figure, t shows a time when some machining process are finished, and some resource holons and job holons change their status from 'machining' to 'idling'. At the time t, all the 'idling' holons have to select their machining schedules in the next time period. The following procedure is newly proposed for the individual holons to select their machining schedules.

(1) Retrieval of Status Data

The individual 'idling' holons firstly get the status data from the other holons which are 'operating' or 'idling'. The 'idling' holons can start the machining operation in the next time period.

(2) Selection of Candidate Holons

The individual 'idling' holons select all the candidate holons for the machining operations in the next time period. For instances, the job holon i selects the resource holons which can carry out the next machining process M_{ik}. On the other hand, the resource holon j select all the candidate job holons which can be machined by the resource holon j.

(3) Determination of Effectiveness Values

The individual holons determine the effectiveness values for the individual candidates selected in the second step. For example, the job holon determines the effectiveness values, based on its own criteria for all the candidate resource holons which can carry out the next machining process. The effectiveness values are given as follows.

JVEi(j) ($0 \leq$ JVEi(j) ≤ 1): Effectiveness value of the candidate resource j for the job holon i.
RVEj(i) ($0 \leq$ RVEj(i) ≤ 1): Effectiveness value of the candidate job i for the resource holon j.

(4) Coordination

All the 'idling' holons send the selected candidates and the effectiveness values of the candidates to the coordination holon. The coordination holon determine a suitable combination of 'idling' job holons and 'idling' resource holons which carry out the next machining processes in the next time period, based on the effectiveness values. The decision-criteria of the coordinate holon is to maximize the total effectiveness values of all the 'idling' holons.

3. EFFECTIVENESS VALUES AND COORDINATION

3.1 Effectiveness Values

The effectiveness values are evaluated based on the criteria of the individual holons, and various criteria are considered for the holons. Therefore, it is assumed that the individual holons have one of the objective functions shown in Table 1 for evaluating the effectiveness values, in the present research. They are:

(1) Availability factors of resource holons: ME

ME is the ratio of the total operating time of a resource holon and the total time after the resource holon starts the operations. The total time

Table 1. Objective functions of holons

Objective		Objective Functions
Resource Holon	Availability Factor	Σ Machining Time / Total Time
	Machining Accuracy	Σ (Machining Accuracy of Resources – Required Machining Accuracy of Jobs)
Job Holon	Total Processing Time	Σ (Machining Time + Idling Time)
	Machining Cost	Σ (Machining Cost of Resources)

includes both the operating time and the idling time of the resource holon.

(2) Machining accuracy of resource holons: MC
MC is the difference between the level of machining accuracy of the resources and the required level of accuracy of the machining process.

(3) Total processing time of job holons: JT
JT is the total time after the job is inputted to the HMS. JT includes the operating time and the idling time of the job.

(4) Machining cost of job holons: JC
JC is the sum of the machining cost of the job holon, which are obtained from the machining costs of the resource holons.

The following procedures are provided for the resource holons to evaluate the effectiveness values.

Let us consider a resource holon j at a time t. The total time after the resource holon j starts its operations, the availability factor, and the evaluated value of machining accuracy of the resource holon j are assumed to equal to $TT_{j.t}$, $ME_{j.t}$, and $MC_{j.t}$, respectively. If the resource holon j selects a candidate job holon i for carrying out the machining process M_{ik}, the availability factor and the evaluated value of the machining accuracy are estimated by the following equations.

$$ME_{j.t+1}(i) = (ME_{j.t} \cdot TT_{j.t} + T_{ikm})/(TT_{j.t} + T_{ikm}) \quad (1)$$

$$MC_{j.t+1}(i) = MC_{j.t} + (MA_{ikm} - A_{ik}) \quad (2)$$

Where, the resource holon j can carry out the machining process M_{ik} of job holon i. ($j = R_{ikm}$)

As regards the job holons, the following equations are applied to evaluate the total processing time and the machining costs, for the case where a job holon i selects a candidate resource holon j ($= R_{ikm}$) for carrying out the machining process M_{ik}. It is assumed that the total time after the job holon i is inputted to the HMS and the machining cost equal to $JT_{i.t}$ and $JC_{i.t}$, respectively.

$$JT_{i.t+1}(j) = JT_{i.t} + T_{ikm} \tag{3}$$

$$JC_{i.t+1}(j) = JC_{i.t} + MC_{ikm} \tag{4}$$

The objective functions mentioned above have different units. Some of them shall be maximized and others shall be minimized. Therefore, the effectiveness values are normalized from 0 to 1, by applying the following equations.

(1) Availability factors of resource holons:

$$RVE_j(i) = 1-(max(ME_{j.t+1}(i))-ME_{j.t+1}(i))/(max(ME_{j.t+1}(i))-min(ME_{j.t+1}(i))) \tag{5}$$

(2) Machining accuracy of resource holons:

$$RVE_j(i) = (max(MC_{j.t+1}(i))-MC_{j.t+1}(i))/(max(MC_{j.t+1}(i))-min(MC_{j.t+1}(i))) \tag{6}$$

(3) Total processing time of job holons:

$$JVE_i(j) = (max(JT_{i.t+1}(j))-JT_{i.t+1}(j))/(max(JT_{i.t+1}(j))-min(JT_{i.t+1}(j))) \tag{7}$$

(4) Machining cost of job holons:

$$JVE_i(j) = (max(JC_{i.t+1}(j))-JC_{i.t+1}(j))/(max(JC_{i.t+1}(j))-min(JC_{i.t+1}(j))) \tag{8}$$

Where, *max*(f(k)) and *min*(f(k)) give the maximum value and the minimum value of f(x) evaluated for all candidates k, respectively.

3.2 Coordination Process

After evaluating the effectiveness values, all the 'idling' holons send all the candidates and their effectiveness values to the coordination holon, and the coordination holon select a most suitable combination of the resource holons and the job holons, which execute the machining processes in the next time period. The coordination process is summarized in the following, for the case where the coordination holon determine a suitable combination of the job holon Job-i ($i = 1,2,..,\alpha$) and the resource holons Resource-j ($j=1,2,..,\gamma$).

The effectiveness value δ_{ij} of the combination of Job-i and Resource-j is given by the following equation, as shown in Table 2.

$$\delta_{ij}= RVE_i(j) + JVE_j(i) \tag{9}$$

The problem to be solved by the coordination holon is to select a combination of job holons and resource holons which maximize the total of the effectiveness value, as shown in the following equation.

Table 2. Effectiveness values of holons

	Resource-1	Resource-2	...	Resource- γ
Job-1	δ_{11}	δ_{12}	...	$\delta_{1\gamma}$
Job-2	δ_{21}	δ_{22}	...	$\delta_{2\gamma}$
...	...	...	...	...
Job- α	$\delta_{\alpha 1}$	$\delta_{\alpha 2}$	...	$\delta_{\alpha\gamma}$

Table 3. Combination of resource and job holons

	Resource-1	Resource-2	...	Resource- γ
Job-1	a_{11}	a_{12}	...	$a_{1\gamma}$
Job-2	a_{21}	a_{22}	...	$a_{2\gamma}$
...	...	...	...	...
Job- α	$a_{\alpha 1}$	$a_{\alpha 2}$	...	$a_{\alpha\gamma}$

$$\max\left(\sum_{i=1}^{\alpha}\sum_{j=1}^{\gamma} a_{ij} * \delta_{ij}\right) \tag{10}$$

where, a_{ij} (= 0 or 1) are the decision parameters, as shown in Table 3. If a_{ij} =1, the job holon i is machined by the resource holon j in the next time period. Otherwise, job holon i is not machined by the resource holon j. Only one job holon is machined by one resource holon, therefore, the following equation shall be satisfied.

$$\sum_{i=1}^{\alpha} a_{ij} \leq 1, \sum_{j=1}^{\gamma} a_{ij} \leq 1 \tag{11}$$

The branch-and-bound method is applied to determine the decision parameters a_{ij}, which maximize Eq. (10) under the constraints given by Eq. (11).

4. CASE STUDY

A prototype of real-time scheduling system for HMS has been developed based on the proposed methods to evaluate the effectiveness values and to carry out the coordination. Some case studies have been carried out to verify the proposed methods.

A holonic manufacturing system model consisting of 10 machining centers (MC) is considered for the case study. The machining centers are classified into 4 types, which have different machining functions, capacities and accuracy. The individual machining center holons have different objectives to evaluate the effectiveness values. The characteristics of the

Table 4. Combination of resource and job holons

Machine Tool Holon	Objective		Machine Type
	Case A	Case B	
MC1	Availability	Accuracy	Type1
MC2	Accuracy	Availability	
MC3	Availability	Accuracy	
MC4	Accuracy	Availability	Type2
MC5	Availability	Accuracy	
MC6	Accuracy	Availability	
MC7	Availability	Accuracy	Type3
MC8	Accuracy	Availability	
MC9	Availability	Accuracy	Type4
MC10	Accuracy	Availability	

Table 5. Objectives and machining processes of products

Job ID	objective	Processes	
		Case 1	Case 2
Job 1	Time	Type2→Type1→Type3→Type4	Type3→Type1→Type4→Type2
Job 2	Cost		
Job 3	Time	Type1→Type3→Type2→Type4	Type1→Type4→Type2→Type3
Job 4	Cost		
Job 5	Time	Type3→Type2→Type1→Type4	Type4→Type2→Type3→Type1
Job 6	Cost		
Job 7	Time	Type2→Type3→Type1	Type2→Type3→Type1
Job 8	Cost		
Job 9	Time	Type1→Type2→Type3	Type1→Type2→Type3
Job 10	Cost		
Job 11	Time	Type3→Type1→Type2	Type3→Type1→Type2
Job 12	Cost		

machining centers are summarized in Table 4. Cases A and B, in the table, mean that tow cases are considered for the case study from the viewpoints of the objectives of the machining center holons. As regards the job holons, the characteristics of the individual job holons are given in Table 5. The table gives the objectives to calculate the effectiveness values and the machining process of the jobs. Two cases are also considered from the viewpoints of the machining processes, as shown in the table.

An example of the machining schedules obtained in the case study is shown in Fig. 3. The obtained results were compared with the results determined by the rule-based method proposed in the previous paper [13], from the viewpoints of the objectives of the individual holons. Figure 4 summarizes the comparison of the proposed method and the previous method. In the figure, horizontal axis gives the objective functions of the individual holons, and the vertical axis shows the number of holons, the objective functions of which are improved by the proposed methods. As shown in the figure, the proposed methods improve the objective functions of the individual holons.

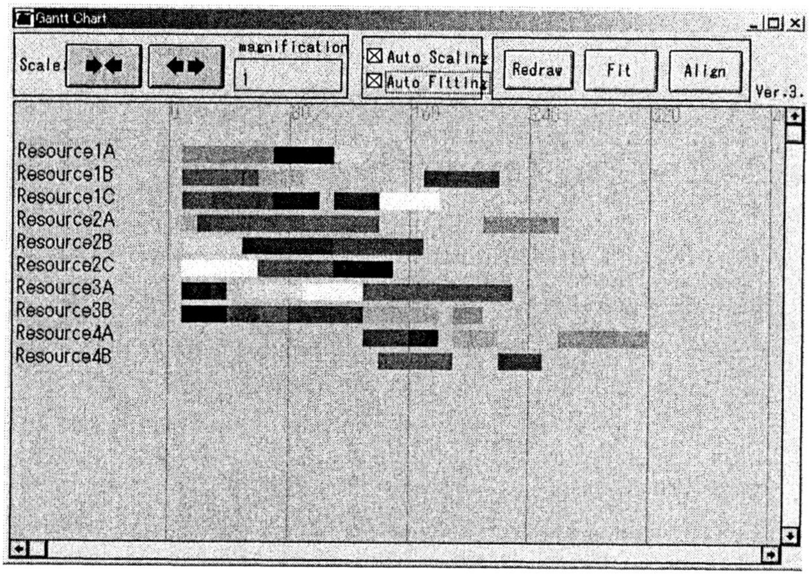

Figure 3. Machining schedules obtained

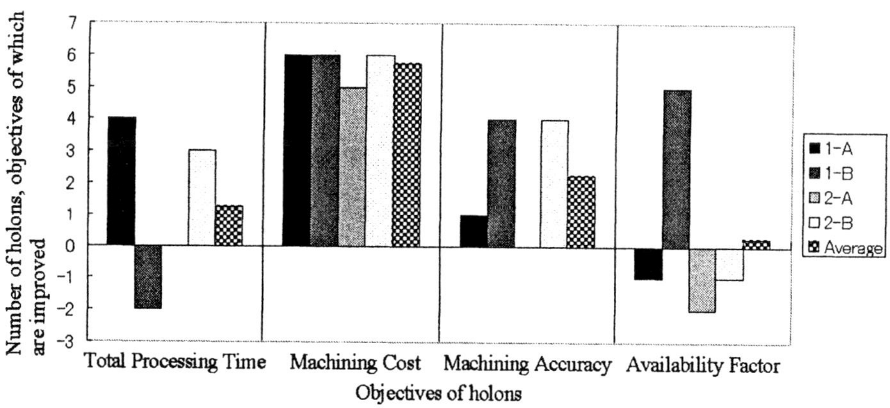

Figure 4. Comparison of proposed method and rule-based method

5. CONCLUSIONS

New real-time scheduling methods are proposed to select a suitable combination of the resource holons and job holons which carry out the machining process. The following remarks are concluded.

(1) A distributed decision-making procedure is proposed to select a suitable combinations of the resource holons and the job holons for the next machining processes, based on the effectiveness values for the candidates.
(2) A systematic method is proposed for the resource holons and the job holons to calculate the effectiveness values of the candidate holons. A coordination method is also developed for the coordination holon to determine a suitable combination of the resource holons and the job holons based on the effectiveness values of the individual holons.
(3) Some case studies of the real-time scheduling have been carried out to verify the proposed methods in comparison with the previous method. It was shown, through case studies, that the proposed methods are effective to improve the objective functions of the individual holons.

REFERENCES

1. ISO (1994). ISO 10303 Industrial automation systems and integration – Product data representation and exchange.
2. ISO (2001). ISO 14649 Industrial automation systems and integration – Physical device control - Data model for computerized numerical control.
3. Okino, N. (1992). A prototyping of bionic manufacturing system, Proc. of the ICOOMS '92, pp.297-302.
4. Ueda, K. (1992). An approach to bionic manufacturing systems based on DNA-type information, Proc. of the ICOOMS '92, pp.303-308.
5. Moriwaki, T. and Sugimura, N. (1992). Object-oriented modeling of autonomous distributed manufacturing system and its application to real-time scheduling, Proc. of the ICOOMS '92, pp.207-212.
6. Warnecke, H.J. (1993). The fractal enterprise, Springer-Verlag, New York.
7. Iwata, K., et al. (1994). Random manufacturing system: A new concept of manufacturing systems for production to order, Annals of the CIRP, Vol. 43, No.1, pp.379-384.
8. Wiendahl, H.P. and Garlichs, R. (1994). Decentral production scheduling of assembly systems with genetic algorithm, Annals of the CIRP, Vol.43, No.1, pp.389-396
9. Wyns, J., et. al. (1996). Workstation architecture in holonic manufacturing systems, Proc. of the 28th Int. Seminar on Manufacturing Systems, pp.220-231.
10. Sugimura, N., et. al. (1996). Modeling of holonic manufacturing system and its application to real-time scheduling, Manufacturing Systems, Vol. 25, No. 4, pp.1-8.
11. Sugimura, N. et. al. (1998). A Study on object-oriented modeling of holonic manufacturing system, Manufacturing Systems, Vol. 27, No. 3, pp.253-258.
12. Sugimura, N. et. al. (2001). Integrated process planning and scheduling in holonic manufacturing systems, Proc. of the 2001 ISATP, pp.250-255.
13. Sugimura, N., et.al (2002), A study on real-time scheduling for holonic manufacturing system, Proc. of CIPR Int. Seminar on Manufacturing Systems, pp.213-218.

36

SENSITIVITY ANALYSIS OF CRITICAL PATH AND ITS VISUALIZATION IN JOB SHOP SCHEDULING

Ryosuke Tsutsumi and Yasutaka Fujimoto
Department of Electrical and Computer Engineering, Yokohama National University, Japan

Abstract: A visualization of a critical path in a job shop scheduling for manufacturing system is presented. The visualization is based on influence factor of the jobs in a given job shop scheduling. The influence factor is defined for each job as quantity of contribution to the critical path that is given by a sensitivity analysis of the path. The information of the sensitivity helps operator to improve the makespan of existing schedule easily.

Key words: Job Shop Scheduling, Tabu Search, Sensitivity Analysis, Critical Path

1. INTRODUCTION

In recent years, as scheduling problem getting more and more large-scaled and complicated, approximation algorithms such as Tabu Search (TS), Simulated Annealing (SA) and Genetic Algorithm (GA) are used for such problems[1][2][3]. In general, the job shop scheduling problem (JSSP) is well known to be one of the particularly hard optimization problems among the NP-hard combinatorial problems. It is difficult to find the optimal solution to the large-scale, practical problem within practical time. Thus the human interaction in combination with a heuristic algorithm is one of the important factors in the practical job shop scheduling systems. In combination with human expertise, the appropriate solution in the practical field is obtained.

In this paper, we propose sensitivity visualization tool to improve solution by human interaction. The sensitivity in this paper means a degree of influence of each operation on the whole schedule. In this system, a feasible schedule is obtained first by use of tabu search scheduling algorithm. The schedule manager can interactively tune the schedule based on visual information.

The structure of this paper is described as follows. In section 2, the model of job shop scheduling and conventional TS is defined. Also the new neighborhood selection in use of sensitivity is proposed. In section 3, the idea of sensitivity analysis and its algorithm are proposed. In section 4, the simulation of TS by conventional method and by proposed method is performed, and the visualization is presented.

2. PROBLEM DEFINITION

2.1 Job Shop Scheduling Problem (JSSP)

The JSSP has complexly intertwined two set of constrains, one is for jobs and the other is for machines. The problem is defined as finding a sequence of operations on each machine that minimizes an objective function, e.g., makespan, under a given sequence of operations for each job. The problem is considered as a particular hard combinatorial optimization problem. The JSSP is briefly described as follows. There are a set of jobs and a set of machines. Each jobs consists of a sequence of operations, and each of them needs fixed duration on one of the machines. Each machine can process at most one operation at one time, each operation is not interrupted if once it has started. The purpose of this problem is to find a minimal makespan schedule.

The model of the JSSP is described as follows. Consider a manufacturing system that consists of m-machine and n-jobs. The sequence of operations is defined for both jobs and machines. Let a set of jobs be $J = \{1,...,n\}$, a set of machines be $M = \{1,...,m\}$, a set of sequence of operations for jobs be $P = \{p_1,...p_n\}$, and a set of sequence of operations for machines $Q = \{q_1,...,q_n\}$. The sequence $p_i = \{f_{i,1},.....f_{i,u_i}\}$, $f_{i,k} \in M$ for job i consists of a finite sequence of machines, which means the job is processed on machines in order of the sequence. The sequence $q_j = \{g_{j,1},...,g_{j,v_j}\}$, $g_{j,l} \in J$ for machine j consists of a finite sequence of jobs, that means the machine processes the jobs in order of the sequence, as well. The relation between P and Q must be feasible. Then the start time of the kth operation for job i is denoted by variable $s_{i,k}$. Also the start time of the lth operation for machine j is denoted by variable $\tau_{j,l}$. If $s_{i,k} = \tau_{j,l}$ then $f_{i,k} = j$ and $g_{j,l} = i$ is satisfied. The uninterrupted processing time for job i on machine j is denoted by $\mu_{i,j}$. The setup times are ignored in this model. The constraints are described by

$$s_{i,k} + \mu_{i,f_{i,k}} \leq s_{i,k+1} \quad (1)$$

$$\tau_{j,l} + \mu_{g_{j,l},j} \le \tau_{j,l+1}. \qquad (2)$$

Fig. 1 and Fig. 2 show an example of Gantt-Chart of a feasible solution.

A Critical Path (CP) is defined as a set of operations that affect the makespan of its schedule directly. In other words, the delay of the operations on the critical path means the delay of whole schedule. Thus the manager of schedule has to pay the most attention to the operations on the critical path. A series of operations on the CP that are processed on same machine successively is called critical block (CB). The critical path for the initial sequence P and Q is obtained as shown in *Fig. 3* by solving the earliest schedule for $s_{i,k}$ and $\tau_{j,l}$ and checking the activity of constraints.

Figure 1. Gantt-Chart for each job

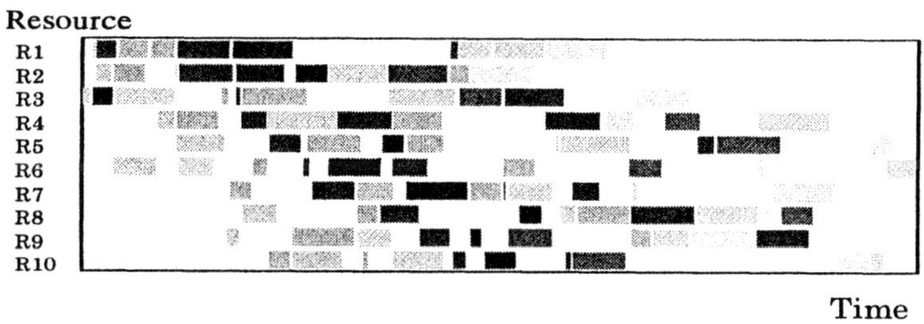

Figure 2. Gantt-Chart for each resource

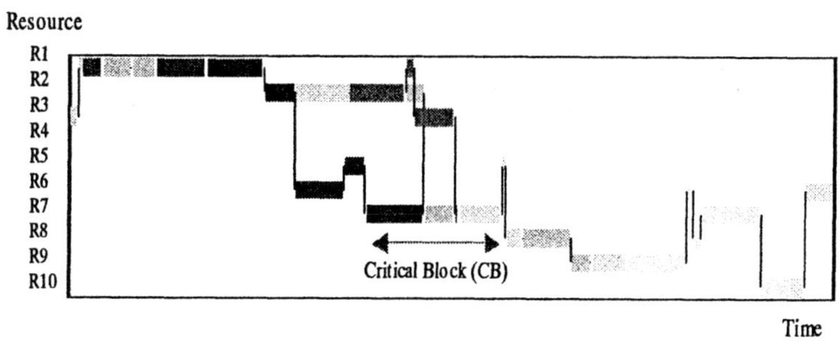

Figure 3. Critical path and critical block

2.2 Tabu Search (TS)

The TS is a heuristic search framework for solving hard combinatorial problems such as job shop scheduling, graph coloring (related), the Traveling Salesman Problem (TSP), the capacitated arc routing problem, etc. The TS iterates a swap of a specific neighborhood to improve the objective function. The TS has two lists to realize efficient convergence to quasi-minimal point. One is Tabu List, another is Superior Solution List[1].

2.2.1 Tabu List

As the TS consists of iterations of neighborhood swapping, it is very important to escape from locally optimal but not globally optimal solutions efficiently. It is also important to prevent the swap cycling. Movements in a certain past term are stored in Tabu List to make the backwards moves taboo.

2.2.2 Superior Solution List

When a superior solution is found, the current state of the solution is stored in the Superior Solution List. The state consists of the processing sequence of the solution, current tabu list, and neighborhood except next candidate. If the tentative best solution is not improved after enough iteration, the state of the past superior solution is restored according to the list that realizes Backtrack. This handling makes the convergence speed fast by searching the neighbor of the past good solutions.

2.2.3 New Neighborhood

A set of schedules generated from current schedule by interchange of a pair of the operations is called neighborhood of current schedule. As mentioned above, one feasible schedule should have the critical path that determines its makespan. In other word, the improvement of sequence on the critical path is necessary to improve the whole solution. Thus the interchange of the operations on the CB is used to improve the solution. The full neighborhood is generally too large to search completely. In a conventional TS method, neighborhood means an interchange of an adjacent pair of the operations on the CB. However, it is afraid that it takes large computation time to improve the solution because only a very narrow range of neighborhood is searched in the conventional method.

Thus the new neighborhood based on sensitivity analysis is introduced in this paper. The most sensitive operation, in other words the operation that should be paid attention extremely, on each of CB is detected first. Then the schedules generated by swapping the most sensitive operation for the other operations on the same CB is defined as new neighborhood. By use of this method, large range of neighborhood can be searched and the calculation cost is reduced.

3. SENSITIVITY ANALYSIS

The structure of schedule at any given time is paid attention. The influence of each operation on the critical path on whole schedule is defined as sensitivity. The constraints that satisfy the equality (1) (2) are defined as *active constraints*. The sensitivity of the operations on critical path is obtained by use of this active concept. The algorithm is described as follows.

1. Select one operation O^* on the critical path. On forwardly justified (earliest) schedule, let the operations whose start time is later than the finish time of O^* be a set A. Let the other operations except O^* be a set B. O^* belong to neither A nor B.
2. Then perform a next modification for an arbitrary operation $a \in A$, where a and $^{\exists}b \in B$ have a same active constrains.
   $$A := A - \{a\},\ B := B + \{a\}$$
   This modification is iterated until a becomes empty.
3. Find a minimal slack in the constrains between A and B, and let it be $l_{\min}$. Then reduce the start time of the operation that belongs to A by $l_{\min}$.
4. Iterate procedure 2, 3 until a becomes an element of the critical path. The sensitivity of O^* is defined as $\sum l_{\min}$.
   Perform above procedures for all operations on the critical path.

The sensitivity expresses quantity of contribution to the makespan of the given schedule. Therefore, improvements of high sensitivity operations have possibility of yielding better solutions.

4. SIMULATIONS AND VISUALIZATION TOOL

4.1 Simulation (Tabu Search Scheduling)

Simulation for typical 10 machines 10 jobs (10x10) benchmark problem designed by Muth and Thompson[4] is performed to compare proposed method with conventional method. The procedure is described as follows.

1. Make initial schedule by use of activation algorithms (see appendix)
2. Detect the critical path by PERT
3. Search the neighbors (by conventional method and by proposed method).
4. Estimate the neighbor and move to one of them.
5. In case the best solution is not improved after enough iterations of move, restore the state of a past superior solution according to the list.

4.2 Simulation Results

Fig. 4 shows the convergence tendency of both by proposed method and by conventional method in the process of search. Conventional method needs a lot of iterations to escape from local optimal solution, however the proposed method by use of sensitivity analysis escapes from local optima solutions efficiently as shown in *Fig. 4*. The average number of evaluation of the objective function is reduced to 60% as shown in *Table. 1*.

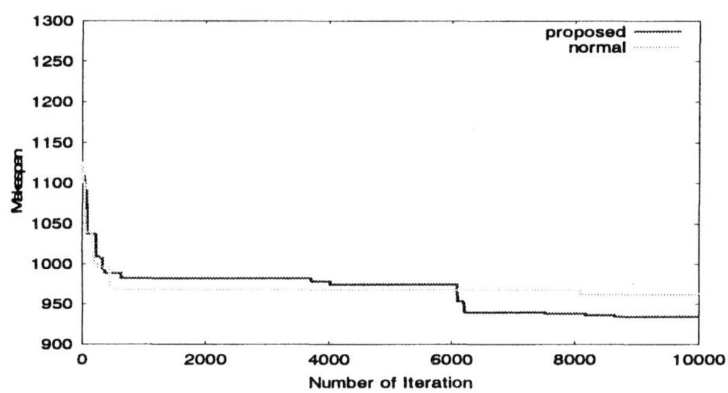

Figure 4. A convergent tendency

Table 1. The average and maximum number of evaluation of the objective function

	Average	Maximum
Conventional Method	11.9	29
Proposed Method	7.1	22

4.3 Visualization

Scheduling system with 3-D Gantt-Chart is developed (*Fig. 5*). This visualization tool aids the user to make the schedule as he wished. As the user can understand the schedule visually and can modify the schedule by use of this tool, the interaction between human and a heuristic algorithm is realized. The function of the system is described as follows. When the scheduling problem is given, the tentative quasi-optimal schedule is searched by use of aforementioned TS. Then the critical path and sensitivities are displayed to the user. The large influence factor (sensitivity) for job *i* expresses the easiness to improve the makespan by modifying the sequence of the neighborhood of the job on the CB. The paths are visualized by density plots on the Gantt-Chart as shown in *Fig. 6*.

This tool provides some of the quasi-optimal schedules that searched as candidates. The user can choice the most convenient schedule from them and also can modify the schedule freely by clicking the operations in the window. The system notifies the user whether the modifications is possible and suggests the idea of better schedule. The tool also can make the modified schedule active by the activation algorithm. (See appendix)

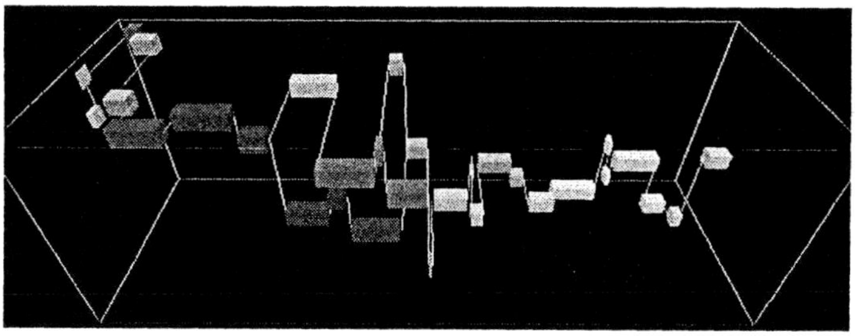

Figure 5. 3-D Gantt-Chart (Critical Path Mode)

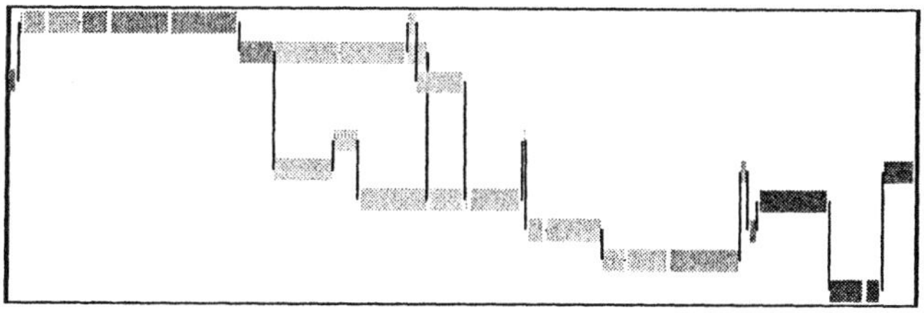

Figure 6. 2-D Visualization of sensitivity (density plot; The dark colored jobs are sensitive. The light colored jobs are insensitive)

5. CONCLUSION

It is very difficult to find the optimal solution to the large-scale for hard combinatorial problems such as JSSP within practical time. Thus the human interaction is considered as one of the important factors to solve such a hard problems.

In this paper, we presented an idea of sensitivity analysis and proposed a visualization system based on the TS with new neighborhood. Proposed system is considered as the useful interactive tool between human expertise and a heuristic algorithm. This system enables the user to improve the schedule effectively. For the future work, it is necessary to consider what kind of function is required for the interface.

6. REFERENCES

1. E. Nowicki, C. Smutnicki (1996). A Fast Taboo Search Algorithm for the Job Shop Problem, *Management Science*, **42**(6), 797-813
2. Takeshi Yamada, Ryohei Nakano (1996). Job-Shop Scheduling by Simulated Annealing Combined with Deterministic Local Search, *Journal of Information Processing Society of Japan,* **37**(4), 597-604. (in Japanese)
3. Takeshi Yamada, Ryohei Nakano (1997). Job-Shop Scheduling by Genetic Local Search, *Journal of Information Processing Society of Japan,* **38**(6), 1126-1138. (in Japanese)
4. J. E. Beasley (1990). OR-Library: Distributing Test Problems by Electronic Mail, *J. Opl. Res. Soc*, **41**(11), 1069-1072
5. B. Giffler, G. L. Thompson (1960). Algorithms for Solving Production Scheduling Problems, *Operations research*, **8**, 487-503
6. Muth, J. F., G. L. Thompson (1963). Industrial Scheduling, Prentice-Hall

APPENDIX

[Active Schedule] To find an initial solution or to improve a given solution, an idea of active schedule proposed by B. Giffler and G. L. Thompson is used[5][6]. An active schedule is defined as a feasible schedule that has a property where no operation can be made to start sooner by permissive left shifting. Algorithm to get active schedule is described as follows.

<Activation Algorithm>

1: Let $ES(O)$ and $EC(O)$ be the earliest start time of operation and the earliest finish time of operation. Both of these are the earliest time without upsetting the schedule that is already decided. Let a set of operations that is both not yet attended and feasible from the viewpoint of technical procedure be X . Let one of the operations whose $EC(O)$ is the earliest in $O \in X$ be O^* and the machine that process O^* be F^* .

2: In the operations $O \in X$ which is processed on the machine F^* , if $ES(O) \leq EC(O^*)$ then add the operation O to the set Y .

3: select one of the operations $O \in (Y \cup O^*)$ for the next processed operation on machine F^* . Iterate this procedure until X become the empty set.

The optimal schedule is obviously active schedule. Thus a better solution is obtained if the activation algorithm is applied to a given schedule.

37

ENTERPRISE INTEGRATION OF MANAGEMENT AND AUTOMATION IN A REFINERY[1]

Chengen Wang
Shenyang Institute of Automation, Chinese Academy of Sciences, P. R. China
e-mail: wangc@sia.ac.cn

Abstract: Traditionally, problems in a petroleum refinery were separately modeled and solved with respect to disciplines. The segregated implementations of various disciplinary technologies resulted in considerable barriers impeding the pursuit of global optimal performance. It is recognized that enterprise-wide integration of the managerial and automation systems is of fundamental significance for refineries to promptly respond to global market requirements. In this paper, the technical implementations are disciplinarily categorized into managerial and automatic systems. Then, typical managerial and automatic implementations in a refinery are depicted to give an insight perception of the heterogeneous data sources manipulated by these systems. Finally, an integration approach based on data reconciliation techniques is proposed to link up the heterogeneous data sources.

Key words: managerial systems, automation systems, enterprise integration, data reconciliation.

1. INTORDUCTION

In the past, problems in a refinery were usually tackled separately in accordance with specific technical or managerial disciplines. In practice each discipline is often initiated by different departments: SPC (statistical process control) by the quality department and production; TPM (total quality maintenance) by the maintenance department; APC (automated process control) by the engineering department [1].

[1] The research work is supported by the National Science Foundation of China, under grant No. 79931000.

However, it has been increasingly identified that many problems are difficult to be decoupled into simple technical or managerial sub-problems, and solutions to these complicated problems require synergic efforts from multiple domains. The disciplinary isolation of managerial and technical implementations gives a rise to the obstacles for refineries to achieve the global performance optimization. Enterprise integration becomes a significant way to accommodate the rapid environmental change, emerging new technologies and diversified customer requirements. Refineries are forced to breakthrough the disciplinary segregations by integrating strategies, human resources, production processes and control techniques.

Despite numerous understandings of integration, it is argued that the basic goal of integration is to improve overall system efficiency by linking its elements by means of communication networks and thereby obtaining a higher responsiveness and effectiveness of the whole system compared with the isolated operation of its components [2]. This paper presents an intermediate result of an on-going research project for developing and integrating the managerial and automation systems in a petroleum refinery.

2. REQUIREMENTS FOR MANAGEMENT AND AUTOMATION INTEGRATION

A refinery manufactures from crude oils, through a series of physical and chemical reactions, over a hundred kinds of products such as petrol, kerosene, diesels, naphtha, petroleum cokes, acids, polypropylenes and paraffin. Survival and prosperity in the turbulent environment require a refinery to concurrently tackle a wide range of challenges in many perspectives. For instance, a refinery has to efficiently collect and process the information concerning market demands, oil procurements and supplies, product distributions, and inventories. It has to promptly respond to demand change by optimally adjusting oil processing programs and product structures. It must keep balanced production flows among the pipelined facilities, and, at the same time, to maintain maximal throughputs of the critical facilities. Pursuits of high product qualities demand tight monitoring and control of a huge amount of coupled process variables.

The multiple disciplinary systems developed in a refinery can be viewed from four levels according to scientific disciplines. 1) ***Operation Control*** includes the basic automation techniques, such as data sampling, feedback control, optimal control and predictive control, which are directly used for monitoring and controlling various transmitters, regulators, instruments and facilities on production sites. 2) ***Process Coordination*** includes process modeling and simulation, process supervision, fault-shooting and prediction techniques. These techniques are applied to coordinate operations of an equipment complex and a group of units in the shop floors. 3) ***Production Optimization*** mainly concentrates on production planning, scheduling, quality control, inventory control, equipment maintenance and coordination of resources. 4) ***Enterprise Management*** basically deals with marketing, resources management, purchases, supplies and sales. Approaches and technologies that support business management and decision-making are used for finding solutions to the company-wide strategic problems.

According to business nature, these systems are classified into two categories: managerial systems and automation systems. Generally speaking, systems dealing with problems of production optimization and enterprise management are attributed to managerial systems. While the systems developed to manipulate problems of operation control and process coordination are attributed to automation systems. To achieve optimal operations, a refinery has to fill up the gap between managerial and automation systems.

3. FUNCTIONALITIES OF MANAGERIAL AND AUTOMATION SYSTEMS

3.1 Typical Managerial Systems In A Refinery

Managerial systems mainly handle structured data and support decision-making on business planning, marketing, finance accounting, human resource management, procurement and inventory control, equipment management, production planning and scheduling (as shown in figure 1).

Marketing and Sales: An important function of marketing is to calculate market demands with regard to actual customer orders and estimations obtained by applying forecasting techniques. Other routine functions of marketing and sales are collecting and processing data concerning customers, orders, market fluctuations, distributors and services. It maintains customer master-files, handles product price-quoting, controls customer credits, and manages bills of lading and invoicing.

Human Resource Management: It manipulates data concerning organization structures, employees, innovative culture, core competencies, and performance appraisals. Optimal business performance requires that human resources be efficiently managed with help of an organization model that explicitly represents organization structures, human attributes, roles and behaviors. Organization structures are used to specify the functional relationships among human resources, and to confine human behavior and activity space.

Procurement Management: Decisions on oil procurements must be made with full considerations of oil compositions and properties, in addition to other marketing factors such as prices. An important task of oil procurement management is efficiently analyzing the compositions, properties and special processing requirements of various crude oils. The procurement management also has to choose oil transportation means: such as by ship, trains, can-trucks and pipelines. It keeps close track of all purchase information, such as the en-route fleets.

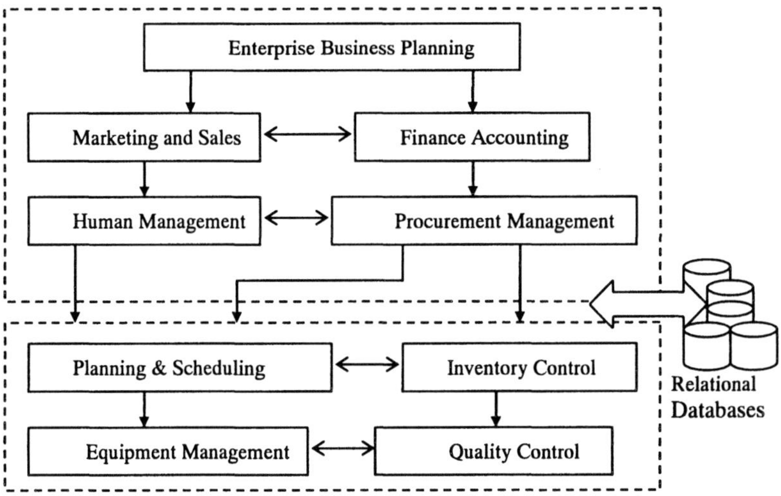

Figure 1. Typical managerial systems in an oil refinery

Production Planning: As crude oil compositions and properties vary in accordance with the yield locations, multiple manufacturing programs, production plans and schedules have to be developed to accommodate change in crude oils. The essence of production planning is to select the appropriate oil manufacturing programs and to optimally decide the volumes of different oil compositions at inlets and outlets of equipment settings.

Production Scheduling: Production scheduling takes as input short-term production plans, dynamically sequences production batches by allocating resources to them in specific time intervals. Production scheduling is to maintain stability in manufacturing processes by adjusting material flows to keep fluctuations in reasonable ranges. Multi-agent systems are very promising approaches for scheduling these decentralized manufacturing processes. Functionally, a scheduling system allows user to define process flows by graphical diagrams and helps users to determine compositions mix ratios of material fluxes. It sets the operation sequences and process parameters, aiming at increasing high market value oil products.

Product Quality Control: Qualities of oil products are basically determined by their compositions, chemical and physical properties. This means that petroleum product qualities depend on interactions and microstructures at molecular level. In practice, it is impossible to directly control product qualities because of lack of sensors that can on-line detect oil compositions and properties. Therefore, product quality models have to be built to associate product qualities (or molecular structures) with process parameters that can be simply sampled, such as flux, temperature and pressure. Product quality control system analyzes the on-line collected data and manually input records about manufacturing process statuses. It generates a variety product quality reports and suggests measures to be taken against quality deviations. In addition, the product quality control system manages quality standards, inspection routines, and problem-handling procedures.

3.2 Automation Systems In A Refinery

Automation systems are those that directly deal with operations of equipment settings. In a refinery, all facilities must operate in equilibrium states defined by precise values of the process variables. Nevertheless, external disturbances and frequent transitions of manufacturing programs for coping with the diversity of oil compositions and properties tend to cause fluctuations in the pipelined production facilities. Amplifications of these fluctuations increase the instability of manufacturing processes and production costs, and reduce product qualities. The main task of automation systems is to monitor, analyze and control the process variables, in order to retain optimal operations in all the production facilities.

Material Balance: The main objective of material balance is to get the demanded products by optimally adjusting the raw materials and the intermediary substances. Basically, there are three functions of a material balance system. First, a variety of manufacturing programs and production flows are applicable to the same crude oils, hence, yielding different product structures, quantities and profits. To obtain the desired products and to decrease invaluable products and wastes, a material balance system needs to determine what raw materials and how many are the inputs, and what products and how many are the outputs. Second, it accurately calculates and adjusts the flux compositions and volumes in different production lines. In another word, the inbound material flows and the outbound material flows in an equipment unit, a production zone and/or a line must be chemically balanced. Thirdly, it keeps material balances in multiple production zones to avoid amplifications of disturbances. Material balances reduce the risks that disturbances in a production zone be transferred to the adjacent zones.

Real-Time Monitoring: Most manufacturing processes in a refinery are partitions of compounds on basis of molecular movements. Product qualities are highly sensitive to change in process parameters, which need to be closely observed. The main functions of a real-time monitoring system are as follows. Configuration utilities provide visual development tools for configuring manufacturing processes, which permit human staffs to observe on-line process evolutions on screens. It provides I/O drivers that support the commonly used communication protocols, such as RS232/245, Ethernet and PROFITBUS. Thus, the equipment and devices supplied by different vendors can be easily interconnected with the monitoring system. The monitoring system also has interfaces with the commonly used database management systems. And the Alarm and Event function generates alarms on disturbances, deviations and errors in production processes. It also keeps historical records of alarms and events to support the after-event analysis.

Process Optimization and Advanced Control Systems: Operations of the petrochemical facilities involve a large number of tightly coupled variables. Disturbances to a control variable can easily influence other variable, exhibiting bullwhip-effects. The objective of operation control is to keep all variables within desired ranges and to maintain optimal performance of the facilities. The operation control is conducted by two systems: namely process optimization and advanced control.

Process optimization creates process models and based on which calculate the optimal control parameters subject to economical, technical and managerial constraints. A process optimization system, working in an off-line mode, aims at increasing qualities and volumes of desired products, while decreasing material and energy consumptions. To this end, it determines which variables are to be controlled, and which variables are to be manipulated. It decides what process parameters are measured and what control strategies or laws are applied during operations. It calculates optimal or near-optimal process parameters specifying the target trajectories, and transfers them to the advanced process control (APC) system. Objectives of advanced process control are reducing disturbances, increasing equipment adaptability and stability and gaining maximum profits, subject to process constraint conditions. An APC system evaluates the real-time sampling data of outputs collected by various sensors against the target values. Once, drifts or errors are detected, an APC system will apply control strategies, such as PID and multi-variable predictive control, to force unit operations returning to target trajectories.

4. INTEGRATION OF HETEROGENEOUS DATA SOURCES

In refineries, data sources handled by applications are managed in different databases on basis of data processing frequencies and influencing scopes. As shown in figure 2, the essential foundation of enterprise integration is information integration across different hardware platforms.

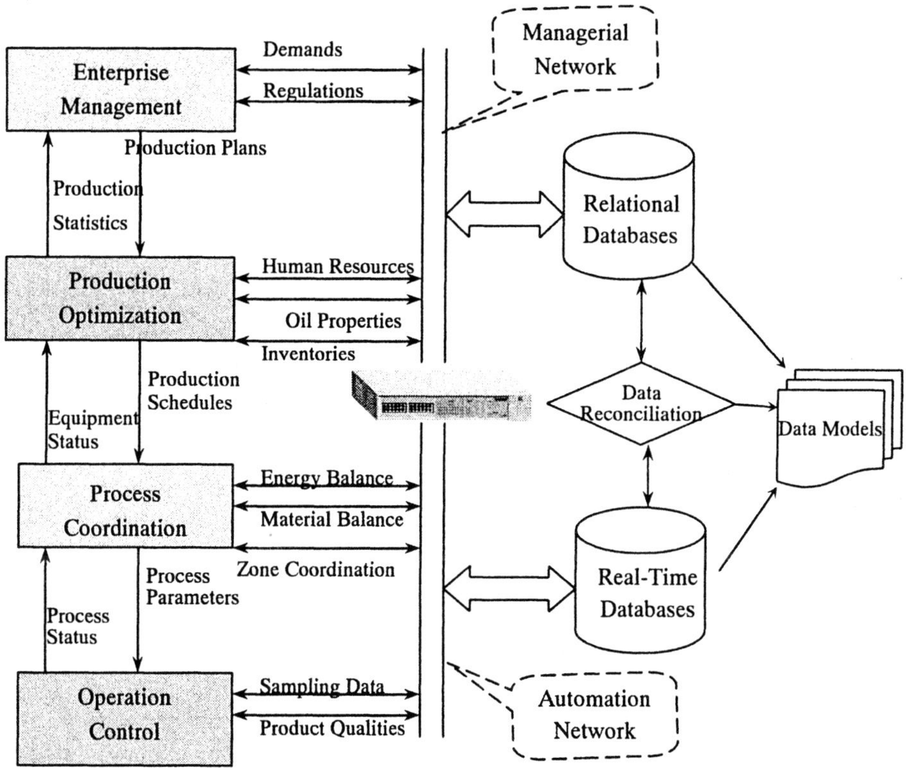

Figure 2. A framework for management and automation integration

In perspectives of process optimization and control, real-time databases are built to include real-time collected on-site data, graphical system configurations, interfaces for control systems and alarm sets. Real-time database management systems are a disciplinary conjunction of the real-time system and database management, which are required to process data and transactions in limited time intervals. Cost-effective technologies of time scheduling, data compression and resource allocations are core competencies of real-time database management systems. Real-time database systems also provide configuration utilities, APIs, and equipment interfaces.

Integration of the above-mentioned heterogeneous data sources is attained by data reconciliation technologies in respect to enterprise data models. Here, data reconciliation has two basic functions. First, data reconciliation maintains information consistency among the data sources. In manufacturing environments, there are many reasons that will cause information inconsistencies or mismatches, which include lower sensor precisions, inaccurate device calibrations, misinterpretations, signal drifting, duplications and operation errors. Data reconciliation evaluates the nominal values against enterprise data models, and detects inconsistencies. According to intrinsic data relationships, data reconciliation trims off errors and smoothes the data to actual values. Second, data reconciliation provides an interoperation mechanism among the databases and binary files. It creates data snapshots from an information source and copies them to target sources. For instance, data reconciliation continuously duplicates the changing real-time data and sends the duplications to update existing data in relational databases. Data reconciliation provides a data subscription service for notifying related applications and data sources about change in each data source. To this end, data reconciliation continually filters and mergers data objects from multiple data sources to present specific data views.

5. CONCLUSIONS

Comprehensively integrating the managerial and automation systems in a refinery is a tremendously challenging task that requires synergic efforts of professionals from multiple disciplines. It is identified that seamless integration of heterogeneous data sources is the basic concern of integrating the managerial and control applications. Consequently, the managerial and automation systems are presented to unveil what types of data and how they are handled. In practice, numerous data sources are built on basis of intrinsic characteristics of business functions that uses these data. It is proposed, in this paper, that information integration can be attained by data reconciliation approach with regard to global data models.

REFERENCES

1. Schippers W. A. J. (2001). An Integrated Approach to Process Control, *International Journal of Production Economics*, 69, 93-105.
2. Kosanke K., Vernadat F., Zelm M. (1999). CIMOSA Enterprise Engineering And Integration, *Computer in Industry*, 40, 83-97.

PART IV

Man-System Collaboration

38

CAI SYSTEM WITH MULTI-MEDIA TEXT THROUGH WEB BROWSER FOR NC LATHE PROGRAMMING

Yoshio Mizugaki[1], Koichi Kikkawa[1], Masahiko Mizui[2], and Keisuke Kamijo[2]
[1]Dept. of Mechanical And Control Eng., Kyushu Institute of Technology, Japan
[2]Dept. of Mechanical Eng., Kyushu Kyoritsu Univ., Japan
e-mail: mizugaki@mech.kyutech.ac.jp

Abstract: A new Computer Aided Instruction (CAI) system for NC lathe programming has been developed with use of multi-media texts including movies, animations, pictures, sound and texts through Web browser. Although many CAI systems developed previously for NC programming consist of text-based instructions, it is difficult for beginners to learn NC programming with use of them. In the developed CAI system, multi-media texts are adopted for the help of users' understanding, and it is available through Web browser anytime and anywhere. Also the error log is automatically recorded for the future references. According to the NC programming coded by a user, the movement of the NC lathe is animated and shown in the monitor screen in front of the user. If its movement causes the collision between a cutting tool and the lathe, some sound and the caution remark are generated. If the user makes mistakes some times at a certain stage in learning NC, the corresponding suggestion is shown in the form of movies, animations, and so forth. By using the multi-media texts, users' attention is kept concentrated during a training course. In this paper, the configuration of the CAI system is explained and the actual procedures for users to learn the NC programming are also explained too. Some beginners tested this CAI system and their results are illustrated and discussed from the viewpoint of the efficiency and usefulness of this CAI system. A brief conclusion is also mentioned.

Key words: CAI, NC lathe programming, multi-media, Web browser, CMI

1. INTRODUCTION

NC (Numerical Control) programming is the basic and fundamental technology in manufacturing. NC lathes are the most popular machine tool and widely used in actual machining processes. Its programming seems to be easier than that of 3-axis or higher order machine tools, but it is still difficult for beginners to learn it by oneself. The reason is that it needs the spatial perception on the collision avoidance between a cutter and a work. So CAI (Computer Aided Instruction) systems on NC lathe programming have been strongly requested especially in industries. The objective of this study is to develop a user-friendly CAI system for NC lathe programming in the Internet era.

2. DEVELOPMENT OF CAI SYSTEM

2.1 NC lathe programming and procedures of CAI development

Fig. 1 shows the specimen of the exercise in the CAI system developed. As easily understood, the NC lathe programming requires not only the geometric aspect including the avoidance of collision, the calibration of the position of cutting edge and so forth, but also the information of machining conditions such as the cutting speed, the feed per revolution, the depth of cut and so forth. The specimen needs the operations of normal turning, tapping, grooving, chamfering and so forth. Therefore its programming should be written with some fundamental NC codes such as G, F, M functions as shown in Fig. 2.

The CAI system has been developed in the procedures shown in Fig. 3. The procedures consist of three stages. The first is about making of the CAI text based on the education theory. The second is about the determination of the CAI system configuration based on the CAI theory. The third is about the coding of the CAI system and its uploading in the Web site based on the computer theory. In the below, the main two stages are explained according to the actual procedure of the development.

2.2 Content of CAI text

In the first stage of CAI system development, the content of NC programming training course held in 2000 through 2001 were surveyed through Web search engines that were google, excite and lycos. The keywords of search were “NC, lathe and training course.” As the result, ten case studies were collected. Considering the contents collected and surveyed, the following text contents were adopted in the CAI system developed.

1. General guidance of NC lathe
   Fundamental configuration of NC lathe, Features of turning, Automatic Tool Changer, Coding process of NC programming.

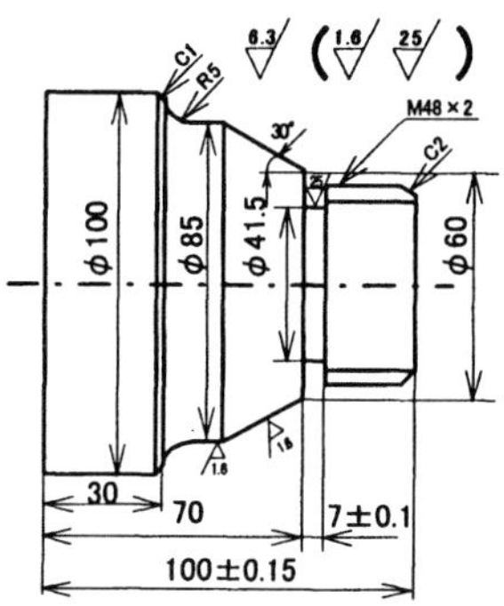

Figure 1. Specimen of work For NC lathe programming.

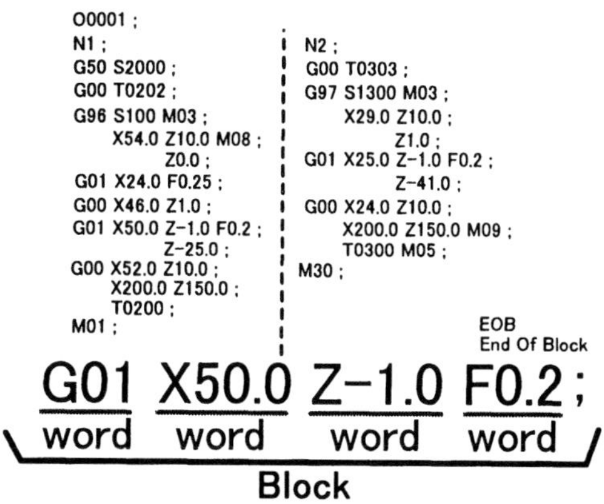

Figure 2. Sample codes of NC programming.

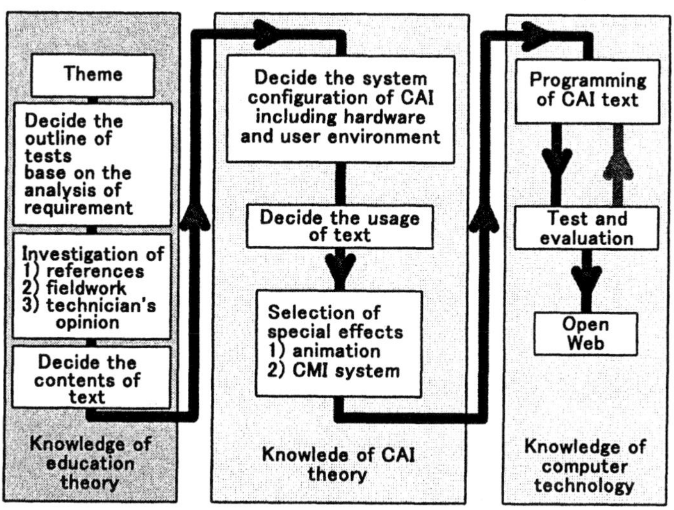

Figure 3. Procedures for the development of the CAI system.

2. Fundamental terms of NC lathe programming
   Coordinate systems of machine tool and workpiece, zero offset, absolute and incremental programming, absolute and incremental dimension, etc.
3. NC programming
   Program start, end of program, Block; word; address; character, Sequence number, dimension word (X, Y, Z, U, V, W, P, Q, R, A, B, C, D, E), F , G, S, T and M function words.
4. Instruction code of G, S, F and T function series
   Rapid traverse, linear interpolation; constant surface speed control; feed rate, override, ATC tool registration

The above content is equivalent to the content of ordinary NC training course in 2 to 3 hours. The CAI system was designed so that a trainee could finish and master the above content generally within one and a half hours.

2.3 Configuration of CAI system and its feature

In the second stage, the requirements for the CAI system are below.

1. Trainees as users of the CAI system are university students of the department of Mechanical Engineering.
2. The expected goal is that trainees understand the outline of NC lathe programming and have the experience of generating NC programs through a simple exercise.
3. All operation in the CAI system can be executed through a Web browser.
4. The number of trainees is 6 to 8 persons per class.
5. The time for CAI training is expected 90 minutes.

Fig. 4 illustrates the hardware configuration of the CAI system with using PCs for CAD/CAM facilities in a classroom of Kyushu Kyoritsu University. (The specification of CPU is Pentium III, 667 MHz, 128 MB of memory with a 17" monitor.)

The fundamental policy of giving suggestions in this study is the tutorial type in the CAI system category. In this tutorial type, a proper suggestion is shown in another window of Web browser when a trainee requests a suggestion by oneself and/or makes a mistake in answering a question. Usually the questions are displayed with a drawing of the specimen and the answer columns follow, as shown in Fig. 5.

The button for the request of suggestion is prepared as shown in Fig. 6. In this suggestion, some visual assistance including drawing, animation and video would be shown too. When the tool path was programmed and answered by the trainee, its motion would be displayed as the animation in the window. At the same time, the collision detection between the cutter and the workpiece would be carried out as the 2-dimensional simulation and the occurred collision would be animated with using the software of FLASH 5 developed by Micromedia Co. (See Fig. 7.) In the context of the CAI system, the various animations are prepared and embedded in 17 scenes totally.

One of significant features of the CAI system developed is that this system can be accessed from anywhere and anytime through a Web browser. This feature offers many advantages for trainees as mentioned below.

1. A trainee can learn and master the subject according to his/her own speed of understanding.
2. A trainee can learn the subject whenever and at any number of times he/she wants.
3. Visual and sound aids with the multi-media text can encourage the attentiveness and concentration of a trainee and the efficiency of his/her learning.
4. The quick response of the CAI system offers the interactive and real-time effects in giving marks of questions.
5. The self-learning mechanism is supported based on the Computer Managed Instruction (CMI) system and automatically the proper CAI text would be shown to a trainee due to the state of learning.

The CMI system deals with the following three parts. The first one is to manage the trainees' records. The second one is to mark the answer of trainees. The third one is to analyze the trainees' records and their state of answer. Based on the analysis of the trainee's state, proper suggestions would be presented or shown in another window. Fig. 8 shows the total configuration of the CAI system developed. This CAI system has been improved through the trainee's execution.

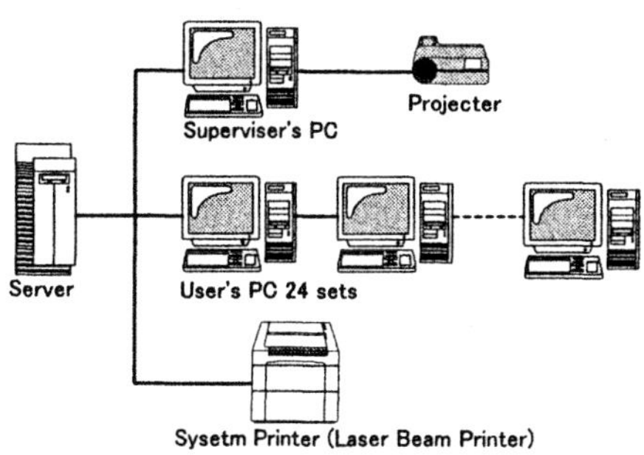

Figure 4. Hardware network in a classroom equipped with the CAI system.

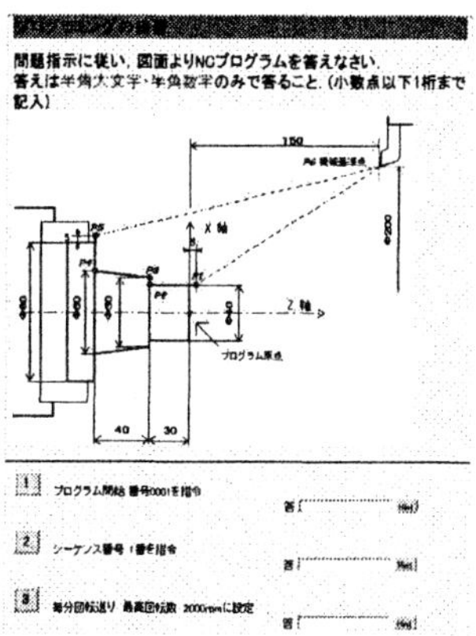

Figure 5. Sample of exercise system in the CAI.

3. EXPERIMENTS

3.1 Outline of experiments

In order to evaluate the accomplishment of the CAI system developed, the comparative experiments had been done. Four different texts of NC lathe programming were handed to four groups of students respectively. Then the result of learning in each group was compared with each other. Each group consists of three students and total twelve students were cooperated. Every student joins only one group. The four different texts and assistance to each group are listed below.

A: The text is the multi-media text provided in the CAI system. The organized assistance by a teacher has been applied progressively when a student faces to some difficulty in solving a question. Before starting the CAI experiment, the additional explanation how to use the CAI system was delivered to students for five minutes. Five minutes are excluded in the total time of experiment.

B: A printed text including the same content of the multi-media text in the CAI system has been used. The organized assistance by a teacher has been done only based on the request of a student.

C: The text is the multi-media text provided in the CAI system as well as in A. There is no additional explanation for students how to use the CAI system. The organized assistance by a teacher has been done only based on the request of a student.

D: The text is the revised multi-media text for the revised CAI system. The revised CAI system has been improved according to the feedback information of students. No organized assistance by a teacher.

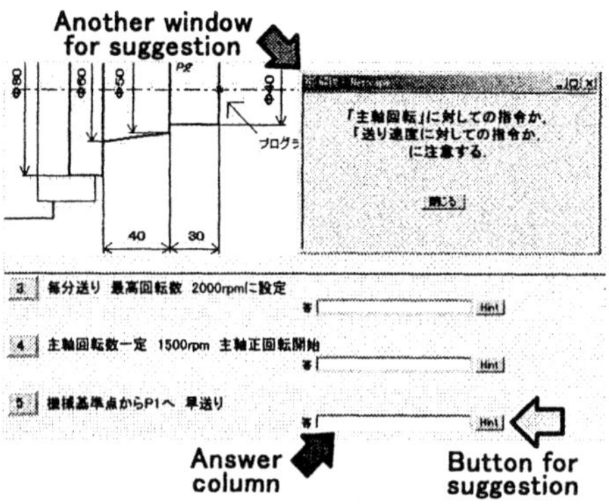

Figure 6. Helpful suggestion in another Window.

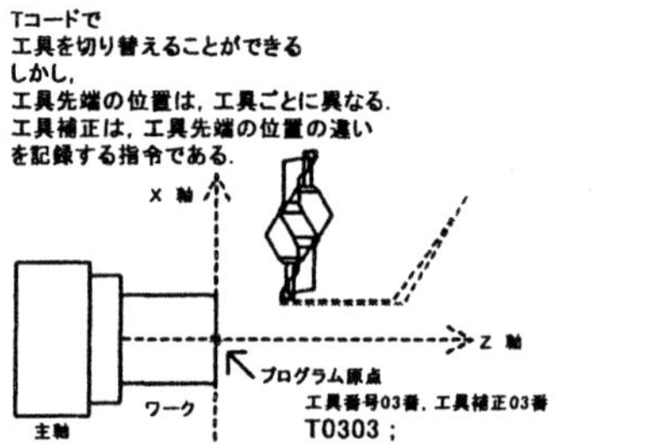

Figure 7. Advised suggestion for the trainee based on one's request or error.

The results of the exercise by the four groups have been compared with each other and summarized in Fig. 9 (a) and (b). The exercise consists of 13 questions. Students reply a question to get the right answer, and then proceed the next question. Fig. 9 (b) summarizes the time required from the start of self-learning to the end of self-learning.

3.2 Results

The CAI system seems to be more useful and helpful compared with the printed text as shown in Fig. 9(a). The result of Group B shows that the printed text is unsuitable for beginners to master the NC programming.

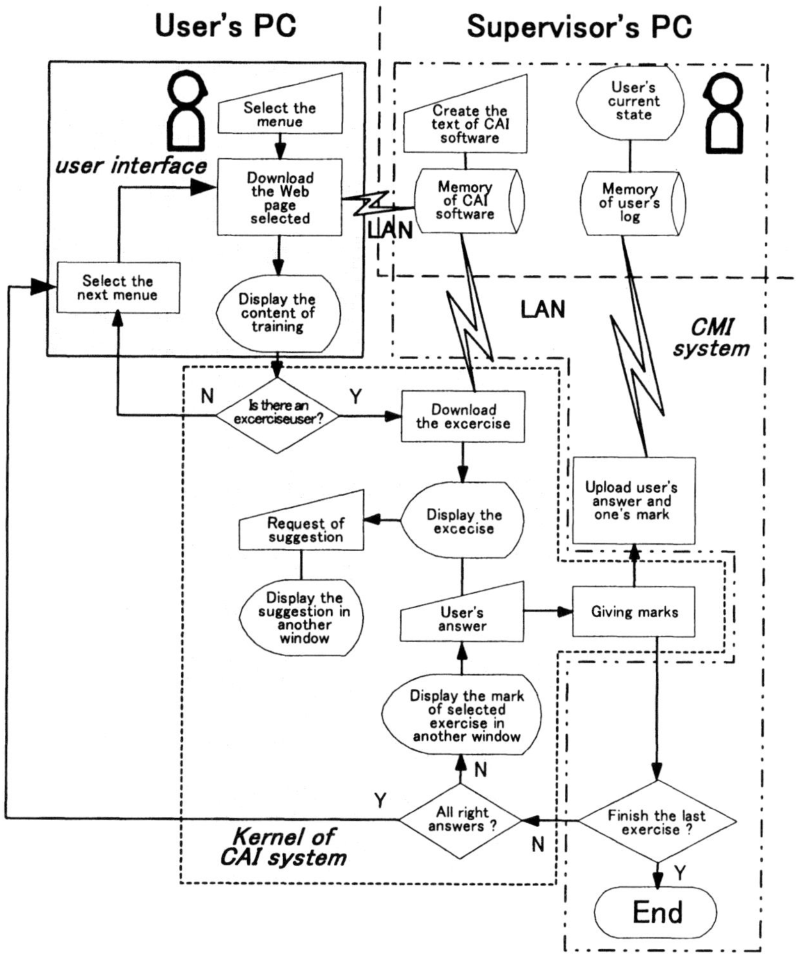

Figure 8. The schematic configuration of the CAI system.

After the interview to the members of Group B, it was found that they skipped the important explanation of NC programming in the printed text.

The printed text has a few educational effects in the latter trial except the first trial. Fig. 9 (b) shows that the combination of CAI system and organized assistance is very useful in learning. It suggests that the organized assistance had better be embedded as the improvement of the CAI system in future.

4. CONCLUSION

In order to give a CAI system the independence on the locality and on the time-restriction, the web-based multi-media-assisted CAI system has been developed. Through its development and the comparative experiments with other learning texts, its usefulness has been verified. Especially trainees could keep attention during the training. The CAI system is essential and very promising in the learning of NC programming. The well-organized assistance should be investigated deeply and embedded in it as the kernel of the machine intelligence. It is the further study the future CAI system.

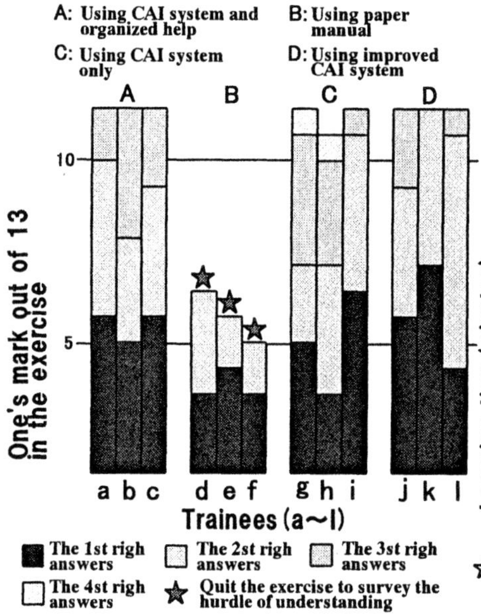

Figure 9(a) Experimental results on the trainees' marks.

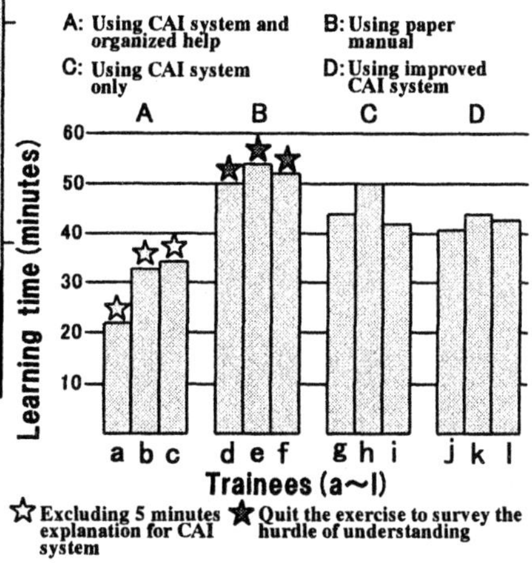

Figure 9(b) Experimental results of trainees based on categories of used texts.

REFERENCES

1. IWATA K. (1991). Exercise of CAD/CAM/CAE/CIM, Kyoritsu Press.
2. FUJIMOTO H. (1996). Computer Integrated Manufacturing System, Corona Press.

39

WEB BASED OPERATION INSTRUCTION SYSTEM USING WEARABLE COMPUTER

Yoshiro Fukuda[1], Takao Kurahashi[2], and Yoichi Kamio[3]

[1] *Hosei University, Japan*

[2]*NTT Cyber Lobaoratory,Japan*

[3] *Toyo Engineering Corporation, Japan*

e-mail: fukuda@k.hosei.ac.jp

Abstract: In our research, the Web-based Instruction System using the wearable computer is developed. The system consists of a wearable computer, a wireless communication system and web-applications. Web-applications are included to a time estimation system, a simulation system, an active instruction manual system, a scheduling system and a statues acquisition system.

In this paper, a system configuration is described, each application is explained, and the result of application is summarised. And in this paper, we pointed out the issues for using the system in the assembly line.

This system is developing under the GLOBEMEN project in the IMS program.

Key words: Wearable Computer, Operation Instruction, Simulation, Web Application

1. INTRODUCTION

Recently, manufacturing systems have begun change from mass-production to small-batch and frequent change of production volume. And also, manufacturing systems change from full-automation to manual operation. In order to operate the system, we should consider instructing effectively the operation to operators in the system [1]. Especially, in global manufacturing system, operation planning division and operation site are at long distance. It is necessary to direct operations based on operation situations.

In order to solve the issues, the Web-based Instruction System using the wearable computer is developed. The system consists of a wearable computer, a wire-less communication system and web-applications. Web-applications include a time estimation system, a simulation system, an active instruction manual system, a scheduling system and a status acquisition system.

The web-applications are located in the server in the network. The operators wear the computer and operators can get the necessary information from the server. The wearable computer is equipped with a camera and a bar-code reader. Operation planning division staff can get the operation status through the camera and bar-code reader. Operators and Operation planning division staff can communicate through the system and network.

In this paper, a system configuration is described, each application is explained, and the result of application is summarized. And in this paper, we point out the issues for using the system in the assembly line.

This system has been developed under the GLOBEMEN project in the IMS program.

2. OPERATION INSTRUCTION SYSTEM

Recently, assembly systems are installed a cell production system, because production volume becomes smaller than ten years ago. The cell production system is a small shop that consists of a few operators [2]. Operators in the Cell have many operations to be assembled. Operators have responsibility to complete the product. Typical Cell system is shown in Figure 1. However, operators can not share training time for new model to be assembled. Therefore, we are necessary to use instruction system for assemble operations in real time.

Figure 1. Typical Cell Production System

We analyzed operations in the cell production system, and then we extract the requirements for a real time operation instruction system. These issues are the following;

1. To indicate the information for need to know.
2. To indicate the newest information.
3. To inform to operator on moving to station.
4. To indicate the easy understandable information.

To satisfy the requirements, we are developing the web-based instruction system by wearable computer (WBIS).

The system consists of the simulation system on the web-application, wearable computer and the wireless LAN. The system configuration is shown in Figure 2. This wearable computer is XYBERNAUT Mobile Assistant IV.

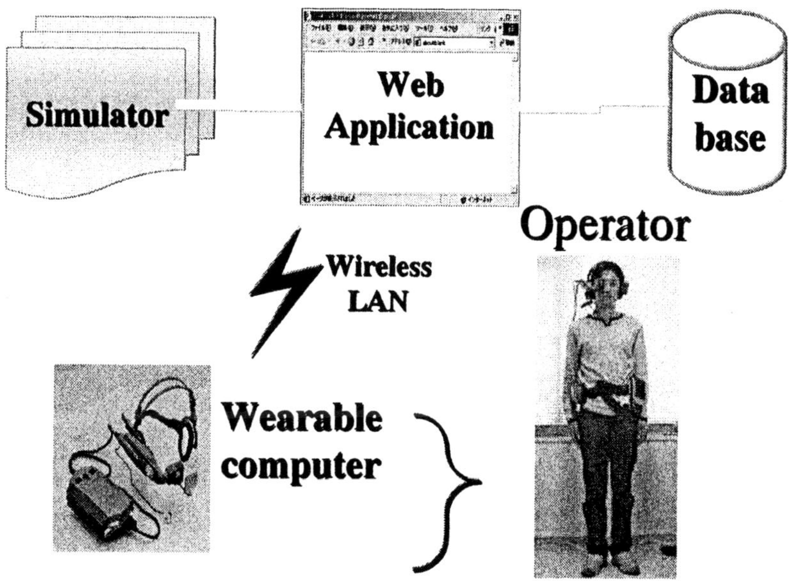

Figure 2. The System Configuration

Web application is included the instruction manual, simulation and scheduling based on the progressing operation. The web-application is called "web-instruction simulation system (WISS)".

Wearable computer can indicate the operation instruction on working the operation [3] [4]. Wearable computer is equipped the wireless LAN. Therefore the operator can get the context for instruction from web-server computer. Also wearable computer has a video-camera near by operator's eye. The camera can take a view and send the visual data to server.

3. WEB-INSTRUCTION SIMULATION SYSTEM

3.1 System Configuration

The Web-Instruction Simulation System (WISS) is a core system in the Web based operation instruction system. WISS provides the contents to the wearable computer. This system configuration is shown in Fig 3.

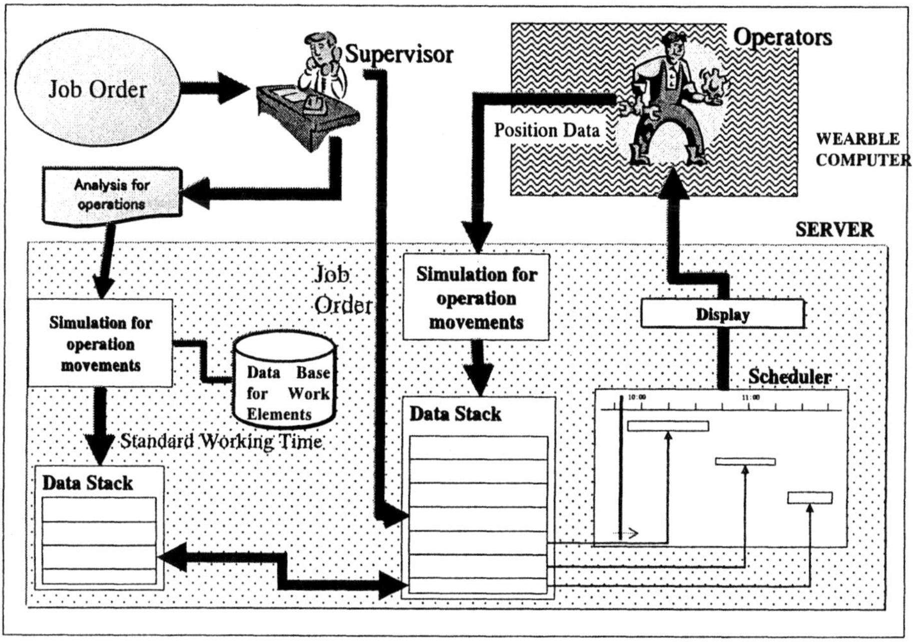

Figure 3. WISS System *Configuration*

WISS has a simulation for work time estimation. This simulation decides the work time and work movements depend on the work order from the supervisor. The simulator has a data base for work elements in the assembly shop. The data base provides the work element for assembly job to simulator. The simulator can calculate the working time. However, the calculated time is not included the moving time from present position to working position. Operator input the position data by wearable computer on finishing pervious work. Simulator will indicate the next work and the next position on the screen in the wearable computer. And also, if operator requests the working procedure, wearable computer can provide the manual to operator. The operator manual is edited by the simulators.

3.2 Simulator for time estimation

This simulator is for estimating the assembly time and the movement for the assembly work. Assembly work breaks down into the elements of work, such as handling, walking and waiting. The element also breaks down into motion elements such as Reach, Grasp, Move, Position and Release with MTM method. The system calculates the assembly time based on the motion elements. The system has the formulated functions for estimating the time [5]. Example of the function is followings;

Time of operation to hold (Grasp)

S : The size of the parts to hold (path)

$2<S<8$ (mm)

In the case of perpendicular supply

$$T= 10^{0.94-0.3S} \times 1/100$$

In the case of level supply

$$T= 25.12\, S^{0.94-0.56} \times 1/100 \text{(Unit 1/100sec)}$$

And then, each motion elements are aggregated to assembly work by the simulator. The simulator can consider the effective sequence and can display the result, as figure 4. The human model in the simulator can be animated. And the result is also in the manual for assembly operator. The human model can be abatable based on the individual physical size.

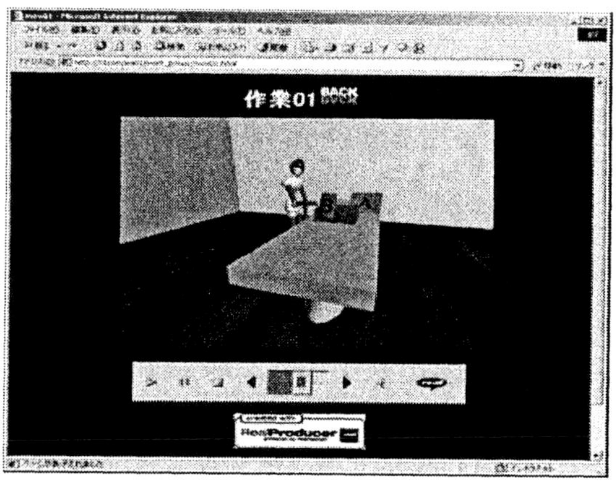

Figure 4. Simulation Output

3.3 WISS human interface

WISS is equipped some applications, such as moving analysis, scheduling, position information, assembly manual, notification, and animation for assembly work. And also, WISS has a database for standard time and a database for work assignment.

However, operators do not need much information on working. Therefore, WISS is separate the information which is processed by supervisor and operator. Supervisor side prepares the moving motion and the standard time for assembly by simulator and analyzer. And Supervisor side makes a planning and a schedule by scheduler, and makes manuals and notifications for assembly work by the simulator. This information is stored in the web-server by HTML format.

The operator side inputs the individual physical data before working. And the operator inputs signal after working. If the operator needs the information, they request the data on the screen in the wearable computer. Software configuration is shown in the Figure 5.The software is running on the Table 1 specifications

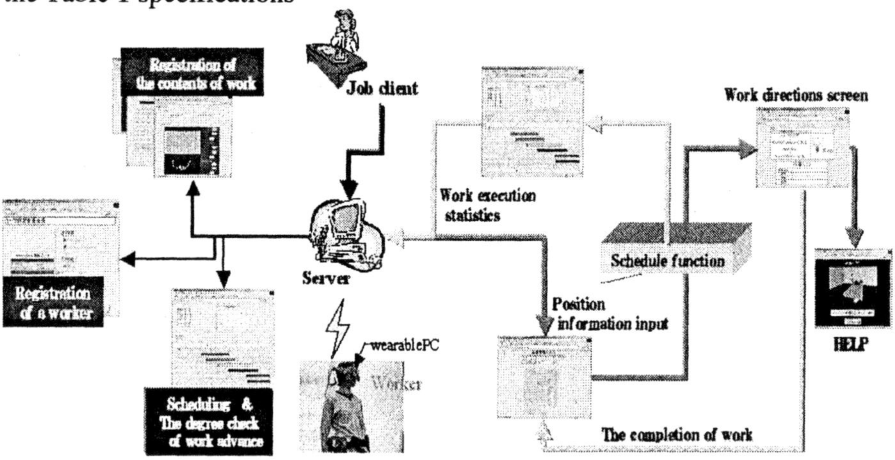

Figure 5. Software Configuration in WISS

Table 1. Software Specification

Operation System	Windows2000 (server), Windows CE(Client)
Web server	IIS5.0 (Internet Information Server)
Database Server	SQL 7.0 & Microsoft Access 2000
Programming	ASP, SQL, Java, HTML, JavaScript
Browser	IE5.5

4. AN IMPLEMENTATION EXAMPLE

For evaluating the proposed system, we have done a simple experiment. The experiment environment is U-shape assembly system. The operator equipped a wearable computer. Operator assemble three parts into a product depend on the web-instruction. The experience scene is in Figure 6.

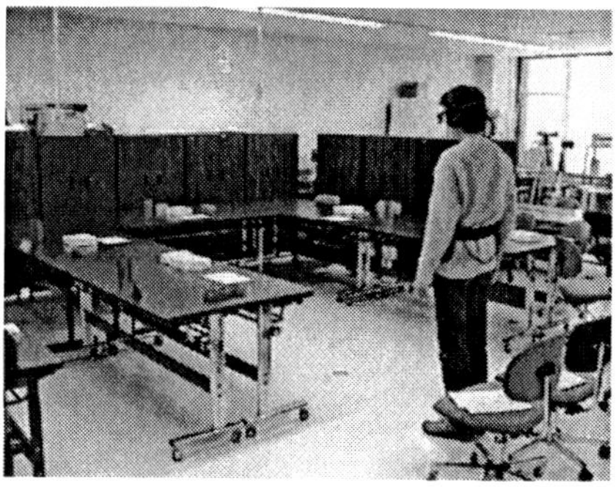

Figure 6. Experience Scene

We have two experiences, one is with wearable computer, and another one is without wearable computer. In each experience, 8 students are as testee.

The results are Figure 7 and Figure 8.

According to the results, we can recognize that the WISS squeezes the amplitude of the working time.

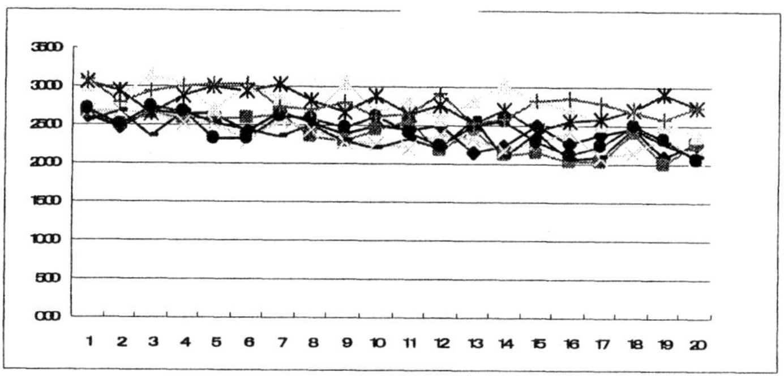

Figure 7. The Assembly Time without WISS

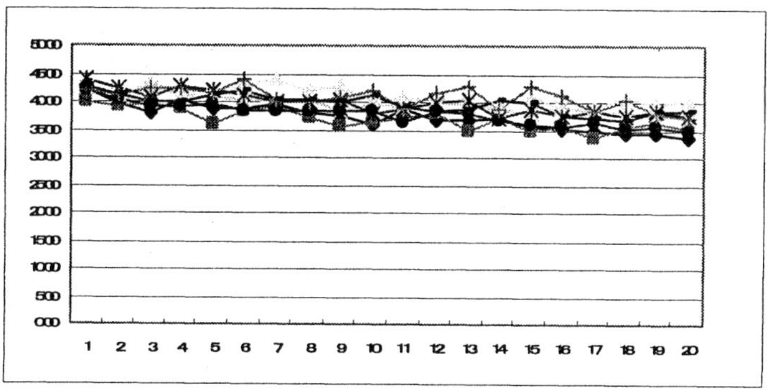

Figure 8. The Assembly Time with WISS

However, the assembly times with WISS are bigger than the one without WISS. The testees said that WISS is helpful for the instruction, but WISS is disturbed the operation. We are necessary to modify the interface of WISS.

5. CONCLUSION

We have developed the Web-based Operation Instruction System using a wearable computer. And we applied the system to the assembly system. The system can communicate the data on working the job. But some problems are clarified by our experience. Wearable computer has a poor display function on working, has a poor input function on wearing.

Simple job is not necessary to use the system; however we can believe the usefulness tools for maintenance and repair field. Web-manual and web-application is powerful tool for instructing the complex work. And we can change easily the content in the web-server from assembly to maintenance.

ACKNOWLEDGEMENT

This research has been carried out under GROBEMEN project, one of the IMS International Research Program. The authors would like to acknowledge all members of GLOBEMEN project for variable support.

REFERENCES

1. Morita K, Kawashima K and Fukuda Y. (2000). Web-based Maintenance Manual with Tree-Dimensional Simulation Model, Global Engineering Manufacturing and Enterprise Networks, Kluwer Academic Publishers, 212-219..
2. Mori K, Yoshikawa N,. et.al. (2000). Neo-Kaizen Application on the Generic Operations Support and Renewal, Global Engineering Manufacturing and Enterprise Networks, Kluwer Academic Publishers, 236-243...
3. Barfield W. and Caudell T. (2001). Fundamentals of Wearble Computer and Augmented Reality, LEA
4. Yanagihara Y, Muto S. and Kakizaki T(2001). Evaluating User Interface Multimodal Teaching Advisor Implemented on a Wearable Personal Computer, Journal of Intelligent and Robotic Systems, Vol.31. No.4. 423-438
5. do M, (1998) Research on Functional Type Working Time Estimation System, Doctor Thesis, Hosei University
6. Baber C, (2001) . Wearable Computers: A Human Factor Review, International Journal of Human-Computer Interaction, Vol.13. No.3. 123-145.

40

MODEL-BASED DESCRIPTION OF HUMAN BODY MOTIONS FOR ERGONOMICS EVALUATION

Sayaka Imai
Dep of Computer Science, Gunma University, Japan
e-mail: sayaka@cs.gunma-u.ac.jp

Abstract: This paper presents modeling of Working Process and Working Simulation factory works. I focus on an example work (motion), its actual work(motion) and reference between them. An example work and its actual work can be analyzed and described as a sequence of atomic action. In order to describe workers' motion, some concepts of Atomic Unit, Model Events and Mediator are introduced. By using these concepts, we can analyze a workers' action and evaluate their works. Also, we consider it as a possible way for unifying all the data used in various applications (CAD/CAM, etc) during the design process and evaluating all subsystems in a virtual Factory.

Key words: Motion Modeling, Info-Ergonomics, Model Event, Mediator

1. INTRODUCTION

Recently some new applications supporting the designing process of a virtual factory, simulating its work are gaining popularity in areas like process planning and work scheduling and any other evaluation. In addition to the CG, they involve many new other types of data, such as video and X-ray images, sensing information, etc. High complexity of those data causes a lot of difficulties to the researchers and developers. Here we must point out two problems which should be methods for solved for the future development. First, as those applications have been using mainly application-oriented data, methods for sharing data with other applications are quite difficult and very often --- impossible. In our opinion, a flexible automation system requires data about a given product to be used in all applications. Second, we think that human factors are usually overlooked or underestimated in eco-factory. Human labor has an important role in the manufacturing process. It is examined from the viewpoint of industrial engineering, ergonomics, etc. and results are used in product planning, working analysis, and work environment design. But aspects like efficiency of works and comfort of the workers are not explored completely.

As a promising solution to the above problems we have proposed the use of Real World Database (RWDB) and ergonomic simulation[1]. RWDB is able to capture various types of data, namely characters, video images, 3D graphics, and shapes of 3D objects. We call all those *Real world data*. The data of all types are unified and stored in a Multimedia Database (MMDB).

On the other hand, we consider it necessary to focus on human-machine cooperation, especially for employees in the factories and analyze their work evaluation, environment, and amenities against this background. We proposed the human body modeling method from two viewpoints in the framework of Info-Ergonomics Concept[2][3][4]. One proposed modeling method is description of precise human body and by using the model, motion evaluation system from medical point of view is offered[2]. As the other modeling method, we also introduced mediator-based modeling of workers' (body/structure and motion) in a factory. The mediator-based model is transformed from measured data, and standardized to common models of body/structure and motion. However, from another point of view, measured data from real world have low-level semantics such as a worker's hand is touching to drill[3][4].

In this paper, we offer a model for analyzing employee's work evaluation and using the results in the manufacturing process design. We also propose a methodology for modeling of the human body and its motions in order to store and query them on a database. For storing all available data about human motion, we need a precise human body model as well as method for representing its motions.

2. DESCRIPTION OF HIERARCHICAL PROCESS AND ACTUAL WORK

In factory work, a process is designed with hierarchical structure divided into such classes of a process, element works, and element operations[5]t is usually carried as shown in Figure 1. An aim of such design is the increase of efficiency by schematizing the workers' work, and making manuals for them.

From a worker's point of view, a process and element of work have targets of each work. It can be called intention of the work. However, in actual work, even the intention is the same, the sequence of work operation changes with individual, time, degrees of skill and/or fatigue, etc. The sequence is also different from the process as it was design to be performed. For instance, if two products are manufactured simultaneously in parallel in the same process, the efficiency of work can be increased. Identically a difference can appear between the experts and the unskilled operators. Only an individual evaluation of the difference to process can support an evaluation of this difference in their efficiency and the improvement of the performance of the work (process).

Figure 1 show a concrete example of process hierarchy in a real factory. In this process with ID "B24" is decomposed into five phases, i.e., element works. In the example, the process is dedicated to the assembling of a transmission box named "kbt024" in this factory. Its first element work is to bring the material object (named "kbt024-1" which is an exterior part of "kbt024") and to set it on the workbench. The second element work is to put a gear into the material. Then the third is to fix it up, the forth is to verify and to confirm mobility, and the fifth is to put the complete object in a specific place. Each element work is described via some typical operation, i.e., element operation.

From another view, a designer can design a process hierarchical as a manual like Figure 1. As opposed to this, in actual works, all workers do not work in the same operation. For example, right side of Figure 2 shows an example of element operation description for the element work "B24-3" of left side of Figure1. In the example, stretching arm to pick up a screw bolt from the box of screws, twisting the bolts, and pushing an electric screwdriver to tighten the bolts are described as element operations. It is thought that this process description is the general description applied by each worker. It is what described the intention of work to each worker.

In order to compare Workers' motion in the same elementary operation and evaluate their actions, some methods which can describe actions different from designed elementary operation (action standardization) and difference in workers' action in same elementary operation (action description and modeling) are required. For description and modeling of action, we introduce Event Graph, which actualize comparing motions in same action by querying motions of a part of workers' body. On the other hand, for schematization of workers' motion, we introduce Mediator based modeling, which actualize comparing actions in same operation by querying kinds of operation.

3. SCHEMATIZATION OF FACTORY WORKERS BEHAVIOUR

3.1 Atomic Units

Here we define the basic elements in our model. We introduce atomic units as simple primitives, which at the same time contain enough information for deriving the higher-level semantics.

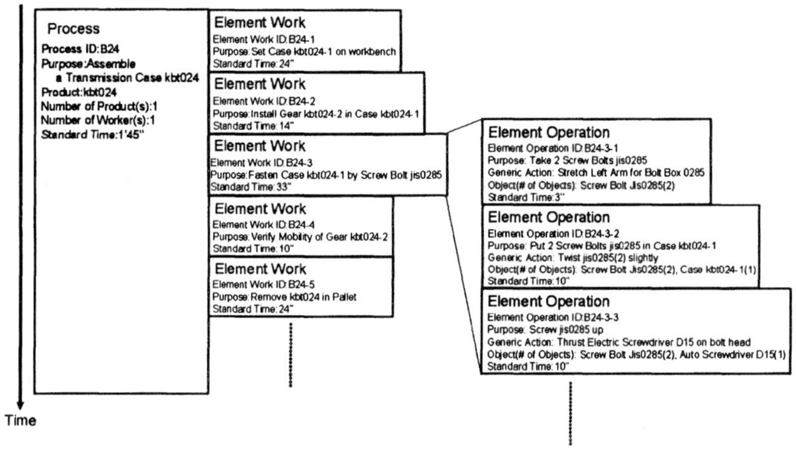

Figure 1. Example of Hierarchy of process, Element Work and Element Operation

Definition: We call an atomic unit (or AU) the tuple (*T, P, O, S, R*) where *T* is a time instant, *P* is a point in 3D space, *O* is a 3D object located in point *P* at the time instant *T*, *S* is a subject in relation *R* with the object *O*.

R is a relation that can be observed and detected in a given point of time, independently from the preceding and succeeding time instances. Also *R* is independent from the view point in the 3D space. Examples of such relations are "touches", "is inside", "is close to", etc. Right part of Figure 2 shows an example of an AU instance. Following from the definition, AUs don't have temporal duration. Those basic atomic Unit of motions can be detected automatically by analyzing video image[6]. The reason for choosing such representation is that we assume all media to be in digital format, therefore (in case of time-dependent media) we have some sampling rate with clearly defined time instants and they determine the highest resolution at which media can be observed. In addition, the existing methods for automatic detection of video features are mainly frame-based. Since evolving in time is an essential characteristic of events, AUs cannot express it and can be regarded as semantics-free in the context of event semantics. In some cases, some of the components of an AU instance may be omitted. For instance, if we want to specify just the existence of a subject at a given point of time and at a given location, then no values need to be specified for the object and relation components.

3.2 Event Graph

The basic idea of introducing event graph is to form a part of the database that contains knowledge about the typical features of action and to use it for extracting action instances from the raw video data on the basis of the detected AUs. An event graph is defined as a directed graph $G_{ME} = \{N, TOR, TDR\}$, where

$$N = \{N = i\},\ N_i \in \{S_i \times O_i \times R_i\}$$
$$TOR \subseteq \{N_i \times N_j\}$$
$$TDR \subseteq \{N_i \times N_j\}$$

N represents a set of nodes. Each node consists of a subject S_i object O_i, and relation R_i, between them. The meaning they represent is similar to the AUs, but in case of event graph nodes, subject and object are generalized and refer to a class of subjects/objects in the AU's meaning. For instance, while in case of an event graph node an example of subject value can be 'Auto Screwdriver' in the AU it would be 'Auto Screwdriver D15(1)'

TOR is a set of one-way edges, representing temporal relations between nodes. The tuple $<N_i, N_i>$, representing the edge from N_i to Nj means that is before Nj in time. If both $< N_i, N_j >$ and $< N_j, N_i >$ are presented, that means N_i and Nj are simultaneous. Since nodes, like AUs, are instantaneous, all possible temporal relations between them are 'before' 'after' and 'simultaneous'. *TDR* is a set of two-way edges representing temporal distance between nodes. In many cases it is not enough to define only the order of AUs, since the time between their occurrence is important, too. Temporal distance relations are labeled with minimum and maximum time limit values expressing these requirements. *TOR* and *TDR* are semantically independent. It is possible, for instance, to have some temporal distance requirement without relation to the order in which AUs are happening. An example of event graph is shown in Figure 2. It represents the action "screwing of a screw", which consist of taking a screw from the box with screws, fixing the screw in some position, and then applying the screwdriver in the same position. Here we assume that if the worker's hand is inside the box of screws, then a screw is taken. Also, the requirement for a very small temporal distance between the second and the third node represents approximately the moment when the screw is released.

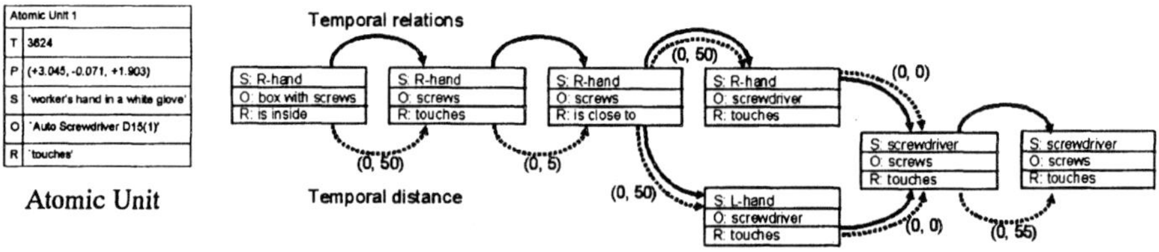

Figure 2. Example of Atomic Unit and Event Graph

3.3 Correspondence between Atomic Units and Event Graph

We introduce here action instances in order to represent the correspondence between AUs and event graph. An event instance represents the occurrence of an action in the real world and its order/causality is expressed by some event graph. The event instance can be visualized by playing the respective part of the video data or by CG Simulation.

Each action instance is determined by a set of AUs in the following way. Each event graph node corresponds to exactly one AU from the set that matches the node components and satisfied all temporal relations included in the event graph. Matching between AU and event graph node holds if and only if: (i) the AU's subject/object values are elements of the class, represented by the respective subject/object values of the node: (ii) both the AU and event graph node have the same value of their relation component. Temporal relations can be verified easily by using the values of time and place components of AUs. Action instances develop in time and occupy some time interval that includes all AUs involved, i.e. the temporal closure of all time instants. There may be various algorithms for calculating it, depending on the specifics of the application field, but the topic is beyond the scope of this paper.

4. SEMANTICAL CORRESPONDENCE BETWEEN EVENT GRAPH AND WORKERS' MOTION

In order to descript a worker's body/motion data, and to store all data into database, we offered "Mediator" Concept about human body/motion in our previous research[3][4]. Our Mediator-based modeling is a model-based modeling which has link with intensions of action, which we call it 'Hub', and actual motion which are detectable from video images. In this section, we mention reference between Mediator data and event graph. We suppose the human's motions are sequence of postures, consequently we define the human's Motion Mediator as collection of characteristic internal postures, namely, relative time and joint angle(value) of each Shape/Structure Mediator in this time, in which characteristic internal postures of a motion of a human. On the other hand, Motion Mediator is made for each motion of a human, and represent differences each motion of a human.

A Hub is the information for expressing the essence in common and standardizing it about human motions. *Motion Hub* is expressed with the intermediate posture along which it surely passes in case it appears in common with the motion is performed using the mutual position relation of the human body parts in shape and a structure Hub.

It can be said that the Hub shows the minimum "item" for describing the characteristics about shape, structure and motion of the human. When Motion Mediator is created from a Motion Hub, it is important what intermediate postures are chosen as a Motion Hub. At a present stage, the designer has to design an intermediate posture for every motion. For example, a characteristic intermediate posture sequence which generally raises Motion which tightens a screw with a drill in a factory can be show as follows.

The postures of tightening a screw using a drill:

- I: Starting Motion - The person is touching the drill
- II: Motion
  - II-1: The person located the drill at the tip of a screw.
  - II-2: The person finished tightening a screw with a drill
  - II-3: Next II-1, repeat sometimes
- III: Ending Motion -- T the person turned the drill.

Generally, it is very difficult to extract human's posture and its semantics precisely from videos by using stereo video cameras etc. But, if it restricts to the specific postures specified by the Motion Hub, the analysis of postures (joint angles) can be carry out easily by means of detection of corresponding points of characteristic points on the human body. Therefore, the Motion Mediator can be created from the Motion Hub. Figure 3 shows a relation of Hub-Mediator-Real World data about a motion of tightening a screw using a drill of worker in factories. The posture sequence about Hub, Mediator and Real World Data is shown in Figure 3, which displays joint value data of motion for convenience as a shape of a posture.

5. CONCLUSION

In the present paper we have reported the description method of detectable factory workers' action in the framework of Info-Ergonomics. For the purpose of representing their actions precisely, we have introduced

Atomic Unit, Event Graph Concept and Mediator Concept. By using these concepts, we can compare their motions and evaluate it. Such approach will benefit to Manufacturing Process Design.

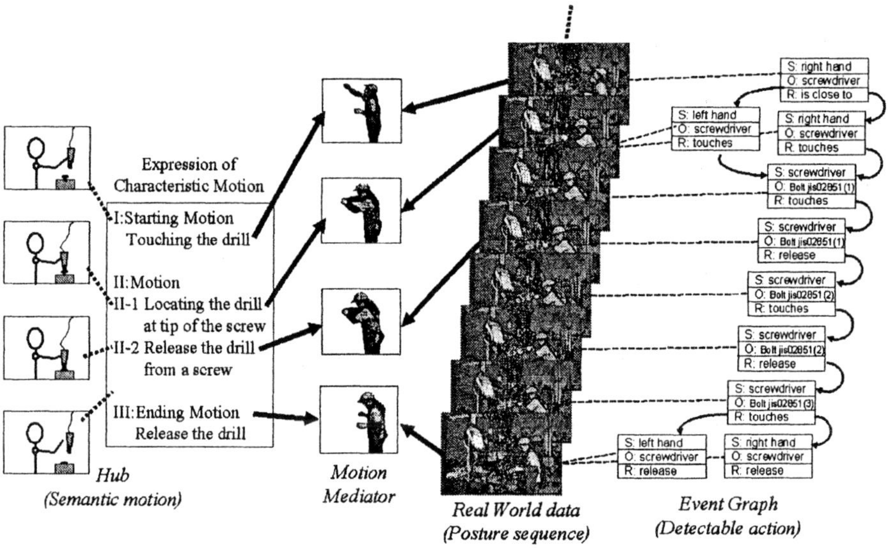

Figure 3. Correspondence of Motion and Event Graph

REFERENCES

1. S. Imai, K. Salev, T. Tomii and H. Arisawa (1998). Modeling of Working Processes and Working Simulation based on Info-Ergonomics and Real World Database Concept, Proc. of 1998 Japan-U.S.A Symposium on Flexible Automation (JUSFA'98) 147-154
2. H. Arisawa, T. Sato (2001). Human-Body Motion Simulation using Bone-based Human Model and Posture Database, Proc. Fourth International Scientific Conference on Prevention of Work-Related Musculoskeletal Disorders PREMUS2001, pp.128
3. T. Tomii, S. Varga, S. Imai, H. Arisawa (1999). Design of Video Scene Databases with Mapping to Virtual CG Space, Proc. of the 1999 IEEE International Conference on Multimedia Computing & Systems ICMCS'99, pp.741-746
4. S. Imai, T. Tomii, H. Arisawa (2001). Motion Modeling and Simulation of Human Workers based on Info-Ergonomics Concept, Proc. of the International Workshops on Modeling of Human/Organizational/Social Aspects of Manufacturing Activities (HUMACS2001), pp.106-114
5. K.Salev, T.Tomii and H.Arisawa (1999). Extracting Event Semantics from Video Based on Real World Database, Lecture Notes in Computer Science -- Advances in Database Technologies, Spatio-Temporal Data Management, Springer, pp.554-567
6. K. Sakaki, T. Sato, H. Arisawa (2002). Motion and Posture Detection by Multi-perspective Video Cameras and Load Analysis on the Precise Human Mock-up, Proc. of the 5th International Conference on DIISM2002.

41

MODEL-BASED MOTION ANALYSIS OF FACTORY WORKERS USING MULTI-PERSPECTIVE VIDEO CAMERAS

Kazunori Sakaki, Takako Sato and Hiroshi Arisawa
Yokohama National University

Abstract: Motion simulation of factory workers is one of the core technologies to achieve optimum Human-Machine Co-existing Environment in modern factory design. For the purpose of posture and motion detection evaluation, we introduced Model based Motion Capture with a limited number of markers and Bone-Based Human-body Mockup (BBHM) which can represent bones, skin, muscles, etc. Also it is required that in any environment the motion capturing method can extract the feature points with sufficient accuracy in any environment. Then, we solved by performing a strict calibration using a simple calibration board and Light Emitting Diode (LED).

Key words: Load analysis, precise human mock-up, Calibration, Motion capture, Info-Ergonomics

1. INTRODUCTION

In order to achieve optimum Human-Machine Co-existing Environment in modern factory design, motion simulation of factory workers is one of the core technologies. We have proposed "Info-Ergonomics" as an integration of simulation/evaluation technologies of human workers' motion, in conjunction with database technologies which provides storing/retrieving functions [1]. We think that human centered factors are often overlooked or underestimated although human labor has an important role in the manufacturing process design. It is usually examined from the viewpoint of

Industrial Engineering, Ergonomics, etc. and results are used in production planning, working analysis, and work environment design. Info-Ergonomics is a promising solution to provide total environment of work simulation and evaluation[2]. It consists of a number of component technologies like Human body modeling, Customizing human body model to specific workers, Motion description, Capturing motions from video images, Simulation, Storing time-dependent data into databases, and Spatio-temporal retrieval on databases.

This paper first outlines the Info-Ergonomics concept, next we will be focusing on motion detection and precise human modeling for Info-Ergonomics simulation. Especially we will also discuss requirements of precision of motion detection and propose the motion detection method by multi-perspective video cameras.

2. MOTION ANALYSIS USING HUMAN BODY MOCKUP

2.1 Info-Ergonomics

Info-Ergonomics is a framework of information integration aiming for conceptual and total design of Manufacturing Systems. Recently modeling the machines in the factory, creating virtual machines by using CG, simulating their work and evaluating it are coming into use gradually [3]. But modeling human beings and creating "virtual employee" as a 3D object on computers still has very limited use, because of the human body's high complexity and the limits imposed by the computer techniques. We focus on human body modeling and cooperative work with machines, especially from the viewpoint of comfortability, safety and efficiency.

In Info-Ergonomics approach, we construct a precise human body mock-up which can represent the differences between individuals in body-part sizes, flexibility and other properties. Giving a movement to the human mock-up model, one can estimate physical load and comfortability on each part of that body.

However, due to the complexity of human body, physical-model based evaluation/estimation is applicable only in highly-restricted cases even on "precise" human body mock-up. For example, we cannot solve force balances on musculoskeletal models because infinite solutions exist depending on muscle tension assignment in each part.

On the other hand, in the traditional Ergonomics field, evaluation/estimation of fatigue on a specific part of human body has been done based on questionnaire studies. Also, in Medical Science field, many such discussions have been done from a variety of viewpoints depending on deep knowledge provided by medical experts.

Following the above discussions, we decided the optimal solution should be "hybrid" evaluating systems standing on the three different methodologies. This idea is shown in Fig. 1.

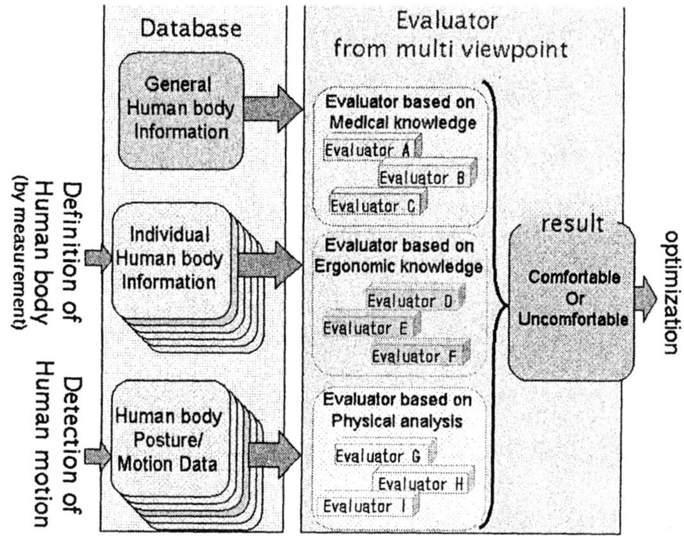

Figure 1. Concept of Info-Ergonomics

2.2 Conceptual Architecture of Info-Ergonomics Simulation System

In order to achieve precise simulation/evaluation, we must dctect the postures and motions automatically from specific individuals. We are developing a prototype system on which whole functions are realized in a uniform way. The conceptual architecture and data flow of this system is shown in Fig. 2.

The left-side part of Fig. 2 shows database and data capturing functions. A Video Capture and Motion Analysis System captures several image sequences by multi-perspective video cameras and detects posture sequence of a specific human body (factory worker) by 3D image analysis method. The technical issues on this method will be discussed later.

On the other hand, to realize a precise simulation, the human body itself must be modeled precisely. For this purpose Human Body Measurement

System detect individual body data. All data described above must be stored in a uniform way into databases.

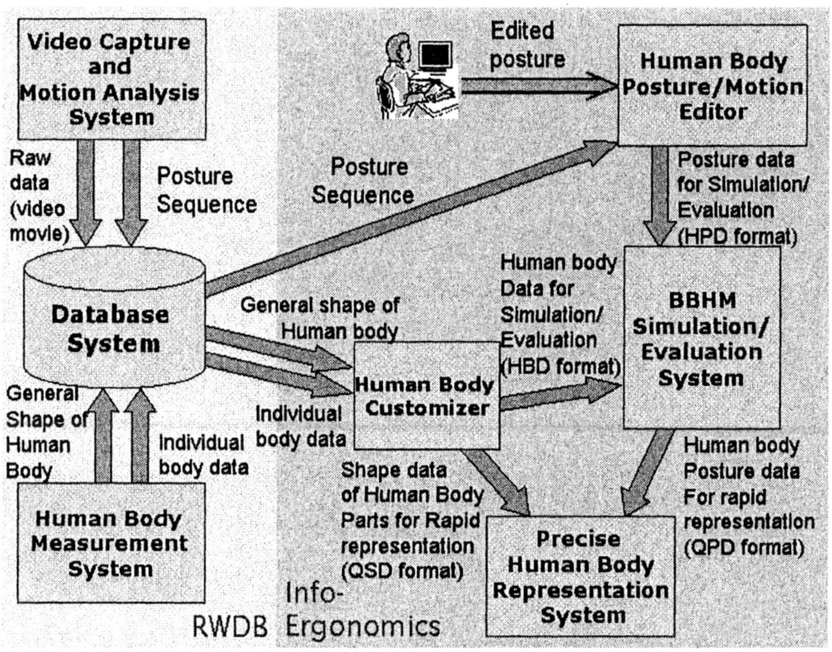

Figure 2. Conceptual Architecture of Info-Ergonomics Simulation System

The right-side part of Fig. 2 shows simulator/evaluator and result viewer. “Human Body Customizer” constructs precise human model from individual body data stored in the database. For the purpose of simulation/evaluation, the module constructs precise human model which reflects the “real” human body size, flexibility, physical strength, and other parameters. The module also constructs human body surface and bones’ shape for the purpose of visualizing simulation results. On the other hand, “Human Body Posture/Motion Editor” retrieves a posture sequence in databases, and converts to posture sequence on the precise human model. From those data, “BBHM Simulation/Evaluation System” does hybrid evaluation mentioned above. “Precise Human Body Representation System” displays simulation/evaluation results graphically.

3. MOTION ANALYSIS BASED ON A FEW MARKERS AND HUMAN MOCKUP

3.1 Method of motion analysis

As discussed in previous sections, one of the most important issues is the method of motion analysis.

Especially in motion analysis of factory workers, reduction of testee's load and flexibility of motion capturing environment are quite essential. However, existing systems do not pay attention to such restrictions because most of them force the users to prepare specific environments.

In our approach we have proposed a new method for motion capturing. First of all, we considered again the role of "markers." In the existing systems, the testee must wear 40 or more markers on his/her body. Each camera of the system can detect only markers positions by hardware algorithm. By calculating spatial positions and connecting markers to each other, you can re-construct the posture on the simplified stick model of the human body.

Our first target is to reduce the number of markers, preserving precision. For instance, if we can reduce the markers to less than 6, testee's load will be remarkably decreased, and time for analysis will be also decreased. To solve this problem, we introduced "model based analysis" for posture detection. The outline of Model Based Motion Capturing is shown in Fig.3 and the brief algorithm in Fig. 4.

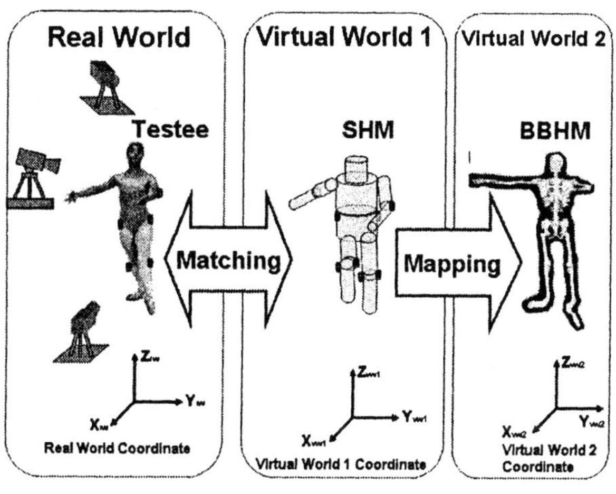

Figure 3. Outline of Model Based Motion Capturing

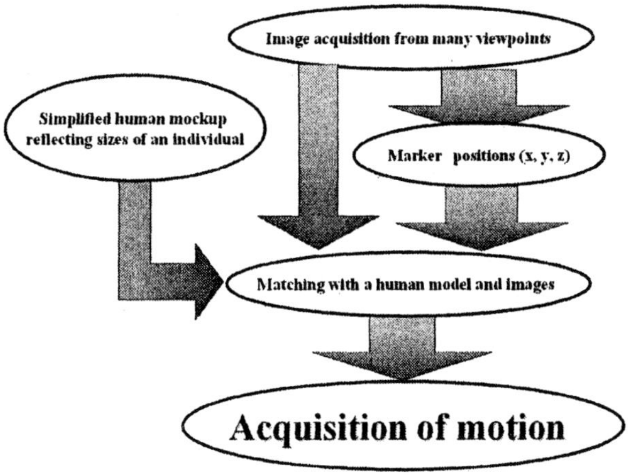

Figure 4. Procedure of Model Based Motion Capturing

In this procedure, after calculating spatial (x, y, z) positions of each marker, the system assigns a tentative posture to the SHM which represents the testee's body. Here we use inverse kinematics to decide the positions of several parts because markers are placed on the limited number of parts. Then using the multi-perspective camera images, we can adjust the posture precisely in a heuristic way.

3.2 Requirements for markers

In the method described in the previous sections, the most important technology is to detect precise positions of markers. Also, the small numbers of markers must always specify the correct posture of a SHM continuously. Generally those are very hard problems to solve. In our experimental system, we introduced three ideas to overcome these problems.

- Camera calibration method using a "standard" object

  To guarantee the accuracy of marker position, a simple standard object are taken from each video camera simultaneously. Then we can calibrate the required camera parameters beforehand.

- Adjusting the markers brightness on the environment

  The brightness of marker points always changes in the environment in which we actually take pictures. We use a light emitting diode (LED) as a marker and choose the brightness depending on the environment.

- Model based Analysis
  We proposed analyzing method which uses a human mockup depending on personal information. This contributes to reduce the number of markers.

In our on-going development, each technology has almost fully been established and contributed to significant results.

4. INVESTIGATION ON DETECTING MARKERS

As discussed in previous sections, markers play the most important role in the motion capturing system. This paper, from now on, will be focusing on marker-related issues. That is:

- Accuracy of the 3-D position detection of the markers
  We confirmed that 3-D positions can be detected with sufficient accuracy by using a standard object.
- Traceability of the marker points in various environments
  We checked that a marker can be pursued by using LED marker, even if brightness changes.

The following subsections describe the verification method concretely.

4.1 Verification of the accuracy of the 3-D position of marker points

In order to detect 3-D positions from images, we have to compute the camera position, camera posture (pan and tilt), and focal length strictly. That is, we have to calibrate the camera parameters.

Generally this problem is called PnP problem, and various discussions has been done. Tsai's algorithm [6] which considers distortion of a camera lens is quoted by many references.

The Tsai's algorithm inputs a set of points on the picture coordinate system and converts them to those of the world coordinate systems ones. Then, it computes a rotation procession and a parallel translation vector as an external parameter, and computes a focal length, lens distortion, a scale coefficient, and the picture starting point as an internal parameter.

We verified whether it can be extracted the marker points with sufficient accuracy by this calibration. As the technique, after calibrating the camera by Tsai's algorithm we take pictures of the pendulum which is attached the LED (light emitting diode) downward, and compute the position of this LED.

External appearance of experimental equipment is shown in Fig. 5.

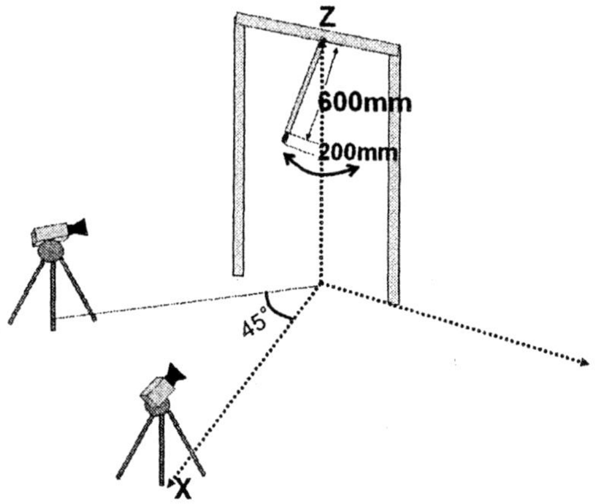

Figure 5. Experiment environment

We used 2 cameras. One is in front position and the other is left. Then, we took an image of the standard object for calibrations.

We compute the 3-D position of a LED using the camera parameters and compare the Actual measurement values.

4.2 Traceability of the marker point in various environment

Since we were able to extract 3-D positions, next problem is how to extract them in the arbitrary photographical conditions. The background brightness is always changing violently in actual capturing environment. Therefore, a marker must be extracted independent from background color changes. Then, we used the LED as a marker which hardly changes hue, even if luminosity value changes.

As for verification method, we used the pendulum attached LED used in the previous section. Since the LED has shone itself, its hue should seldom change even if surrounding brightness changes. We took image sequences in which the pendulum is swaying by 2 cameras. Then, we performed calibration and pursued a light emitting diodes by image analysis.

Now we will explain the procedure of this LED tracing. First, we performed the following procedures in the 1st frame.

1. Edge-preserving smoothing
2. Extract the domain of the color which resembled the specified pixel position mostly.
3. Calculation of each maximum and minimum of HSV in the domain obtained by 2.
4. Calculation of the middle of the domain extracted by 2.

To the successive frames we trace LEDs by following procedure.

1. It leaves only the pixel which is contained in each maximum and minimum of HSV calculated with the 1st frame.
2. Dilation and erosion processing
3. Labeling processing
4. Extracting the domain near the position of the middle of the pixel in the 1st frame.
5. Extracting the position of the middle of the domain extracted by 4.

With the above procedure, we extracted and pursued the 3-D positions of an object from the pixel positions in each extracted picture.

5. AN EXPERIMENT RESULT AND EVALUATION

5.1 The accuracy experiment of the 3-D position of the feature point

We show a calibration result with 2 cameras in Table 1.

Table 1. Calibration result

	Left Image		Center Image	
	Measured value[mm]	Result value[mm]	Measured value [mm]	Result value[mm]
x	2300	2303.2	0	-38.7
y	1329	1351.6	2234	2379.8
z	1041	1041.1	1055	1056.5
x	90	94	90	89
y	-60	-59.4	0	1.2
z	-180	-175.1	180	179.4
f	30	30	25	25.8

Based on this calibration result, we show the result of the 3-D coordinates values of the center of the stationary LED of a pendulum in Table 2.

Table 2. Calibration result

	Measured value[mm]	Result value[mm]
X	0	-46.73
Y	0	-10.83
Z	945	946.65

5.2 The pursuit experiment of the feature point in various environment

We show the result of tracing without shining LED in Fig. 6, and show the result of tracing with shining LED in Fig. 7.

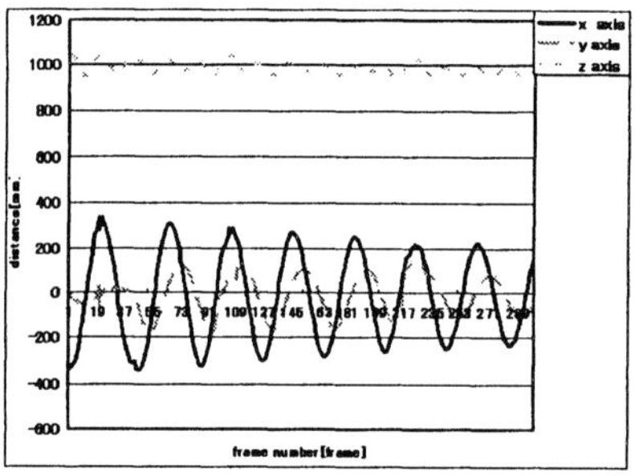

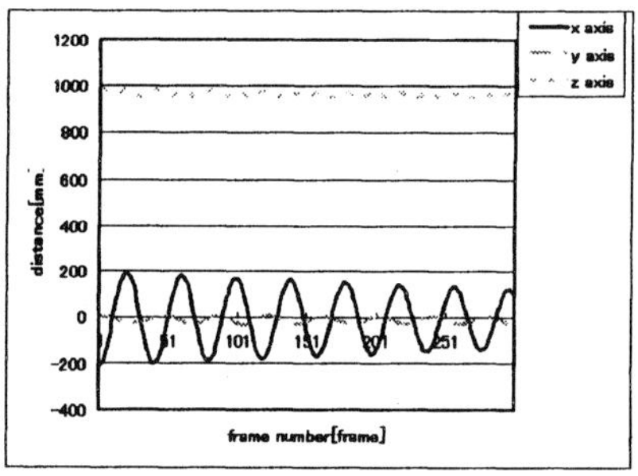

Figure 6. The result when turning off a LED *Figure 7.* The result when turning on a LED

From these two tables, it can be said that the result with a LED turned on was better. Especially, in fig.6 there is a large difference of the value of y-axis.

6. CONCLUSION

In this paper we described Info-Ergonomics concept which makes it possible to simulate/evaluate human posture/motion with precise human model called BBHM. We also described the conceptual architecture of a simulation system.

Also, we proposed a motion capturing method based on a few markers and human mockup. Then we discussed the motion capturing method which can extract the feature points required for motion analysis with sufficient accuracy in arbitrary environment. The problems are performing calibration correctness and the extraction method of the feature point. We solved them by performing a connect calibration using a standard object, and using Light Emitting Diode as markers.

Many problems are left for the future. By modeling the other components of the human body such as tendon and muscle, it becomes possible to evaluate human motions more precisely and to obtain a lead to understand diseases and factors which restrict the range of motion of joints.

REFERENCES

1. H. Arisawa, S. Imai (2001). Mediator-Based Modeling of Factory Workers and Their Motions in the Framework of Info-Ergonomics, *Human Friendly Mechatoronics (Selected Papers of ICMA2000)*, 395-400
2. Kageyu Noro (1995). Illustrated Ergonomics. JIS
3. H. Arisawa, T. Sato, T. Tomii (2002). Human-Body Motion Simulation Using Bone-Based Human Model and Construction of Motion Database, *Conceptual Modeling for New Information Systems Technologies (ER2001 Workshops, HUMACS, DASWIS, ECOMO, and DAMA, Yokohama Japan, November 27-30, 2001. Revised Papers)*, 115-26
4. I. A. Kapandji (1986). Physiologie Articulaire, Maloine S. A. Editeur (1980)
5. T. Sato, S. Nagano, T. Tomii, H. Arisawa, N. Sakai (2001). Design and Construction of Human-Body Motion Database Using Bone-Based Human Model, *IPSJ Journal Transaction of Database*, 42(SIG1(TOD8)), 92-102 (In Japanese)
6. ROGER Y. TSAI,"A Versatile Camera Calibration Technique for High-Accuracy 3 D Machine Vision Metrology Using Off-the-Shelf TV Cameras and Lenses", *IEEE Journal of Robotics and Automation*, Vol. RA.3, No.4, August 1987

42

HUMAN FACTOR AND ITS IDENTIFICATION IN MANUFACTURING PREDICTION

Yang Jianhua and Yasutaka Fujimoto
Dept. of Electrical and Computer Engineering, Yokohama National Univ. Japan
e-mail: yangjh@fujilab.dnj.ynu.ac.jp, fujimoto@ynu.ac.jp

Abstract: A decision model, stemmed from Bayesian thinking, is proposed to predict the operator's behavior in manufacturing system. The decision model is addressed using non-parametric distribution where a binary division method is proposed to reduce the complexity of the model, eliminating irrelevant features.

Key words: human factor, prediction, Bayesian theorem, non-parametric model

1. INTRODUCTION

In general a manufacturing plan should be set up to meet the time constrains of orders, while the delivery dates are also determined by prediction of complete time of orders based on the capacity of manufacturing system. In many cases, we find that a shop floor is controlled by a group of operators who make their decisions according to some rules usually given by a guide, their experience obtained from manufacturing history, and probably their mood of that day. Therefore, the human interference should be included if we hope to predict the future of manufacturing. Errors might be decreased if we can correctly identify the behaviors of operators using the past decision data.

The identification of human behavior in manufacturing system differs from general pattern recognition[1] in which sampling data, the manufacturing history, have been given and random sampling is not applicable. Therefore sufficient manufacturing history data are needed to make it possible to identify the human behavior. The model used in this paper is primarily developed and originated from Bayesian thinking, where some special transformations are introduced for constituting feature vector from parameters of factors. On the other hand, a full non-parametric model is proposed to dispose of both continuous and discrete variables with irregular distributions. To solve the complexity problem where a non-parametric model might result in an explosion of data storage[2] and the relevance selection problem where irrelevant factors are redundant[3], we propose a binary division method in this paper.

2. MODELLING

With respect to the problem we consider, it is unnecessary to obtain a general model of human's brain. In fact, it is also unrealistic until today even if we have received some clues about its operation mechanism. In order to clearly illustrate the essence of the problem, we formulate the process of human's decision making as follows. Let the operator's surroundings be C, the operator's status be M, the operator's decision mechanism be D, and the final decision be Q. The process of human's decision making can be represented by $D:C\times M\rightarrow Q$. Unfortunately we often only know the partial surrounding information I around the operator and try to employ $D:I\rightarrow Q$ to represent the decision process of the manufacturing system. In other words, the decision recognition can be expressed by $R:(I,Q)\rightarrow D$, where R stands for recognition mechanism. Errors are inevitable if $(C-I)\times M\neq\phi$. Therefore the topic about error decreasing in this paper is always discussed over R.

Bayesian thinking[4] is often employed in pattern recognition. Let Ω be sampling space, which is composed of n independent hypotheses, noted by $\{B_1,B_2,\cdots,B_n\}$. The probability of occurrence for result x can be computed by following equation:

$$p(B_k\mid x)=p(x\mid B_k)p(B_k)/p(x) \tag{1}$$

For identification of operator's behavior, we let hypotheses be operator's decision D, let results be history system data I and history decision data Q. According to equation (1), we get

$$p(D\mid(I,Q))=p((I,Q)\mid D)p(D)/p((I,Q)) \tag{2}$$

Furthermore, suppose that an operator always select no more than one job based on the current status of surroundings. The surroundings data at that time are addressed by a feature vector X, then (2) can be expressed by

$$p(D\mid X)=p(X\mid D)p(D)/p(X) \tag{3}$$

From manufacturing history, distribution $p(X\mid D)$ and $p(X)$ can be obtained although sometimes it is hard and complex to do so. For prior distribution $p(D)$, suppose that it is a uniform distribution. As a result, the posterior distribution $p(D|X)$ is fundamentally determined by $p(X \mid D)/p(X)$. Let

$$\eta(X \mid D) = p(X \mid D)/p(X). \tag{4}$$

The $\eta(X \mid D)$ can be regarded as a force, the operator's decision, driving the prior probability distribution $p(X)$ to the posterior probability distribution $p(X|D)$. Finally for the future status, we predict that operator will select a job $j^* \in J$ such that $p(D|X_{j^*}^F) = \max_j p(D|X_j^F)$, i.e.,

$$\eta(X_{j^*}^F \mid D) = \max_j \eta(X_j^F \mid D) \tag{5}$$

3. DECISION ACQUISITION

As illustrated in Section 2.1, the decision recognition distribution $\eta(X \mid D)$ naturally describes the operator's decision mechanism and shows how much information is obtained.

To simply illustrate it, let $[\bullet]$ be logarithm of probability variable $p(\bullet)$, then we have

$$[X \mid D] = \log p(X \mid D),\ [X] = \log p(X). \tag{6}$$

And we define

$$[\eta(X \mid D)] = [X \mid D] - [X] \tag{7}$$
$$F(D) = \int |[X \mid D] - [X]| dX \tag{8}$$

where $F(D)$ is referred to as intensity of operator's decision and $\int$ stands for a generalized integral operator which can compute over both discrete and continuous variables.

In general, the higher the intensity means the stronger operator's decision. If the $p(X)$ have the same distribution with $p(X|D)$, we get $F(D) = 0$. It means that we learn nothing from manufacturing history, i.e., one might select jobs randomly. Therefore we certainly cannot predict the future. But for an effective identification, we always have $F(D) > 0$. For instance, only a feature x is considered. Given $p(x = 0) = 0.6$ and $p(x = 1) = 0.4$. After operator's decision, we get $p(x = 0|D) = 0.8$ and $p(x = 1|D) = 0.2$. Then how much can we learn from history? Or $F(D) = ?$.

Here the $\int$ is substituted by the $\sum$, then we get

$$F(D) = \sum |[X \mid D] - [X]|$$
$$= \sum |\log p(X \mid D) - \log p(X)| = |\log 0.8 - \log 0.6| + |\log 0.2 - \log 0.4| = 0.426$$

Note that definition of $\eta[X \mid D]$ is invalid if $p(X \mid D) = 0$ or $p(X) = 0$. Therefore integral $F(D)$ defined in formula (8) does not always exist. To strictly define it, we discuss some properties over so-called valid sampling space. Let Ω_R stand for valid sampling space of $p(X)$ such that $p(X) > 0$,

Ω_D for valid sampling space of $p(X \mid D)$ such that $p(X \mid D) > 0$, respectively. We have following conclusions.

[Property 1] $\Omega_R \geq \Omega_D$.

It can be obviously proved because a sampling point of the operator's decision should be one belonging to original sampling space. Particularly, $\Omega_R = \Omega_D$ means that no more than one job waits in the buffer at any time hence the operator has no choice but select the only one.

[Property 2] $p(X) > 0$ if $p(X \mid D) > 0$.

It can be induced from *property 1* and can be regarded as another description of *property 1.*

Based on *property 2*, we can revise formula (8) as

$$F(D) = \int_{\Omega_D} \| [X \mid D] - [X] \| dX \tag{9}$$

whose definition always exists.

[Property 3] let $\Delta\Omega = \Omega_R - \Omega_D$, we have

$$p(X) > 0 \ and \ p(X \mid D) = 0 \ for \ \forall X \in \Delta\Omega.$$

It can be concluded from *property 2* and the definition of valid sampling space. The domain $\Delta\Omega$ is also referred to as deterministic decision space, implying that a sampling point in $\Delta\Omega$, which is also referred to as deterministic decision point, will be surely recognized because zero is the smallest value. Generally the larger the domain $\Delta\Omega$ is, the stronger the decision mechanism is.

4. NON-PARAMETRIC DISTRIBUTION

The simplest way to describe distribution of $\eta(X \mid D)$ is utilization of parametric model such as normal distribution, beta distribution etc, where the distribution can be completely represented by some parameters such as average value, variance, etc. However the distribution type should be known before we employ parametric model. Thus to obtain general description of $\eta(X \mid D)$, non-parametric model is usually a possible choice.

A non-parametric distribution model is generally described by dividing sampling space into many tiny domains, where distribution density $p(X)$ is almost constant. Let a domain be S, corresponding volume be V. The probability of feature vector in S can be calculated by $p(S) \cong p(X)V$. According to Monte Carlo simulation[5], given m sampling data, if among them k data fall in the domain S, the probability of feature vector in S can be obtained by $p(S) = k/m$. Thus the distribution density in domain S can be determined by

$$p(X) = k/mV. \tag{10}$$

The basic two methods for modeling non-parametric distribution are kernel density method[6][7] and k-nearest neighbors method[8]. For the kernel density method, the probability density of a domain can be calculated by fixing the volume of the domain, counting the data that fall in it. For k-nearest neighbor method, the probability density of a domain can be calculated by fixing the number of data that fall in the domain, changing the volume of the domain. A main drawback of the kernel density method is that a large domain division might result in low smooth while a small domain division might result in low reliability due to limited history data. Moreover, sometimes its implementation is almost infeasible. The k-nearest neighbors method emphasizes that the volume of domain is changeable, fixing the counts of data that fall in the domain. But it is often hard to get such a domain. In fact, no matter what kind of method, the fundamental problem is the division of sampling space. In next section, a binary division method is proposed to provide such a solution, where both the volume and counts are changeable.

5. BINARY DIVISION METHODOLOGY

Noticed that an effective decision means that decision distribution $p(D|X)$ is not a uniform distribution. The larger difference among domains generally implies the more effective decision. So we should emphasize the feature with less variance and consider how to divide it firstly. Here a binary division method is one of possible choices.

Let Ω be the sampling space, $X = [x_1 \ \ x_2 \ \ \cdots \ \ x_k]$ be a feature vector. At first, a binary division is done along each feature $x_i (i = 1,2,\cdots,k)$, so we get a group of bi-subspaces, i.e., domains, denoted by $S(x_i,L,\Omega)$ and $S(x_i,R,\Omega)$, where L stands for the left domain, R for right domain, respectively. As described previously, instead of computing probability density $p(D|X)$, $\eta(X|D)$ is applied to describe recognition distribution therefore we define $\eta(x_i,L,\Omega|D)$ standing for density distribution of $S(x_i,L,\Omega)$, $\eta(x_i,R,\Omega|D)$ for density distribution of $S(x_i,R,\Omega)$. Among k divisions only one along the feature x_{i^*} $(i^* \in \{1,2,\cdots,k\})$ is really selected to be executed, which is such that

$$\Delta\eta(x_{i^*},\Omega) = \max_i \Delta\eta(x_i,\Omega) \tag{11}$$

where $\Delta\eta(x_i,\Omega) = |\eta(x_i,L,\Omega|D) - \eta(x_i,R,\Omega|D)|$. (12)

Similarly for each subspace $S_u \in \{S(x_{i^*},L,\Omega),\ S(x_{i^*},\ R,\Omega)\}$ we can obtain its furthermore divided subspaces $S(x_i,L,S_u)$ and $S(x_i,R,S_u)$ by binary divisions. And the really executed division along the feature x_{i^*} $(i^* \in \{1,2,\cdots,k\})$ at this step is also such that

$$\Delta\eta(x_{i^*},S_u) = \max_i \Delta\eta(x_i,S_u), \tag{13}$$

where $\Delta\eta(x_i,S_u) = |\eta(x_i,L,S_u|D) - \eta(x_i,R,S_u|D)|$. (14)

Apparently such a division might be carried out infinitely, producing countless domains therefore a termination condition should be added.

Hereby, we introduce two thresholds: an integer $\sigma(\geq 0)$ standing for a threshold of sampling points for a subspace S_u and a real number $\delta(\geq 0)$ for a threshold of the difference of density distribution between two subspaces of the subspace S_u. The binary division process will be stopped if

$$C(S_u) \leq \sigma \,\|\, \Delta\eta(x_{i^*}, S_u) \leq \delta \tag{15}$$

where $C(S_u)$ is the sampling points of the subspace S_u and symbol $\|$ represents 'OR' Boolean operator.

The domain division for non-parametric distribution is equivalent to sampling problem in signal processing. An effective technique is that the higher density makes more divisions, vice versa. It is the threshold σ that determines how small a domain should be.

Furthermore, as we consider the problem of division of sampling space, distinguishing relevant and irrelevant features should be also taken in account. Clearly the model will become redundant if an irrelevant feature is involved. Therefore is it possible that irrelevant features can be kicked out when domains are divided?

It is clear that the times of binary division along the each feature x_i, denoted by $\kappa(x_i)$, might be different. And it can be applied to deal with the problem of elimination of irrelevant features. Before some conclusions are induced, the definitions of irrelevant feature are discussed as follows.

[Definition 1] Irrelevant feature in strong sense: A feature x_r is an irrelevant feature if decision distribution $p(D|x_r)$ is a uniform distribution and independent of other features.

Using above definition and the sampling division method, we obtain the following theorem.

[Theorem] The times of binary division along a feature x_r is denoted by $\kappa(x_r)$. $\kappa(x_r) = 0$ if the feature x_r is irrelevant to operator's decision in strong sense.

[Proof]

Based on equations (3), (4), we get

$$\eta(X|D) = p(X|D)/p(X) = p(D|X)/p(D). \tag{16}$$

For the feature x_r, we have

$$\eta(x_r|D) = p(D|x_r)/p(D). \tag{17}$$

The distribution $\eta(x_r|D)$ should be uniform because $p(D|x_r)$ and $p(D)$ are uniform distributions, according to definition and assumption.

The uniform property is kept for all domains if a feature is independent of others, therefore the binary division on x_r for any subspace S_u is always such that

$$\Delta\eta(x_r, S_u) = 0. \tag{18}$$

But according to binary division method, only the binary division such that $\Delta\eta(x_i, S_u) > 0$ is possibly selected and really executed. Thus the binary division will be never really executed on x_r, i.e.,

$$\kappa(x_r) = 0. \tag{19}$$

[End]

However we cannot induce that a feature is irrelevant one in strong sense even if $\kappa(x_r) = 0$ using proposed binary division. Therefore we introduce the definition of irrelevant feature in weak sense as follows.

[Definition 2] Irrelevant feature in weak sense: A feature x_r is an irrelevant one if $\kappa(x_r) = 0$.

That is, we can eliminate irrelevant features in weak sense using binary division method.

6. AN EXAMPLE

Given a set of jobs $J = \{1,2,3,4,5,6,7,8,9,10\}$, waiting in a buffer to be processed on a machine, its corresponding processing time and parts size are represented by $H = \{h(j)\}_{10} = \{12,24,30,28,48,51,61,60,70,66\}$ and $S = \{s(j)\}_{10} = \{20,10,10,20,30,30,30,20,20,10\}$ respectively. Let the parts size of the job before job 1 be 20. Suppose that jobs are mounted according to sequence $1 \rightarrow 2 \rightarrow 3 \rightarrow 4 \rightarrow 5 \rightarrow 6 \rightarrow 7 \rightarrow 8 \rightarrow 9 \rightarrow 10$. Define feature vector $X = [x_1 \quad x_2]^T$ where

$$x_1 = h(i) - \min_j h(j) \quad \text{(Job } i \in J \text{ is the next one to be mounted, } j \neq i\text{)} \tag{20}$$

$$x_2 = \begin{cases} 0 & s(o) = s(i) \\ 1 & s(o) \neq s(i) \end{cases} \quad \text{(Job } o \in J \text{ is the one that just has been processed).} \tag{21}$$

To simplify our example, we suppose that $p(X)$ is approximately a uniform distribution. Therefore $\eta(X \mid D)$ is determined by $p(X \mid D)$. According to above processing sequence, we obtain a set of sampling data

$$\{\begin{matrix}-12\\0\end{matrix}, \begin{matrix}-4\\1\end{matrix}, \begin{matrix}2\\0\end{matrix}, \begin{matrix}-24\\1\end{matrix}, \begin{matrix}-3\\1\end{matrix}, \begin{matrix}-9\\0\end{matrix}, \begin{matrix}1\\0\end{matrix}, \begin{matrix}-6\\1\end{matrix}, \begin{matrix}4\\0\end{matrix}\},$$

where we needn't make a decision for the last job thus only 9 data are generated. Let $\sigma = 1, \delta = 0$. The result of binary division along $X = [x_1 \quad x_2]^T$ is shown in *Fig.1* and obtained histogram is shown in *Fig.2*.

Fig.1 indicates that both of x_1 and x_2 might be related to operators' decision because division times along them, 7 and 4 respectively, are larger than 0. *Fig.2* illustrates the recognition information for operator decision, which is equivalent to $\eta(X \mid D)$ due to our assumption that $p(X)$ is approximately a uniform distribution. It shows that an operator will select a job to be mounted using minimum processing time rule mixed with identical parts size preference. Some domains in Fig.2, whose distribution density equals 0, are invalid because of insufficient data. Therefore generally more past data should be provided if we want to obtain a perfect result.

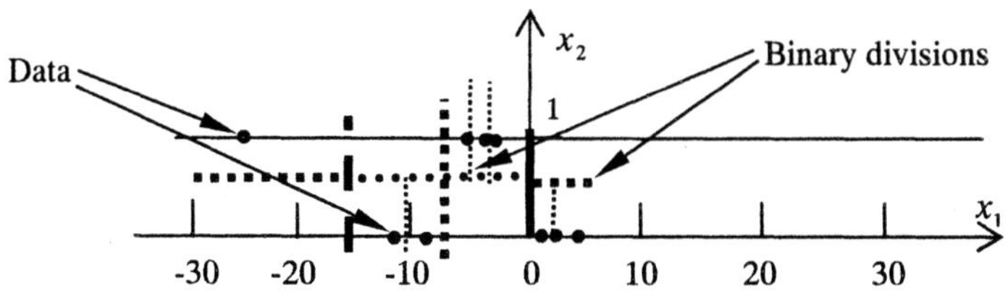

Figure1. Binary divisions performed on 2 dimensions

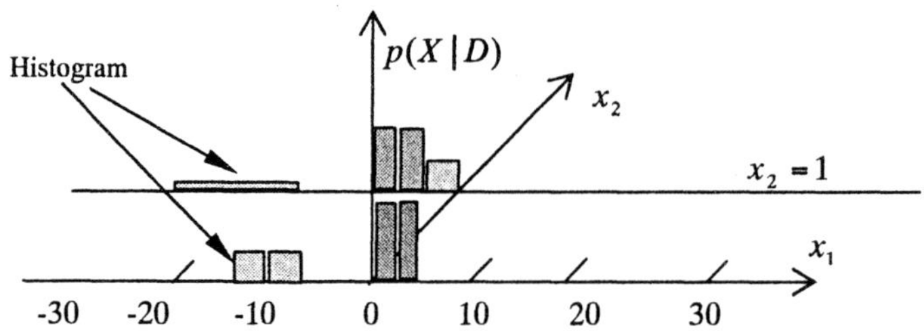

Figure 2. Distribution density after binary division

7. CONCLUSIONS

To recognize operators' decision for a manufacturing system, a model induced from Bayesian thinking is proposed in this paper. We employ non-parametric distribution model to address it and propose a binary division method, whose properties are investigated. It shows that proposed method has the advantage of compressing the model as small as possible, eliminating irrelevant features as well. An example is provided to illustrate the recognition of the operators' decisions.

REFERENCES

1. Anil K. Jain, Robert P. W. Duin, Jianchang Mao, "Statistical Pattern Recognition: A Review", IEEE Trans. On Pattern analysis and machine intelligence, Vol.22, No.1, Jan. 2000, pp4-37.
2. A.G. Gray, A.W. Moore, "'N-Body' Problem in statistical Learning", Advances in Neural Information Processing System 13, May 2001.
3. Avrim L. Blum, Pat Langley, "Select of Relevant Feature and Examples in Machine Learning", Artificial Intelligence, 1997, pp245-271.
4. B. Moghaddam, T. Jebara, A. Pentland, "Bayesian Modeling of Facial Similarity", In Advances in Neural Information Processing System 11, MIT Press, 1999.
5. P.E.Lassila, J.T. Virtamo, "Nearly optimal importance sampling for Monte Carlo simulation of loss system", ACM trans. On modeling and computer simulation, Vol.10, No.4, Oct. 2000, pp326-347.
6. B.W. Silverman, "Density Estimation for Statistic and Data Analysis", Chapman and Hall, London, 1986.
7. G. Terrell, D. Scott, "Variable Kernel Density Estimation", Ann. Statistic, Vol.20, No.3, pp1236-1265, 1992.
8. T.K. Ho, "Nearest Neighbors in Random Subspaces", Lecture Notes in Computer Science: Advances in Pattern Recognition, pp640-648, 1998.

AUTHOR INDEX

KEYWORD INDEX